VIROLOGY

VIROLOGY

Dr. A. Maharajan
M.Sc., M.Phil., Ph.D.
Assistant Professor
PG & Research Department of Zoology
Khadir Mohideen College
Adirampattinam - 614 701
Thanjavur Dist.,
Tamil Nadu, India

2011
DAYA PUBLISHING HOUSE
Delhi - 110 035

© 2011 A. MAHARAJAN (b. 1975–   )
ISBN 9789351242277

Published by   :   **Daya Publishing House**
                          A Division of
                          Astral International Pvt. Ltd.
                          – ISO 9001:2008 Certified Company –
                          4760-61/23, Ansari Road, Darya Ganj,
                          New Delhi-110 002
                          Ph. 011-43549197, 23278134
                          E-mail: info@astralint.com
                          Website: www.astralint.com

Laser Typesetting   :   **Classic Computer Services**
                          Delhi - 110 035

Printed at   :   **Chawla Offset Printers**
                          Delhi - 110 052

PRINTED IN INDIA

Preface

Virology is a fascinating and rapidly developing subject, and is worthy of study purely because viruses are interesting! Furthermore, virology is a branch of science that is of immense relevance to mankind for a host of reasons, not the least of which are the threats to human health caused by viruses, such as HIV. hepatitis B virus, papilloma viruses, measles and influenza viruses, to mention just a few. There is a continuant need for trained virologists, and it is hoped that this book will play a small role to fulfil this need. This book is specially prepared to meet the demands on Zoology, Microbiology and Biotechnology UG and PG students of various University, it brings out the detailed information on introduction, history, classification, architecture, replication, pathogenesis, genetics, immunology, viral host interaction, bacteriophages, plant viruses, animal viruses, interferons, antiviral drugs, prions and viroids, virus genomes, vaccination, diagnostic methods and infection control measures. I am thankful to *Dr. M. Sivalaila., Ph.D*, Principal, Udaya college of Arts and Science and my *Beloved Parents, Sister, Brother in law* and my wife *Mrs.R.Lakshmi sree,* Lecturer, N.M.S.Kamaraj Polytechnic college, Pazhavilai for their cooperation and encouragement.

Dr. A. Maharajan

Contents

Chapter 1
Introduction to Viruses

Definition of a Virus

Viruses are organized associations of macromolecules: *Nucleic acid* (which carries the blueprint for the replication of progeny virions) contained within a protective *shell of protein units.*

A *virus* is a biological agent that reproduces inside the cells of living hosts. When infected by a virus, a host cell is forced to produce many thousands of identical copies of the original virus, at an extraordinary rate. Unlike most living things, viruses do not have cells that divide; new viruses are assembled in the infected host cell. Over 2,000 species of viruses have been discovered.

A virus consists of two or three parts: all viruses have genes made from either DNA or RNA, long molecules that carry the genetic information; all have a protein coat that protects these genes; and some have an envelope of fat that surrounds them when they are not within a cell. Viruses vary in shape from the simple helical and icosahedral to more complex structures. Viruses are about 100 times smaller than bacteria, and it would take 30,000 to 750,000 of them, side by side, to stretch to 1 centimetre (0.39 in).

Viruses spread in many different ways. Plant viruses are often spread from plant to plant by insects and other organisms, known as *vectors*. Some viruses are spread by blood-sucking insects. Each species of virus relies on a different method. Whereas viruses such as influenza are spread through the air by people's coughing and sneezing, others such as norovirus, which are transmitted by the faecal-oral route, contaminate hands, food and water. Rotavirus is often spread by direct contact with infected children. HIV is one of several major viruses that are transmitted during sex. The origins of viruses is unclear: some may have evolved from plasmids–pieces of DNA that can move between cells-while others may have evolved from bacteria.

Viral infections often cause disease in humans and animals; they are usually completely eliminated by the immune system, and this confers lifetime immunity to the host for that virus. Antibiotics have no effect on viruses, but antiviral drugs have been developed to treat life-threatening infections. Vaccines that produce lifelong immunity can prevent some viral infections.

Origins

Viruses are found wherever there is life and have probably existed since living cells first evolved. The origin of viruses is unclear because they do not form fossils, so molecular techniques have been the most useful means of hypothesising how they arose. However, these techniques rely on the availability of ancient viral DNA or RNA, but, unfortunately, most of the viruses that have been preserved and stored in laboratories are less than 90 years old. Molecular methods have only been successful in tracing the ancestry of viruses that evolved in the 20th century.

There are three main theories of the origins of viruses.

Regressive theory

Viruses may have once been small cells that parasitised larger cells. Over time, genes not required by their parasitism were lost. The bacteria rickettsia and chlamydia are living cells that, like viruses, can reproduce only inside host cells. They lend credence to this theory, as their dependence on parasitism is likely to have caused the loss of genes that enabled them to survive outside a cell.

Cellular Origin Theory

Some viruses may have evolved from bits of DNA or RNA that "escaped" from the genes of a larger organism. The escaped DNA could have come from plasmids pieces of DNA that can move between cells while others may have evolved from bacteria.

Coevolution Theory

Viruses may have evolved from complex molecules of protein and DNA at the same time as cells first appeared on earth and would have been dependent on cellular life for many millions of years.

Structure

A virus particle, known as a virion, consists of genes made from DNA or RNA which are surrounded by a protective coat of protein called a capsid. The capsid is made of many smaller, identical protein molecules which are called capsomers. The arrangement of the capsomers can either be icosahedral (20-sided), helical or more complex. There is an inner shell around the DNA or RNA called the nucleocapsid, which is formed by proteins. Some viruses are surrounded by a bubble of lipid (fat) called an envelope (Figure 1).

Viruses are very small in size (20–300 nanometers) and contain *either* DNA *or* RNA (not both as in higher forms of life), The genome (DNA or RNA) codes for the few proteins necessary for replication.

Some proteins are *non-structural*, *e.g.* nucleic acid polymerases and some are *structural*, *i.e.* they become incorporated and form part of the virion.

Protein building blocks are assembled according to general principles of virus architecture to form a tight "shell" *(capsid)* inside which the nucleic acid genome lodges for protection. This shell may take the form of a

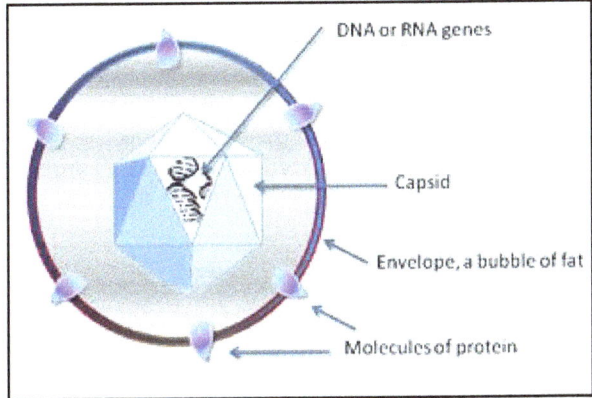

Figure 1. Structure of Virus

polyhedron (usually icosahedral) or it may be spiral (helical symmetry), or it may be more complex. Some viruses acquire an outer lipoprotein coat by "budding" through the host cell membranes (nuclear membrane or cytoplasmic membrane) and are thus called *enveloped* viruses.

All the viral proteins have reactive epitopes which are important for interaction with cellular components during the process of infection and replication. The host's defence mechanisms (cellular and humoral mediated responses) are directed against the viral antigenic epitopes.

Classification

Viruses are broadly classified primarily upon the type of genomic nucleic acid, *e.g.* DNA or RNA and then further by the number of strands of nucleic acid (*e.g.* double-stranded DNA, double-stranded RNA or single-stranded RNA, with a positive or negative "sense" of that single strand). *Retroviruses* are a special category of RNA viruses that require reverse transcription of their RNA to DNA and then integration of that DNA into the host cell genome before replication can take place. They carry a reverse transcriptase enzyme as part of the virion.

Replication

1. Adsorption

Viruses have reactive sites on their surface which interact with specific *receptors* on suitable host cells. This is usually a passive reaction (not requiring energy) and the specificity of the reaction between viral protein and host receptor defines and limits the host species as well as the type of cell that is infected (although transfected nucleic acid can by-pass this limitation and extend the host range). Damage to these binding sites (*e.g.* by disinfectants or heat), or blocking by specific antibodies (neutralizing antibodies) can render virions non-infectious.

2. Uptake

After adsorption, the coat of enveloped viruses may *fuse* with the host cell membrane and release the virus nucleocapsid into the host cytoplasm. Other viruses may enter the cell by a process of *"endocytosis"* which involves invagination of the cell membrane to form vesicles in the cell cytoplasm.

3. Uncoating

Refers to the release of the viral genome from its protective capsid to enable the nucleic acid to be transported within the cell and transcribed to form new progeny virions.

4. Genomic Activation

Messenger RNA (m-RNA) is transcribed from viral DNA (or formed directly from some RNA viruses) and codes for viral proteins that are translated by the host cell." Early" proteins are usually non-structural (*e.g.* DNA or RNA polymerases) and later proteins are structural, *e.g.* capsid proteins, *i.e.* building blocks of the virion. Nucleic acid replication produces new viral genomes for incorporation into progeny virions.

In general, DNA viruses replicate mainly in the nucleus and RNA viruses mainly in the cytoplasm, but there are exceptions, *e.g.* Pox viruses contain DNA but replicate in the cytoplasm of the host cell.

5. Assembly

Assembly of viral nucleocapsids may take place in the nucleus (*e.g.* herpes virus, adenovirus); in the cytoplasm (*e.g.* polio virus); or at the cell surface, *e.g.* "budding" viruses such as influenza. Accumulation of virions at sites of assembly may form "inclusions" that are visible in stained cells with the light microscope.

6. Release

Release of new infectious virions is the final stage of replication.This may occur by *budding* from the cell surface, as occurs with many enveloped viruses. In this case capsid proteins and nucleic acid condense directly adjacent to the cell membrane and viral-coded envelope proteins, introduced into the cell membrane, concentrate in the vicinity of capsid aggregates. The membrane surrounding the nucleocapsid then bulges out and becomes "nipped off" to form the new enveloped virion.

Some viruses utilize the cellular *secretory pathway* to exit the cell. Virus particles enclosed within *Golgi*-derived vesicles are released to the outside of the cell when the transport vesicle fuses with the cell membrane. Disintegration or *lysis* of the infected cell can also result in the release of intact infectious virions.

Methods and Study of Viruses

Diagnosis of Viral Infections

Viruses can be studied in a number of direct and indirect ways and all these methods can be applied in a *diagnostic situation, i.e.* is this patient infected with a particular virus? There are two approaches:

1. Detection and demonstration of the virus itself; and
2. The study of the host's response to that virus

One of the earliest ways of detecting a virus was by inoculating a susceptible host (laboratory) animal with infectious material derived from a patient or sick animal and then observing that animal for signs of disease. Fertile hens eggs proved useful systems for a number of viruses (especially myxoviruses) and are still used for influenza. Today, live animals are rarely used as *"in vitro" cell cultures* have largely replaced them. In recent years "non-cultivable" viruses have been extensively studied by molecular techniques ("genetic engineering").

The structure of different viruses has been elucidated by a range of electron microscopy and x-ray crystallography techniques. Viruses amplified by growth in culture (or in a few special cases, directly from patient specimens without amplification) can be demonstrated by electron microscopy.

Viral antigens can be detected by a wide range of *serological techniques* utilising polyclonal or monoclonal antibodies. Techniques include precipitation, agglutination, immunofluorescence, ELISA, complement fixation and radio immuno assays. These same techniques, utilising purified viral antigens, can be used to detect specific antibodies to those viruses in the patient's serum. Identification of different classes of antibodies (IgG and IgM) can aid in differentiating between a current infection and immunity.

Some viruses (*e.g.* myxo- and paramyxoviruses, including influenza) have the property of *haemagglutination* (causing red blood cells to stick together) which can be used to detect and quantitate the virus (by haemagglutination) or specific antibodies to that virus (haemagglutination inhibition). Similarly, *neutralisation* of viral infectivity by antibodies can be used to detect and quantify either virus or specific antibody to that virus.

Modern *molecular techniques* of both protein chemistry and nucleic acid biochemistry have greatly improved the specificity of virus diagnostic procedures. Methods include:

☆ Polyacrylamide gel electrophoresis (PAGE) of protein fragments

☆ Western blotting, and identification of specific proteins with labelled probes

☆ Polymerase chain reaction (PCR), to amplify specific segments of viral nucleic acid

☆ Southern blotting, and DNA hybridisation with labelled probes

☆ Sequencing of portions of the viral genome

☆ Restriction fragment length polymorphisms of viral nucleic acid

Applications

The application of sophisticated molecular technology has enabled the generation of diagnostic assays for viruses that have not yet been visualized or cultured. *Hepatitis C virus* is the prime example. This RNA virus has never been cultured, but portions of its genome were extracted from blood known to be infectious for hepatitis C. By means of adapted PCR techniques, the nucleotide sequence of the entire viral genome was eventually assembled. Knowing some gene sequences enabled biochemists to synthesise corresponding small portions of proteins (peptides). Some peptides were found to be major antigenic determinants of the virus and these peptides have now been incorporated into commercial ELISA tests designed to detect human antibodies to hepatitis C. The presence of antibodies has been shown to be associated with chronic hepatitis C infection and a high risk of transmitting hepatitis C in blood transfusions. As from 1993, blood transfusion services in South Africa routinely screen all blood donations for hepatitis C antibody.

Molecular biology methods have been used to compare degrees of relatedness of similar organisms and to build *phylogenetic trees* ("family trees" based on genomic similarities). The ability to detect and sequence portions of a viral genome permits genetic markers of specific sub-strains to be identified. This has led to the new science of *genetic epidemiology* (*i.e.* disease tracing).

Sterilisation and Disinfection

Viruses, especially the enveloped viruses, are generally fairly labile and do not survive too well outside their host cells. However, some (*e.g.* hepatitis B virus) are very resistant to inactivation, and healthcare workers need to take special precautions to avoid transmitting such infections. Means of prevention of the spread of infection, and sterilisation and disinfection of viruses, are very similar to those principles that are applied in bacteriology.

Spread may be by

1. Inhalation of aerosolised "droplets";
2. Ingestion;
3. Direct contact (skin/mucous membrane to skin/mucous membrane), or
4. Indirect contact via intermediate "fomites".

Moist heat (autoclaving 120°C x 20 minutes) or *dry heat* (oven, 180°C for 60 minutes) are effective against all viruses–lesser degrees of heat may inactivate many viruses (*e.g.* simple *boiling*) but may not reliably inactivate resistant viruses especially if times of exposure are short.

Chemicals

Halogens, especially chlorine as *hypochlorite* are effective against viruses but corrosive on instruments where *activated gluteraldehyde* ("Cidex") is preferred. *Detergents and lipid solvents* inactivate readily the enveloped viruses which need an intact envelope for effective cell adsorption. *Phenolic disinfectants* damage proteins and thus inactivate bacteria but do not affect nucleic acids. Phenolics are *not recommended* for viral disinfection.

Chapter 2
History of Virology

1796

Edward Jenner (1749-1823) used cowpox to *vaccinate* against smallpox. In 1774, a farmer named Benjamin Jesty had vaccinated his wife and two sons with cowpox taken from the udder of an infected cow and had written about his experience (see 1979). Jenner was the first person to deliberately vaccinate against any infectious disease, *i.e.* to use a preparation containing an antigenic molecule or mixture of such molecules designed to elicit an immune response. Although Jenner is commonly given the credit for vaccination, *variolation*, the practice of deliberately infecting people with smallpox to protect them from the worst type of the disease, had been practised in China at least two thousand years previously.

1885

Louis Pasteur (1822-1895) experimented with rabies vaccination, using the term "virus" (Latin, poison) to describe the agent. Although Pasteur did not discriminate between viruses and other infectious agents, he originated the terms "virus" and "vaccination" (in honour of Jenner) and developed the scientific basis for Jenner's experimental approach to vaccination.

1886

John Buist (a Scottish pathologist) stained lymph from skin lesions of a smallpox patient and saw "elementary bodies" which he thought were the spores of micrococci. These were in fact smallpox virus particles–just large enough to see with the light microscope.

1892

Dmiti Iwanowski (1864-1920) described the first "filterable" infectious agent–tobacco mosaic virus (TMV)–smaller than any known bacteria. Iwanowski was the first person to discriminate between viruses and other infectious agents, although he was not fully aware of the significance of this finding.

1898

Martinus Beijerinick (1851-1931) extended Iwanowski's work with TMV and formed the first clear concept of the virus "contagium vivum fluidum"–soluble living germ. Beijerinick confirmed and extended Iwanowski's work and was the person who developed the concept of the virus as a distinct entity. *Freidrich Loeffler* (1852-1915) and *Paul Frosch* (1860-1928) demonstrated that foot and mouth disease is caused by such "filterable" agents. Loeffler and Frosch were the first to prove that viruses could infect animals as well as plants.

1900

Walter Reed (1851-1902) demonstrated that yellow fever is spread by mosquitoes. Although Reed did not dwell on the nature of the yellow fever agent, he and his coworkers were the first to show that viruses could be spread by insect vectors such as mosquitoes.

1908

Karl Landsteiner (1868-1943) and *Erwin Popper* proved that poliomyelitis was caused by a virus. Landsteiner and Popper were the first to prove that viruses could infect humans as well as animals.

1911

Francis Peyton Rous (1879-1970) demonstrated that a virus (Rous sarcoma virus) can cause cancer in chickens (Nobel Prize, 1966) (see 1981). Rous was the first person to show that a virus could cause cancer.

1915

Frederick Twort (1877-1950) discovered viruses infecting bacteria.

1917

Felix d'Herelle (1873-1949) independently discovered viruses of bacteria and coins the term *bacteriophage*. The discovery of bacteriophages provided an invaluable opportunity to study virus replication at a time prior to the development of tissue culture when the only way to study viruses was by infecting whole organisms.

1935

Wendell Stanley (1887-1955) crystallized TMV and showed that it remained infectious (Nobel Prize, 1946). Stanley's work was the first step towards describing the molecular structure of any virus and helped to further illuminate the nature of viruses.

1938

Max Theiler (1899-1972) developed a live attenuated vaccine against yellow fever (Nobel Prize, 1951). Theiler's vaccine was so safe and effective that it is still in use today! This work saved millions of lives and set the model for the production of many subsequent vaccines.

1939

Emory Ellis (1906-) and *Max Delbruck* (1906-1981) established the concept of the "one step virus growth cycle" essential to the understanding of virus replication (Nobel Prize, 1969). This work laid the basis for the understanding of virus replication—that virus particles do not "grow" but are instead assembled from preformed components.

1940

Helmuth Ruska (1908-1973) used an electron microscope to take the first pictures of virus particles. Along with other physical studies of viruses, direct visualization of *virions* was an important advance in understanding virus structure.

1941

George Hirst demonstrated that influenza virus agglutinates red blood cells. This was the first rapid, quantitative method of measuring eukaryotic viruses.

1945

Salvador Luria (1912-1991) and *Alfred Hershey* (1908-1997) demonstrated that *bacteriophages* mutate (Nobel Prize, 1969). This work proved that similar genetic mechanisms operate in viruses as in cellular organisms and laid the basis for the understanding of antigenic variation in viruses.

1949

John Enders (1897-1985), *Thomas Weller* (1915-) and *Frederick Robbins* (1916-) were able to grow poliovirus in vitro using human tissue culture (Nobel Prize, 1954). This development led to the isolation of many new viruses in tissue culture.

1950

André Lwoff (1902-1994) and colleagues discovered *lysogenic* bacteriophage in *Bacillus megaterium* irradiated with ultra-violet light and coined the term *prophage* (Nobel Prize, 1965). Although the concept of lysogeny had been around since the 1920s, this work clarified the existence of *temperate* and *virulent bacteriophages* and led to subsequent studies concerning the control of gene expression in prokaryotes, resulting ultimately in the operon hypothesis of Jacob and Monod.

1952

Renato Dulbecco (1914-) showed that animal viruses can form plaques in a similar way to *bacteriophages* (Nobel Prize, 1975). Dulbecco's work allowed rapid quantitation of animal viruses using assays which had only previously been possible with bacteriophages.

Alfred Hershey (1908-1997) and *Martha Chase* demonstrated that DNA was the genetic material of a *bacteriophage*. Although the initial evidence for DNA as the molecular basis of genetic inheritance was discovered using a bacteriophage, this principle of course applies to all cellular organisms (though not all viruses!).

1957

Heinz Fraenkel-Conrat (1910-1999) and *R.C. Williams* showed that when mixtures of purified tobacco mosaic virus (TMV) RNA and coat protein were incubated together, virus particles formed spontaneously. The discovery that virus particles could form spontaneously from purified subunits without any extraneous information indicated that the particle was in the free energy minimum state and was therefore the favoured structure of the components. This stability is an important feature of virus particles.

Alick Isaacs and *Jean Lindemann* discovered interferon. Although the initial hopes for interferons as broad spectrum antiviral agents equivalent to antibiotics have faded, interferons were the first cytokines to be studied in detail.

Carleton Gajdusek proposes that a "slow virus" is responsible for the *prion* disease kuru (Nobel Prize, 1976) (see 1982). Gajdusek showed that the course of the kuru is similar to that of scrapie, that kuru can be transmitted to chimpanzees and that the agent responsible is an atypical virus.

1961

Sydney Brenner, Francois Jacob, and *Matthew Meselson* demonstrate that bacteriophage T4 uses host cell ribosomes to direct virus protein synthesis. This discovery revealed the fundamental molecular mechanism of protein translation.

1963

Baruch Blumberg discovered hepatitis B virus (HBV) (Nobel Prize, 1976). Blumberg went on to develop the first vaccine against the HBV, considered by some to be the first vaccine against cancer because of the strong association of hepatitis B with liver cancer.

1967

Mark Ptashne isolates and studies the repressor protein. Repressor proteins as regulatory molecules were first postulated by Jacob and Monod. Together with Walter Gilbert's work on the *E.coli* lac repressor protein, Ptashne's work illustrated how repressor proteins are a key element of gene regulation and control the reactions of genes to environmental signals. *Theodor Diener* discovered *viroids*, agents of plant disease which have no protein capsid. Viroids are infectious agent consisting of a low molecular weight RNA that contains no protein capsid responsible for many plant diseases.

1970

Howard Temin (1934-1994) and *David Baltimore* independently discovered reverse transcriptase in retroviruses (Nobel Prize, 1975). The discovery of reverse transcription established a pathway for genetic information flow from RNA to DNA, refuting the so-call "central dogma" of molecular biology.

1972

Paul Berg created the first recombinant DNA molecules, circular SV40 DNA genomes containing phage genes and the galactose operon of *E.coli* (Nobel prize, 1980). This was the beginning of recombinant DNA technology.

1973

Peter Doherty and *Rolf Zinkernagl* demonstrate the basis of antigenic recognition by the cellular immune system (Nobel Prize, 1996). The demonstration that lymphocytes recognize both virus antigens and major histocompatibility antigens in order to kill virus-infected cells established the specificity of the cellular immune system.

1975

Bernard Moss, Aaron Shatkin and colleagues show that messenger RNA contains a specific nucleotide cap at its 5' end which affects correct processing during translation. These discoveries in reovirus and vaccinia were subsequently found to apply to cellular mRNAs–a fundamental principle.

1976

J. Michael Bishop and *Harold Varmus* determined that the *oncogene* from Rous sarcoma virus that can also be found in the cells of normal animals, including humans (Nobel Prize, 1989). Proto-oncogenes are essential for normal development but can become cancer genes when cellular regulators are damaged or modified, *e.g.* by virus *transduction.*

1977

Richard Roberts, and independently *Phillip Sharp,* show that adenovirus genes are interspersed with non-coding segments that do not specify protein structure (*introns*) (Nobel Prize, 1993). The discovery of gene splicing in adenovirus was subsequently found to apply to cellular genes–a fundamental principle. *Frederick Sanger* and colleagues determine the complete sequence of all 5,375 nucleotides of the bacteriophage fX174 *genome* (Nobel Prize, 1980). This was the first complete genome sequence of any organism to be determined.

1979

Smallpox was officially declared to be eradicated by the World Health Organization (WHO). The last naturally occurring case of smallpox was seen in Somalia in 1977. This was the first microbial disease ever to be completely eliminated.

1981

Yorio Hinuma and colleagues isolated human T-cell leukaemia virus (HTLV) from the patients with adult T-cell leukaemia. Although several viruses are associated with human tumours, HTLV was the first unequivocal human cancer virus to be identified.

1982

Stanley Prusiner demonstrates that infectious proteins he called *prions* cause scrapie, a fatal neurodegenerative disease of sheep (Nobel Prize, 1997). This was the most significant advance in understanding of what were previously called "slow virus" diseases and are now known as transmissible spongiform encepthalopathies (TSEs).

1983

Luc Montaigner and *Robert Gallo* announced the discovery of human immunodeficiency virus (HIV), the causative agent of AIDS. In only 2-3 years since the start of the AIDS epidemic the agent responsible was identified.

1985

U.S. Department of Agriculture (USDA) granted the first ever license to market a genetically-modified organism (GMO)–a modified vaccinia virus to vaccinate against swine herpes (pseudorabies). The first commercial GMO.

1986

Roger Beachy, Rob Fraley and colleagues demonstrated that tobacco plants transformed with the gene for the coat protein of tobacco mosaic virus are resistant to TMV infection. This work resulted in a better understanding of virus resistance in plants, a major goal of plant breeders for many centuries.

1989

Hepatitis C virus (HCV), the source of most cases of nonA, nonB hepatitis, was definitively identified. This was the first infectious agent to be identified by molecular cloning of the genome rather than by more traditional techniques (see 1994).

1990

First (approved) human gene therapy procedure was carried out on a child with severe combined immune deficiency (SCID), using a retrovirus vector. Although not successful, this was the first attempt to correct human genetic disease.

1993

Nucleotide sequence of the smallpox virus genome completed (185,578 bp). Initially, it was intended that destruction of remaining laboratory stocks of smallpox virus would be carried out when the complete genome sequence had been determined. However, this decision has now been postponed (1999).

1994

Yuan Chang, Patrick Moore and their collaborators identified human herpesvirus 8 (HHV-8), the causative agent of Kaposi's sarcoma. Using a polymerase chain reaction (PCR)-based technique, representational difference analysis, this novel pathogen was identified.

1999

Number of confirmed cases of people living with HIV/AIDS worldwide reaches 33 million. The AIDS pandemic continues to grow. Confirmed cases are an underestimate of the true total worldwide. Nucleotide sequence of the largest known virus genome completed: Paramecium bursaria Chlorella virus 1. This 330,742 bp sequence represents the technical advances in sequencing which have occurred since the first genome sequence was completed in 1977. By 1985, *Harald zur Hausen* had shown that two strains of Human papillomavirus (HPV) cause most cases of cervical cancer. Two vaccines protecting against these strains were released in 2006. In 2006 and 2007 it was reported that introducing a small number of specific transcription factor genes into normal skin cells of mice or humans can turn these cells into pluripotent stem cells, known as Induced Pluripotent Stem Cells. The technique uses modified retroviruses to transform the cells; this is a potential problem for human therapy since these viruses integrate their genes at a random location in the host's genome, which can interrupt other genes and potentially causes cancer.

Chapter 3
Virus Classification

Virus classification involves naming and placing viruses into a taxonomic system. Like the relatively consistent classification systems seen for cellular organisms, virus classification is the subject of ongoing debate and proposals. This is largely due to the pseudo-living nature of viruses, which are not yet definitively living or non-living. As such, they do not fit neatly into the established biological classification system in place for cellular organisms, such as plants and animals.

Nomenclature of Virus

Till early 1950s viruses were named according the disease they caused. This system is now out of date. The *International Committee on Taxonomy of Viruses (ICTV)* has recommended to use a set of suffixes to denote different taxonomic ranks in virus classification. According to ICTV, the name of orders should end of *virales*, the family should end in *viridae,* the subfamily should end in *virinae* and of genes should end in *virus*. The species epithet should be formed from the disease caused by the virus and the suffix virus. For instance, the hierarchial system of different taxonomic ranks in placing Mumps virus is given below:

Order	—	Mononegavirales
Family	—	Paramyxoviridae
Subfamily	—	Paramyxovirinae
Genus	—	Rubula virus
Species	—	Mumps virus

Virus classification is based mainly on phenotypic characteristics, including morphology, nucleic acid type, mode of replication, host organisms, and the type of disease they cause. A combination of two main schemes is currently in widespread use for the classification of viruses. David Baltimore, a Nobel Prize-winning biologist, devised the Baltimore classification system, which places viruses into one of seven groups. These groups are designated by Roman numerals and separate viruses based on their mode of replication, and genome type. Accompanying this broad method of classification are specific naming conventions and further classification guidelines set out by the International Committee on Taxonomy of Viruses.

Baltimore Classification

The Baltimore Classification of viruses is based on the method of viral mRNA synthesis. Baltimore classification (first defined in 1971) is a classification system which places viruses into one of seven groups depending on a combination of their nucleic acid (DNA or RNA), strandedness (single-stranded or double-stranded), Sense, and method of replication. Other classifications are determined by the disease caused by the virus or its morphology, neither of which are satisfactory due to different viruses either causing the same disease or looking very similar. In addition, viral structures are often difficult to determine under the microscope. Classifying viruses according to their genome means that those in a given category will all behave in a similar fashion, offering some indication of how to proceed with further research. Viruses can be placed in one of the seven following groups:

☆ I: *dsDNAviruses* (e.g. Adenoviruses, Herpesviruses, Poxviruses)

☆ II: *ssDNA viruses* (+)sense DNA (*e.g.* Parvoviruses)

☆ III: *dsRNA viruses* (*e.g.* Reoviruses)

Sl.No.	Virus Family	Examples (Common Names)	Virion Naked/ Enveloped	Capsid Symmetry	Nucleic Acid Type	Group
1.	Adenoviridae	Adenovirus	Naked	Icosahedral	ds	I
2.	Papillomaviridae	Papillomavirus	Naked	Icosahedral	ds circular	I
3.	Parvoviridae	Parvovirus B19	Naked	Icosahedral	ss	II
4.	Herpesviridae	Herpes simplex virus, varicella-zostervirus, cytomegalovirus, Epstein-Barr virus	Enveloped	Icosahedral	ds	I
5.	Poxviridae	Smallpox virus, vaccinia virus	Complex coats	Complex	ds	I
6.	Hepadnaviridae	Hepatitis B virus	Enveloped	Icosahedral	circular, partially ds	VII
7.	Polyomaviridae	Polyoma virus; JC virus (progressive multifocal leucoencephalopathy)	Naked	Icosahedral	ds circular	I
8.	Circoviridae	Transfusion Transmitted virus	Naked	Icosahedral	ss circular	II

☆ IV: *(+)ssRNA viruses* (+)sense RNA (*e.g.* Picornaviruses, Togaviruses)

☆ V: *(-)ssRNA viruses* (-)sense RNA (*e.g.* Orthomyxoviruses, Rhabdoviruses)

☆ VI: *ssRNA-RT viruses* (+)sense RNA with DNA intermediate in life-cycle (*e.g.* Retroviruses)

☆ VII: *dsDNA-RT viruses* (*e.g.* Hepadnaviruses)

DNA Viruses

☆ *Group I*: viruses possess double-stranded DNA.

☆ *Group II*: viruses possess single-stranded DNA.

RNA Viruses

☆ *Group III*: viruses possess double-stranded RNA genomes, *e.g.* rotavirus. These genomes are always segmented.

☆ *Group IV*: viruses possess positive-sense single-stranded RNA genomes. Many well known viruses are found in this group, including the picornaviruses (which is a family of viruses that includes well-known viruses like Hepatitis A virus, enteroviruses, rhinoviruses, poliovirus, and foot-and-mouth virus), SARS virus, hepatitis C virus, yellow fever virus, and rubella virus.

☆ *Group V*: viruses possess negative-sense single-stranded RNA genomes. The deadly Ebola and Marburg viruses are well known members of this group, along with influenza virus, measles, mumps and rabies.

Reverse Transcribing Viruses

☆ *Group VI*: viruses possess single-stranded RNA genomes and replicate using reverse transcriptase. The retroviruses are included in this group, of which HIV is a member.

☆ *Group VII*: viruses possess double-stranded DNA genomes and replicate using reverse transcriptase. The hepatitis B virus can be found in this group.

ICTV Classification

The International Committee on Taxonomy of Viruses began to devise and implement rules for the naming and classification of viruses early in the 1990s, an effort that continues to the present day. The system shares many features with the classification system of cellular organisms, such as taxon structure. Viral classification starts at the level of order and follows as thus, with the taxon suffixes given in italics:

Order	–	*virales*
Family	–	*viridae*
Subfamily	–	*virinae*
Genus	–	*virus*
Species		

So far, five orders have been established by the ICTV: the *Caudovirales, Herpesvirales, Mononegavirales, Nidovirales,* and *Picornavirales.* These orders span viruses with varying host ranges. *Caudovirales* are tailed dsDNA (group I) bacteriophages, *Herpesvirales* contains large eukaryotic dsDNA viruses, *Mononegavirales* includes non-segmented (-) strand ssRNA (Group V) plant and animal viruses, *Nidovirales* is composed of (+) strand ssRNA (Group IV) viruses with vertebrate hosts, and *Picornavirales* contains small (+) strand ssRNA viruses that infect a variety of plant, insect, and animal hosts. Other variations occur between the orders, for example, Nidovirales are isolated for their differentiation in expressing structural and non-structural proteins separately. However, this system of nomenclature differs from other taxonomic codes on several points. A minor point is that names of orders and families are italicized, as in the ICBN most notably, species names generally take the form of *[Disease] virus.* The establishment of an order is based on the inference that the virus families contained within a single order have most likely evolved from a common ancestor. The majority of virus families remain unplaced. Currently (2008) 82 families and 2,083 species of virus have been defined.

Holmes Classification

Holmes (1948) used Carolus Linnaeus system of binomial nomenclature classification system to viruses into 3 groups under one order, Virales. They are placed as follows:

☆ *Group I: Phaginae* (attacks bacteria)

☆ *Group II: Phytophaginae* (attacks plants)

☆ *Group III: Zoophaginae* (attacks animals)

LHT System of Virus Classification

The LHT System of Virus Classification is based on chemical and physical characters like nucleic acid (DNA or RNA), Symmetry (Helical or Icosahedral or Complex), presence of envelope, diameter of capsid, number of capsomers. This classification was approved by the Provisional Committee on Nomenclature of Virus (PNVC) of the International Association of Microbiological Societies (1962). It is as follows:

☆ *Phylum Vira* (divided into 2 subphyla)

● *Subphylum Deoxyvira* (DNA viruses)

● *Class Deoxybinala* (dual symmetry)

● *Order Urovirales*

● *Family Phagoviridae*

● *Class Deoxyhelica* (Helical symmetry)

● *Order Chitovirales*

● *Family Poxviridae*

● *Class Deoxycubica* (cubical symmetry)

● *Order Peplovirales*

- *Family Herpesviridae* (162 capsomeres)
- *Order Haplovirales* (no envelope)
- *Family Iridoviridae* (812 capsomeres)
- *Family Adenoviridae* (252 capsomeres)
- *Family Papiloviridae* (72 capsomeres)
- *Family Paroviridae* (32 capsomeres)
- *Family Microviridae* (12 capsomeres)
- *Subphylum Ribovira* (RNA viruses)
- *Class Ribocubica*
- *Order Togovirales*
- *Family Arboviridae*
- *Order Lymovirales*
- *Family Napoviridae*
- *Family Reoviridae*
- *Class Ribohelica*
- *Order Sagovirales*
- *Family Stomataviridae*
- *Family Paramyxoviridae*
- *Family Myxoviridae*
- *Order Rbadovirales*
- *Suborder Flexiviridales*
- *Family Mesoviridae*
- *Family Peptoviridae*
- *Suborder Rigidovirales*
- *Family Pachyviridae*
- *Family Protoviridae*
- *Family Polichoviridae*

Casjens and Kings Classification of Virus

Casjens and Kings (1975) classified virus into 4 groups based on type of nucleic acid, presence of envelope, symmetry and site of assembly It is as follows:

☆ *Single Stranded RNA Viruses*
☆ *Double Stranded RNA Viruses*
☆ *Single Stranded DNA Viruses*
☆ *Double Stranded DNA Viruses*

Subviral Agents

The following agents are smaller than viruses but have some of their properties.

Viroids

Family *Pospiviroidae*

Genus *Pospiviroid*; type species: *Potato spindle tuber viroid*

Genus *Hostuviroid*; type species: *Hop stunt viroid*

Genus *Cocadviroid*; type species: *Coconut cadang-cadang viroid*

Genus *Apscaviroid*; type species: *Apple scar skin viroid*

Genus *Coleviroid*; type species: *Coleus blumei viroid 1*

Family *Avsunviroidae*

Genus *Avsunviroid*; type species: *Avocado sunblotch viroid*

Genus *Pelamoviroid*; type species: *Peach latent mosaic viroid*

Satellites

Satellites depend on co-infection of a host cell with a helper virus for productive multiplication. Their nucleic acids have substantially distinct nucleotide sequences from either their helper virus or host. When a satellite subviral agent encodes the coat protein in which it is encapsulated, it's then called a satellite virus.

Satellite Viruses

☆ Single-stranded RNA satellite viruses

● Subgroup 1: *Chronic bee-paralysis satellite virus*

● Subgroup 2: *Tobacco necrosis satellite virus*

Satellite Nucleic Acids

☆ Single-stranded satellite DNAs

☆ Double-stranded satellite RNAs

☆ Single-stranded satellite RNAs

● Subgroup 1: Large satellite RNAs

● Subgroup 2: Small linear satellite RNAs

● Subgroup 3:Circular satellite RNAs (virusoids)

Prions

Prions, named for their description as *"proteinaceous and infectious particles,"* lack any detectable nucleic acids or virus-like particles. They resist inactivation procedures which normally affect nucleic acids.

☆ Mammalian prions: Agents of spongiform encephalopathies

☆ Fungal prions: PSI+ prion of *Saccharomyces cerevisiae*, URE3 prion of *Saccharomyces cerevisiae*, RNQ/PIN+ prion of *Saccharomyces cerevisiae*

Chapter 4
Virus Architecture

Design of the Protein Shell

The complex arrangements of macromolecules in the virus shell are minute marvels of molecular architecture. Specific requirements of each type of virus have resulted in a fascinating apparent diversity of organization and geometrical design. Nevertheless, there are certain common features and general principles of architecture that apply to all viruses.

In 1956, Crick and Watson proposed on theoretical considerations and on the basis of rather flimsy experimental evidence then available, principles of virus structure that have been amply confirmed and universally accepted.

They first pointed out that the nucleic acid in small virions was probably insufficient to code for more than a few sorts of protein molecules of limited size. The only reasonable way to build a protein shell, therefore, was to use the same type of molecule over and over again, hence their theory of *identical subunits*.

The second part of their proposal concerned the way in which the subunits must be packed in the protein shell or capsid. On general grounds it was expected that subunits would be packed so as to provide each with an identical environment. This is possible only if they are packed *symmetrically*. Crick and Watson pointed out that the only way to provide each subunit with an identical environment was by packing them to fit some form of *CUBIC SYMMETRY*. A body with cubic symmetry possesses a number of axes about which it may be rotated to give a number of identical appearances. These predictions were soon confirmed and it became evident that the occurrence of icosahedral features in quite unrelated viruses was not a matter of chance selection but that *icosahedral symmetry* is preferred in virus structure.

An ICOSAHEDRON is composed of 20 facets, each an equilateral triangle, and 12 vertices, and because of the axes of rotational symmetry is said to have *symmetry*.

Axes of Symmetry

There are, in fact, six 5-fold axes of symmetry passing through the vertices, ten 3-fold axes extending through each face and fifteen 2-fold axes passing through the edges of an icosahedron (Figure 2).

Icosahedral symmetry requires definite numbers of structure units to complete a shell. In their discussions, Crick and Watson (1956), thinking in terms of asymmetrical protein subunits packed in such a way that each has an

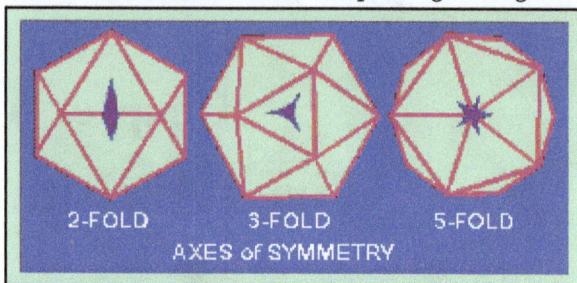

Figure 2. Axes of Symmetry

identical environment, pointed out that a virus with 5:3:2 symmetry required a multiple of 60 subunits to cover the surface completely. Each unit would be related identically and asymmetrically with its neighbours, and none of the units would coincide with an axis of symmetry.

The introduction of *NEGATIVE STAINING* (Brenner and Horne, 1959) revolutionized the field of electron microscopy of viruses. Within just a few years, much new and exciting information about the architecture of virus particles was acquired. Not only were the overall shapes of particles revealed but also the symmetrical arrangement of their components. This led to a need for a new terminology to describe the viral components.

Lwoff, Anderson and Jacob (1959) proposed the terms *"capsid"* and *"capsomers"* to represent, respectively, the protein shell and the units comprising it, and the term*"virion"* to denote the complete infective virus particle (*i.e.* a capsid enclosing the nucleic acid). This terminology was generally accepted although it later proved to be inadequate.

As soon as the first high resolution micrographs of negatively stained icosahedral viruses were obtained (Horne *et al.*, 1959–adenovirus; and Huxley and Zubay, 1960–turnip yellow mosaic virus) it seemed that there was a structural paradox. The number of morphological units observed on the surface of known icosahedral viruses at that time was never 60 or multiples of 60, and was often more than 60. Furthermore, the capsomers themselves appeared to be symmetrical and were located on symmetry axes, *e.g. herpesvirus.*

There was direct evidence that capsomers of herpesvirus were hexagonal and pentagonal in section. It is obvious that five-fold capsomers must be located on axes of five-fold symmetry, and six-fold capsomers may be situated on axes of two-fold or three-fold symmetry, or in indifferent sites where they are suited to hexagonal packing.

It was therefore clear that the capsomers were not equivalent to the subunits of Crick and Watson (1956). An obvious solution to the problem was provided by supposing that the symmetrical capsomers are built from a number of *ASYMMETRICAL SUBUNITS*. In this way it is possible to build a variety of complicated bodies in which 5:3:2 symmetry is preserved and in which the number of subunits is a multiple of 60 as predicted by Crick and Watson.

The theoretical basis for the structure of isometric viruses was put on a firm foundation by Caspar and Klug (1962) with their concept of identical elements in *quasi-equivalent* environments. They defined all possible polyhedra in terms of *structure units*. The icosahedron itself has 20 equilateral triangular facets and therefore $20T$ structure units where T is the *TRIANGULATION NUMBER* given by the rule:

$T=Pf$ where P can be any number of the series 1,3,7,13,19,21,31..(=$h + hK + K$, for all pairs of integers, h and K having no common factor) and f is any integer.

Morphological units can be clustered as $20T$ trimers, $30T$ dimers or separated as $60T$ monomers. The number of morphological units that would be produced by a clustering into hexamers and pentamers can be calculated as follows: There are $10(T-1)$ hexamers plus 12. (and only 12) pentamers.

Caspar and Klug (1962) claimed that most icosahedral viruses fall into 2 classes:- $P=1$ and $P=3$; and that all deltahedra for which $P=>7$ are skew, and therefore exist in right and left- handed forms. One "hand" might be selected by the nucleic acid, but there would still be the chance that mistakes in assembly leading to defective particles might occur. The most probable mistake in assembly would be the formation of tubular forms. Tubular structures which have a diameter and surface structure similar to icosahedral virus particles have been observed associated with polyoma and papilloma viruses.

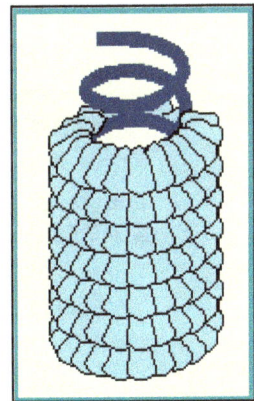

In a review of symmetry in virus architecture, Horne and Wildy (1961) showed that all the viruses then known (with the exception of a few bacteriophages) fell into two main morphological groups:- those with *cubic* symmetry and the others with *HELICAL* symmetry.

"Linear" viral capsids have RNA genomes that are encased in a helix of identical protein subunits. The length of the helical viral nucleocapsid is determined by the length of the nucleic acid. Until 1960, the only known examples of virions with helical symmetry were those of plant viruses, the best studied example being *tobacco mosaic virus* (Figure 3).

Figure 3. Tobacco Mosaic Virus

At that time, the architecture of the *myxoviruses* was poorly understood (Figure 4). Early electron micrographs of shadow-cast preparations revealed particles of varying shape and size but little detail could be reported (Bang, 1948). With the advent of negative staining, it became obvious that the *myxo-* and *paramyxo-viruses* consisted of an inner nucleo-protein component with helical symmetry surrounded

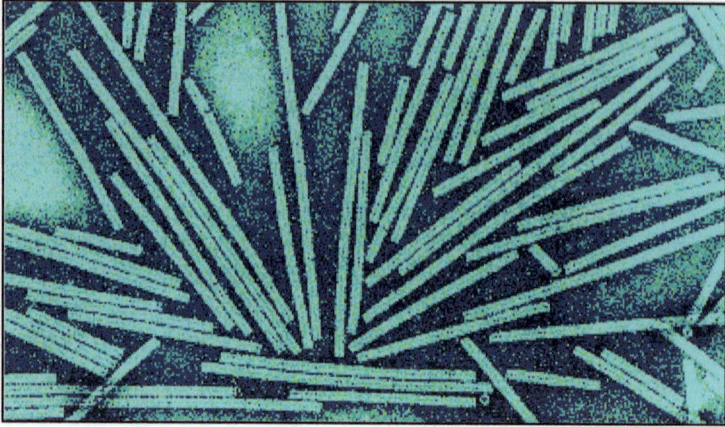

Figure 4. Myxo Virus

by an envelope of characteristic morphology. This realization of the helical symmetry of the myxoviruses laid the foundation for the understanding of the symmetry of other complex groups of viruses such as rabies virus and granulosis virus.

In an attempt to clarify the *terminology for virus components,* Caspar *et al.* (1962) made a number of proposals which were generally accepted. Briefly, the proposals are as follows:

1. The *CAPSID* denotes the protein shell that encloses the nucleic acid. It is built of structure units.

2. *STRUCTURE UNITS* are the smallest functional equivalent building units of the capsid.

3. *CAPSOMERS* are morphological units seen on the surface of particles and represent clusters of structure units.

4. The capsid together with its enclosed nucleic acid is called the *NUCLEOCAPSID*.

5. The nucleocapsid may be invested in an *ENVELOPE* which may contain material of host cell as well as viral origin.

6. The *VIRION* is the complete infective virus particle.

Chapter 5

Replication of Virus

Principal Events Involved in Replication

1. Adsorption

The first step in infection of a cell is attachment to the cell surface. Attachment is via ionic interactions which are temperature-independent. The viral attachment protein recognizes specific receptors, which may be protein, carbohydrate or lipid, on the outside of the cell. Cells without the appropriate receptors are not susceptible to the virus (Figure 5).

2. Penetration

The virus enters the cell in a variety of ways according to the nature of the virus.

Enveloped Viruses

(A) Entry by fusing with the plasma membrane

Some enveloped viruses fuse directly with the plasma membrane. Thus, the internal components of the virion are immediately delivered to the cytoplasm of the cell.

(B) Entry via endosomes at the cell surface

Some enveloped viruses require an acid pH for fusion to occur and are unable to fuse directly with the plasma membrane. These viruses are taken up by invagination of the membrane into endosomes. As the endosomes become acidified, the latent fusion activity of the virus proteins becomes activated by the fall in pH and the virion membrane fuses with the endosome membrane. This results in delivery of the internal components of the virus to the cytoplasm of the cell.

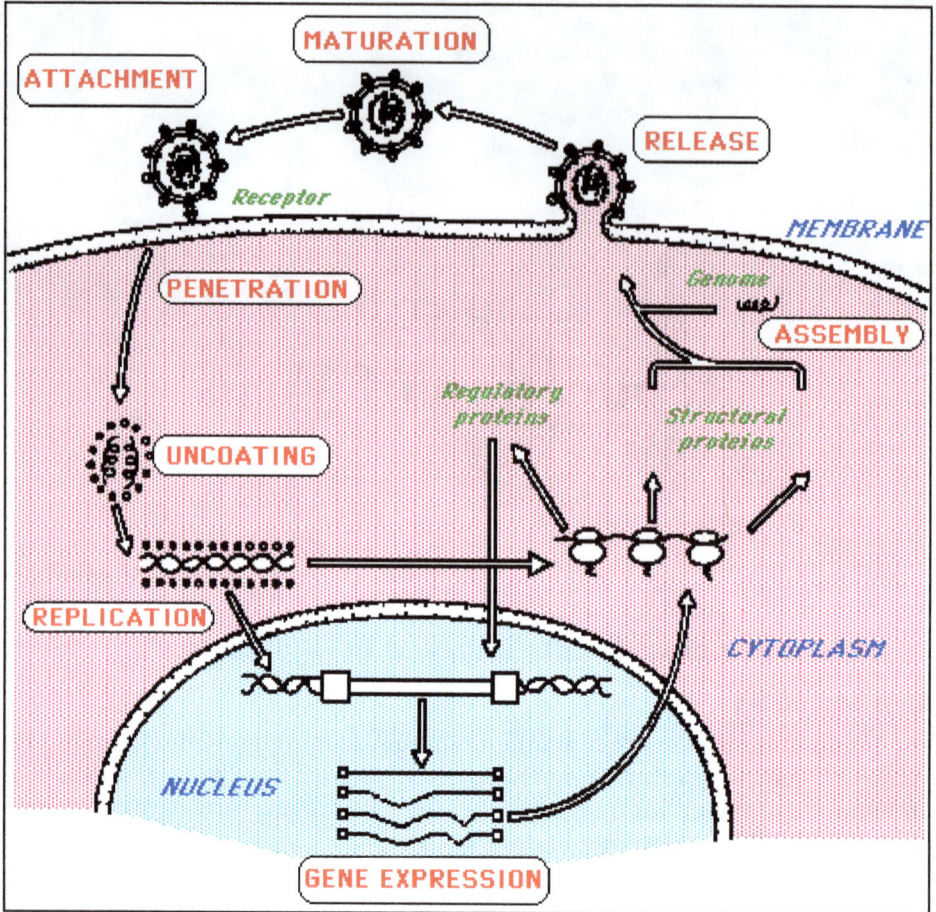

Figure 5. Replication of Virus

Non-enveloped Viruses

Non-enveloped viruses may cross the plasma membrane directly or may be taken up into endosomes. They then cross (or destroy) the endosomal membrane.

3. Uncoating

Nucleic acid has to be sufficiently uncoated that virus replication can begin at this stage. When the nucleic acid is uncoated, infectious virus particles cannot be recovered from the cell–this is the start of the *ECLIPSE phase*–which lasts until new infectious virions are made.

4. Synthesis of Viral Nucleic Acid and Protein

Many strategies are used

5. Assembly/Maturation

New virus particles are assembled. There may be a maturation step that follows the initial assembly process.

6. Release

Virus may be released due to cell lysis, or, if enveloped, may bud from the cell. Budding viruses do not necessarily kill the cell. Thus, some budding viruses may be able to set up persistent infections. Not all released viral particles are infectious. The ratio of non-infectious to infectious particles varies with the virus and the growth conditions.

Structural versus Non-Structural Proteins

All proteins in a mature virus particle are said to be structural proteins–even if they make no contribution to the morphology or rigidity of the virion–non-structural proteins are those viral proteins found in the cell but not packaged into the virion.

Effect of Viruses on Host Macromolecular Synthesis

Many viruses inhibit host RNA, DNA or protein synthesis (or any combination of these). The mechanisms by which the virus does this vary widely.

Cytopathic Effect (CPE)

The presence of the virus often gives rise to morphological changes in the host cell. Any detectable changes in the host cell due to infection are known as a cytopathic effect. Cytopathic effects (CPE) may consist of cell rounding, disorientation, swelling or shrinking, death, detachment from the surface, etc.

Many viruses induce apoptosis (programmed cell death) in infected cells. This can be an important part of the host cell defense against a virus–cell death before the completion of the viral replication cycle may limit the number of progeny and the spread of infection. (Some viruses delay or prevent apoptosis–thus giving themselves a chance to replicate more virions.)

Some viruses affect the regulation of expression of the host cell genes which this can have important results both for the virus's ability to grow, and in terms of the effect on the host cell.

The cytopathic effects produced by different viruses depend on the virus and the cells on which it is grown. This can be used in the clinical virology laboratory to aid in identification of a virus isolate.

Assays for Plaque-Forming Units

The CPE effect can be used to quantitate *infectious* virus particles by the plaque-forming unit assay. Cells are grown on a flat surface until they form a monolayer of cells covering a plastic bottle or dish. They are then infected with the virus. The liquid growth medium is replaced with a semi-solid one so that any virus particles produced as the result of an infection cannot move far from the site of their production. A *plaque* is produced when a virus particle infects a cell, replicates, and then kills that cell. Surrounding cells are infected by the newly replicated virus and they too are killed. This process may repeat several times. The cells are then stained with a dye which stains only living cells. The dead cells in the plaque do not stain and appear as unstained areas on a colored background. Each plaque is the result of infection of one cell by one virus followed by replication and spreading of that virus. However, viruses that do not kill cells may not produce plaques.

Assays for Viruses

Some methods (*e.g.* electron-microscopy) enable every virion to be counted but are not informative about infectivity. Other methods (*e.g.* hemagglutination) are a less sensitive measure of how much virus is present, but again are not informative about infectivity. Other methods, *e.g.* plaque assay, measure the number of infectious virus particles.

DNA Viruses

With animal DNA viruses, transcription and translation are not coupled. Except for poxviruses, transcription occurs in the nucleus and translation in the cytoplasm. Generally, the primary transcripts, generated by RNA polymerase II, are larger than the mRNAs found on ribosomes, and in some cases, as much as 30 per cent of the transcribed RNA remains untranslated in the nucleus. The viral messengers, however, like those of animal cells, are monocistronic. Transcription has a temporal organization, with most DNA viruses only a small fraction of the genome is transcribed into early messengers. The synthesis of early proteins is the key initial step in viral DNA replication. After DNA synthesis, the remainder of the genome is transcribed into late messengers. The complex viruses have immediate early genes, which are expressed in the presence of inhibitors of protein synthesis, and delayed early genes, which require protein synthesis for expression. Regulation is carried out by proteins present in the virions, or specified by viral or cellular genes, interacting with regulatory sequences at the 5' end of the genes. These sequences may respond in *trans* to products produced by other genes and act in *cis* on the associated genes. Different classes of genes may be transcribed from different DNA strands and therefore in opposite directions *e.g.* polyomaviruses. The transcripts may undergo post- transcriptional processing so that nonessential intervening sequences are removed.

DNA Replication

The mode of replication is semiconservative but the nature of the replicative intermediates depends on the manner of replication. Several methods of replication can be recognized.

A. Adenoviruses

Adenoviruses show asymmetric replication, which initiates at the 3' end of one of the strands using a protein primer. The growing strand displaces the preexisting strand of the same polarity and builds a complete duplex molecule. The displaced strand in turn replicates in a similar manner after generating a panhandle structure by pairing the inverted terminal repetitions.

B. Herpesviruses

Herpesviruses have linear genomes with terminal repeats. On reaching the nucleus, the terminal ends undergo limited exonucleotic digestion and then pair to form circles. Replication is thought to take place via a rolling circle mechanism, where concatemers are formed. During maturation, unit-length molecules are cut from the concatemers.

C. Papovaviruses

The DNA of papovaviruses are circular and the replication is bidirectional and symmetrical, via cyclic intermediates.

D. Parvoviruses

The replication of single stranded parvoviruses is initiated when +ve and -ve stranded DNA from different parvovirus particles come together to form a double stranded DNA molecule from which transcription and replication takes place.

E. Poxviruses

The striking feature of poxvirus DNA is that the two complementary strands are joined. The replicative intermediates, present in the cytoplasm, are special concatemers containing pairs of genomes connected either head to head or tail to tail.

F. Hepadnaviruses

Hepatitis B virus employs reverse transcription for replication. The genome consists of a partially double-stranded circular DNA with a complete negative strand and an incomplete positive strand. Upon entering the cell, the positive strand is completed and transcribed. RNA transcripts are in turn reverse-transcribed into DNA by a viral enzyme in several steps, following closely the model of retroviruses, including a jump of the nascent positive strand from one direct repeat (DR) to another. (Figure 6).

RNA Viruses

The replication of RNA viral genomes is dictated by the absence of multiple translation units within the same messenger, a characteristic of all animal cell messengers. To overcome this difficulty, 3 main strategies have developed.

1. The viral mRNA acts directly as the messenger and is translated monocistronically, followed by cleavage to form different proteins.
2. The virion RNA is transcribed to yield various monocistronic mRNAs by initiating transcription at various places.

The genome itself is a collection of separate RNA fragments that are transcribed into monocistronic mRNAs. RNA viruses can be placed into 7 classes, according to the nature of the viral RNA and its relation to the messenger.

Class I (*e.g.* Picornaviruses, Flaviviruses)

The genome, having +ve polarity, itself act as the messenger, specifying information for the synthesis of both structural and nonstructural proteins. The same RNA molecule also initiate replication that requires the expression of proteins first. This format allows little control over replication *e.g.* Poliovirus has no independent mechanism of controlling the numbers of structural proteins made.

Class II (*e.g.* Coronaviruses, Togaviruses)

Many +stranded RNA viruses have subgenomic RNA as part of their cycle. This

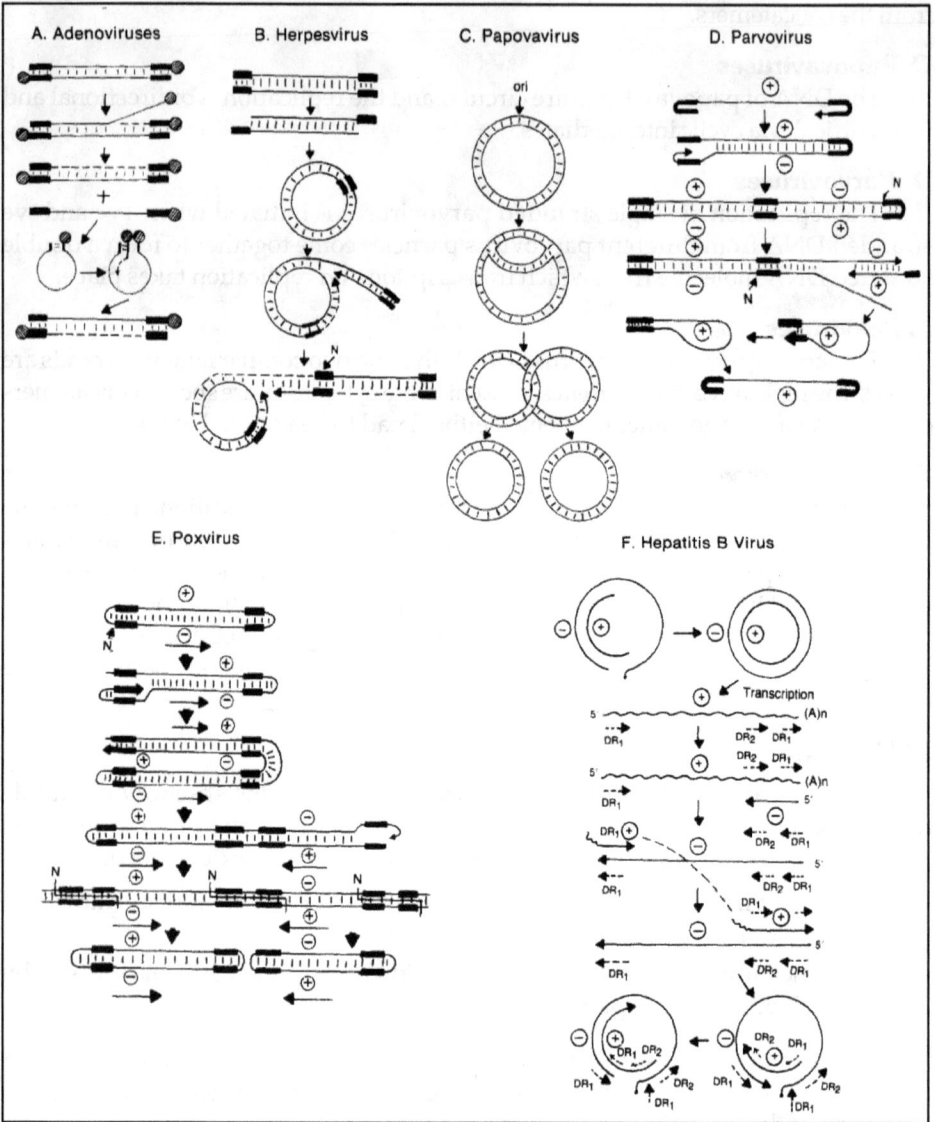

Figure 6. DNA Replication

	Subgenomic	PTC	Nested	Splicing
Picornaviridae	N	Y	N	N
Togaviridae	Y	Y	N	N
Coronaviridae	Y	N	Y	N

would allow a certain amount of control. The subgenomic mRNA cannot be recognized by the RNA polymerase. It can be used solely for the synthesis of structural proteins etc. A second way to get round the problem is to make a nested set of RNAs. The nested set of RNA is the most efficient form of control. They can control which part of their genome to express.

With togaviruses, the 49S genome RNA is first translated into polyprotein that is processed into the nonstructural proteins. The subgenomic 26S mRNA, which is transcribed from the full length -ve RNA, is translated into a smaller polyprotein that is processed into viral structural proteins. With coronaviruses, a nested set of mRNAs is generated in the following manner: the -ve transcript is first generated form the genome, which is then transcribed into monocistronic mRNAs of different sizes. Each begins with an identical short 5' leader sequence that is joined to the transcripts at the start of the various genes and continues to the 3' end of the genome. These mRNAs are not produced by splicing a genomic-size transcript because the virus is able to replicate in enucleated cells.

Class III (*e.g.* Paramyxoviruses, Thabdoviruses)
The genome is of -ve polarity to the messenger. A virion RNA-dependent RNA transcriptase first transcribes the genomes into separate monocistronic messengers initiating at a single promoter. The transcriptase stops and restarts at each juncture between different genes.

Class IV (*e.g.* Orthomyxoviruses, Most Bunyaviruses)
The –ve genome is in several distinct nonoverlapping pieces of ssRNA. The virion transcriptase generate a messenger from each piece. With orthomyxoviruses, most genomic segments contain a single gene but 2 fragments contain 2 overlapping genes: one is expressed by a full-length messenger, the other by a shorter messenger obtained from the former by splicing. The replication of orthomyxoviruses is unusual amongst RNA viruses in that it takes place within the nucleus. The nuclear function it requires is the 5' cap of cellular messengers, which it "pinches" after endonucleotic cleavage of the host messengers. The 5' cap is then used as primers in the synthesis of viral messengers.

Class V (*e.g.* Arenaviruses, Phleboviruses)
Arenaviruses have an ambisense genome in that half the genome is of -ve polarity and is transcribed into a messenger by a virion transcriptase, but the other half, which is of +ve polarity is transcribed twice: first a complete transcript of the genome is made, then the mRNA is transcribed form this transcript. This strategy is seen in the S (small) segment of the genome of phleboviruses. Ambisense genomes are unusual for RNA viruses but not for dsDNA viruses.

Class VI (*e.g.* Reoviruses)
Reoviruses contain distinct nonoverlapping segments of dsRNA, each is transcribed into an independent mRNA by the virion transcriptase. Most messengers are monocistronic, but one is bicistronic and expresses a second protein by initiating at an internal AUG in a different reading frame. Each segment of reovirus RNA is replicated independently. A nascent mRNA strand is first generated by the virion

transcriptase, which then serves as the template for the replicase to make the negative strand. The two strands remain associated in a dsRNA molecule that ends up in a virion. This replication is asymmetric and conservative because (1) the -ve strand of the virion RNA servers as the initial template and (2) the parental RNA does not end up in the progeny.

Class VII (*e.g.* Retroviruses)

Retroviruses are unique in that their genomes are transcribed into DNA and not RNA. They contain two identical ssRNA of +ve polarity, with a poly A tail at the 3' end and a cap at the 5' end. Each is transcribed into DNA by reverse transcriptase that then integrates into the cellular DNA as provirus. Transcription of the provirus by the cellular transcriptase yields the viral molecules that end up in virions.

Since RNA viruses of classes III to VII require a virion transcriptase for synthesizing a messenger, their purified viral RNAs are not infectious. Only those of classes I and II are infectious. With RNA viruses, there is no differentiation between early and late messengers.

Replication of Single-Stranded RNA Viruses (Classes I to V)

In all cases, replication consists of building a template strand complementary to the viral strand of the same length, which then servers as the template for progeny viral strands. These steps are carried out by a collection of enzymes of both viral and cellular origin, in association with the nucleocapsids of the infecting virions. In many instances, replication and transcription interfere with each other: with -ve stranded viruses, both template and transcripts are made from viral strands: with +ve viruses, a viral strand can be used as a messenger or replication template. Initially in the infection, there is no interference as the messenger function is needed to provide proteins needed for replication. Later the supply of these proteins regulate the rate of replication. *e.g.* with poliovirus, replication is initiated when the pVg protein becomes covalently linked to at the 5' ends of the RNA, apparently initiating the formation of a replication complex. Messenger and progeny often differed structurally. *e.g.*, the messengers of influenza virus have capped leader sequences derived from cellular messengers. In addition, they lack 17 to 22 nucleotides at the 3' end. Moreover, replication requires ongoing protein synthesis to provide the required proteins, whereas transcription does not (Figure 7).

In RNA replication, the newly made template strand remains associated with the viral strand on which it is made, forming a double-stranded structure the length of the viral genome, known as the replicative form (RF). Synthesis of new strands occurs by conservative asymmetric synthesis, similar to adenoviruses. An RF with a nascent viral strand is known as RI (replicative intermediate) RF molecules are fairly abundant during replication because after the completion of a new strand, the replicase appear to remain associated for some time with the template before reinitiating synthesis. RFs accumulate at the end of replication, when no more RIs are formed. With the exception of orthomyxoviruses, the viral RNAs replicate in the cytoplasm. The replicase present in infected cells synthesize new viral RNA strands of both polarities. Transcription occurs at the same site as replication. It is unclear whether

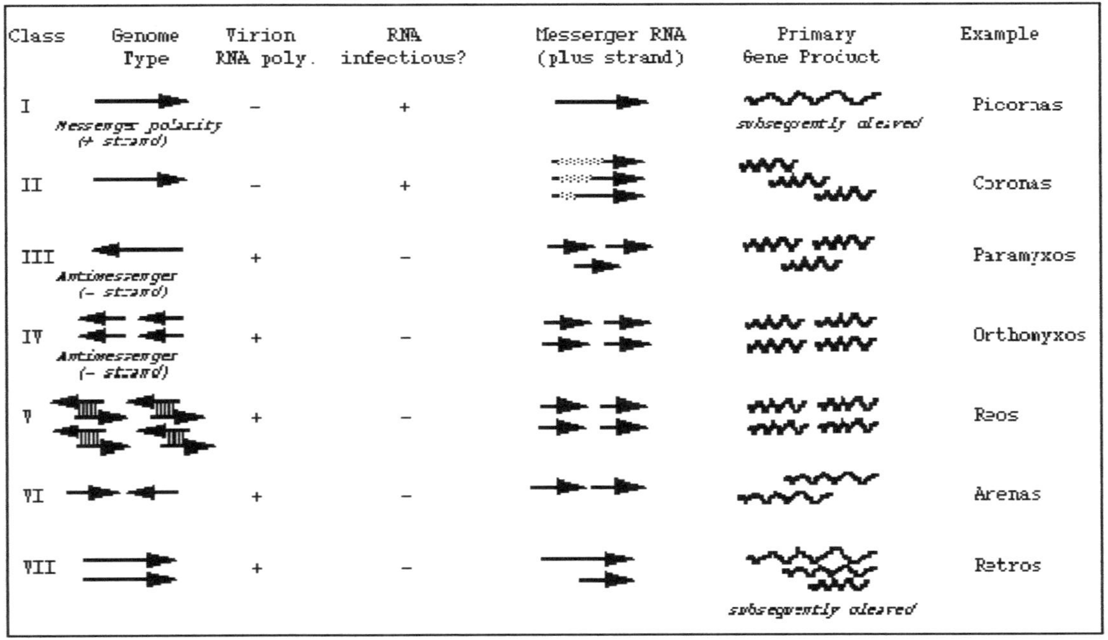

Class	Genome Type	Virion RNA poly.	RNA infectious?	Messenger RNA (plus strand)	Primary Gene Product	Example
I	*Messenger polarity (+ strand)*	−	+		*subsequently altered*	Picornas
II		−	+			Coronas
III	*Antimessenger (− strand)*	+	−			Paramyxos
IV	*Antimessenger (− strand)*	+	−			Orthomyxos
V		+	−			Reos
VI		+	−			Arenas
VII		+	−		*subsequently altered*	Retros

Figure 7. RNA Virus Replication

replication and transcription are carried out by different enzymes or by the same enzyme.

Maturation and Release

Maturation proceeds differently for naked, enveloped, and complex viruses.

Naked Icosahedral Viruses

Preassembled capsomers are joined to form empty capsids (procapsid) which are the precursors of virions. The assembly of capsomers to form the procapsid is often accompanied by extensive reorganization, which is revealed by changes in serological specificity and isoelectric point. *e.g.* picornaviruses and adenoviruses. Naked icosahedral viruses are released from infected cells in different ways. Poliovirus is rapidly released, with death and lysis of infected cells. in contrast, the virions of DNA viruses that tend to mature in the nucleus tend to accumulate within infected cells over a long period and are released when the cell undergoes autolysis, and in some cases, may be extruded without lysis.

Enveloped Viruses

Viral proteins are first associated with the nucleic acid to form the nucleocapsid, which is then surrounded by an envelope. In nucleocapsid formation, the proteins are all synthesized on cytoplasmic polysomes and are rapidly assembled into capsid components. In envelope assembly, virus-specified envelope proteins go directly to the appropriate cell membrane (the plasma membrane, the ER, the Golgi apparatus), displacing host proteins. In contrast, the carbohydrates and the lipids are produced by the host cell. The viral envelope has the lipid constitution of the membrane where its assembly takes place. (*e.g.* the plasma membrane for orthomyxoviruses and paramyxoviruses, the nuclear membrane for herpesviruses). A given virus will differ in its lipids and carbohydrates when grown in different cells, with consequent differences in physical, biological, and antigenic properties.

The envelope glycoproteins are synthesized in the following manner: the polypeptide backbone is first formed on polysomes bound to the ER, which then moves via transport vesicles to the Golgi apparatus where it attains it full glycosylation and fatty acid acylation. The matrix proteins that are present in viral envelope are usually not glycosylated and stick to the cytoplasmic side of the plasma membrane through hydrophobic domains. Matrix proteins connect the cytoplasmic domains of the envelope glycoproteins with the cell's cytoskeleton, and they gather the viral glycoproteins to form the virions. The selection of viral glycoproteins is efficient but not exclusive. *e.g.* rhabdovirus virions contain 10 to 15 per cent of nonviral glycoproteins. They may also contain glycoproteins specified by another virus infecting the same cell. Envelopes are formed around the nucleocapsids by budding of cellular membranes. With orthomyxoviruses and paramyxoviruses, the viral glycoproteins incorporated in the membranes confer on the cell some properties of a giant virion. Thus cells infected by these viruses may bind RBCs (haemadsoption), and paramyxovirus-infected cells may fuse with uninfected cells to form multinucleated syncytia by the fusion of their membranes. This fusion is equivalent to the fusion of the virion's envelope with the plasma membrane of the host cell at the onset of infection.

Complex Viruses

Maturation of the highly organized poxviruses takes place in cytoplasmic foci called "factories" In contrast to simpler viruses, the poxvirus membrane contain newly synthesized lipids that differ in composition to the cellular lipids. The maturation of poxviruses after the precursors have been enclosed within the primitive membranes suggest that poxviruses may be transitional forms towards a cellular organization.

Defective Interfering Particles

Interference may occur during replication by the generation of defective interfering (DI) particles. They are formed during infection with various kinds of RNA viruses, such as rhabdoviruses, togaviruses, orthomyxoviruses, paramyxoviruses, coronaviruses and some DNA viruses (herpesviruses). With some viruses *e.g.* VSV, the DI particles are smaller than regular particles and can therefore be obtained in pure form. They usually contain the normal virion proteins but have a shorter genome. They are replication defective and require the helper functions of a normal virus co-infecting the same cell. In early serial passages, DI particles rapidly increase in titre, then the yield of the infectious virus, and finally the total particle yield is progressively reduced.

The genomes of DI particles are internally deleted but retain both ends, which are essential for the replication of RNA viruses. With DNA viruses, the origin of replication is always conserved and often repeated. Those features show that to cause interference, the DI genomes must replicate. They deprive the regular virus of its replicase by binding to it more effectively. They do not make a replicase of their own because they are always defective in their replicase gene. The formation of DI genomes of RNA viruses is the consequence of high variability of these genomes. The DI genomes are formed by a copy choice mechanism when the replicase, having replicated part of the template, skips to another part of the same or another template. With VSV and other negatively stranded RNA viruses, 4 types of defective genomes are seen;

1. *Deletions*–The polymerase jumps to a site beyond on the same template, skipping a fragment.
2. *Snapbacks*–This occurs when the replicase, having transcribed part of the + strand, switches to the just-made–strand as template. The resultant RNA contain half+ and half- and can produce a hairpin on annealing.
3. *Panhandle*–This is formed by a similar mechanism, when the polymerase carrying a partially made–strand switches back to transcribing the extreme 5' of it, so that on annealing, the strand forms a panhandle.
4. *Compounds*–These genomes are made by a combination of deletions and snapbacks.

The competition of DI genomes with competent genomes depends not only on the structure of the DI genome but on that of the normal competent genome. Different DI genomes may interfere to very different degrees with the same competent genome and competent genomes may acquire mutations that make them resistant to the existing DI genomes. Subsequently, this is overcome by the new types of DI genomes. During viral multiplication, many types of DI genomes are continuously made and they are very heterogeneous.

Viroids

Viroids are responsible for causing serious diseases in many plants. They consist of naked RNA which does not code for any protein, nor is protein associated with it. Essentially, each viroid particle is a circular ssRNA molecule containing 250 to 400 nucleotides. They are highly resistant to enzymatic degradation because they have no free ends and because they have a very tight secondary structure (owing to self-complementary sequences). All viroid strains have similar characteristics. Their genome can be considered a dsRNA, with many unpaired short "bubbles" regions. There is no AUG initiation codon for protein synthesis, or of their complements (in case the RNA is of negative-stranded type). There is no evidence that the RNA is translated. They are replicated in the nucleus of infected cells by host enzymes through double stranded intermediates. Replication is blocked by alpha-amantine, which inhibits RNA polymerase II (the RNA polymerase responsible for generating the transcripts for mRNAs).

The base sequences of viroids have repeats, both direct and inverted, which suggest a relatedness to transposing elements. Moreover, they possess a sequence similar to that used by retroviruses. However, viroids are not transcribed into DNA, and no sequences homologous to viroids are found in the DNA of infected cells. cDNA of the viroid is also infectious and can be transcribed into regular infectious viroid particles. A striking feature of viroid RNA is the presence of sequences highly homologous to some of the small nuclear RNAs U_1 and U_3, which are involved in the splicing of introns in animal cells. This suggests that viroids may have originated from introns and their pathogenecity might be due to interference with the normal splicing of introns in cells. Virusoids are satellites of certain plant viruses that are encapsidated with their helper RNAs in the virions. A candidate for a viroid-like agent in humans is the delta agent which is much larger (1678bp) and is surrounded by a coat.

Chapter 6
Viral Pathogenesis

The consequences of a viral infection depend on a number of viral and host factors that affect pathogenesis. Viral infection was long thought to produce only acute clinical disease but other host responses are being increasingly recognized. These include asymptomatic infections, induction of various cancers, chronic progressive neurological disorders and possible endocrine diseases.

Cellular and Viral Factors in Pathogenesis

Viral virulence, like bacterial virulence, is under polygenic control. It is frequently associated with several characteristics that promote viral multiplication and cell change. The susceptibility of a particular cell to viral infection depends mainly on the presence of cellular receptors. Hence cells resistant to a virus may be susceptible to its extracted nucleic acid. Cultivation may markedly alter the viral susceptibility of cells from that in the original organ. For instance, polioviruses, which multiply in the nervous tissue but not in the kidney of a living monkey, multiply well in culture cells derived from the kidneys, since receptors develop in the cultivated kidneys cells. Marked changes in susceptibility accompany the maturation of animals. Many viruses are much more virulent in newborn animals than in adults *e.g.* coxsackieviruses, HSV or vice-verse *e.g.* polioviruses, hepatitis A. Genetic factors are also thought too play an important role in determining the susceptibility of an animal to a virus.

Cellular Response to Viral Infection

Cells can respond to virus infection in 3 different ways

1. No apparent change
2. CPE and death
3. Loss of growth control (transformation)

Patterns of Disease

In a host, viruses cause 3 basic patterns of infection: localized, disseminated, and inapparent.

Localized

Viral replication remains localized near the site of entry *e.g.* skin, respiratory or the GI tract).

Systemic (Disseminated)

Systemic infections usually take place through several steps:

1. Entry of virus
2. Spread to regional lymph nodes
3. Primary viraemia spread to other susceptible organs such as the liver and spleen
4. More intense secondary viraemia $\rightarrow$ dissemination to other organs, such as the skin.

Immunological and Other Systemic Factors

1. Circulating Antibodies

Abs in the serum and extracellular fluid provide the main protection against primary viral infections *i.e.* at the site of entry into the host. For infections where viraemia is an essential feature of the disease *e.g.* measles, polio, mumps and smallpox, the degree of protection is directly related to the level of neutralizing antibodies in the blood when the virus enters it. In experimental HSV infections, the B-cell response limits the spread of the virus to the CNS and reduces the establishment of latency in the peripheral ganglia. Protection of the respiratory and the GI tract is associated with IgA antibodies that are secreted into the extracellular fluid. Hence, by inducing the secretion of IgA antibodies, natural infections produce specific local as well as systemic immunity. The protective role of Abs is also evident in the prophylactic effectiveness of passive immunization. Administration of immune serum before or early in the incubation period can prevent or modify the disease.

Although humoral antibodies generally develop during recovery from a viral disease, they appear to play a less prominent role in this process than in protection. Most patients with agammaglobulinaemia make a normal recovery from most viral diseases, although some affected children may develop persistent or fatal echovirus infections. Furthermore, even patients with selective IgA deficiency do not develop more prolonged or more severe respiratory or enteric disease. The limited effect of humoral Abs on recovery is not surprising, since they are ineffective against intracellular viruses. Furthermore, many viruses can spread directly to contiguous, uninfected cells and thus remaining inaccessible to Abs. However, Abs do play a role in restricting the dissemination of some viruses *e.g.* polioviruses and togaviruses, the pathogenesis of which depends on the viraemic stage.

Long-lasting immunity, with persistence of circulating Abs, follows infection with a number of viruses, especially those causing viraemia. Second attacks are extremely rare in measles, smallpox, yellow fever or poliomyelitis. In contrast, second infections are common with most acute localized infections without viraemia, particularly respiratory diseases. It is thought that adequate levels of neutralizing antibodies do not persist in respiratory secretions, even though sufficient circulating antibodies are present. Adenoviruses are notable exceptions, perhaps because they frequently terminate in a latent infection of the lymphoid tissue in the respiratory and the GI tract.

2. Cell-Mediated Immunity

Patients who lack immunoglobulins but develop CMI ordinarily recover from viral diseases without difficulty. Whereas patients with defective CMI but normal humoral immunity recover poorly from certain viral infections, such as vaccination with vaccinia. CMI also appears to play a critical role in maintaining latent virus infections. Such infections are frequently reactivated in patients undergoing organ transplants whose CMI is suppressed by therapy. These latent infections include CMV, VZV, HSV, EBV, adenoviruses, measles, human papillomaviruses, JC and BK viruses. Herpesvirus infections are also often activated in patients with extensive burns (HSV and CMV) and in the aged (VZV) owing to diminished CMI. Activation of these viruses is also a common event in AIDS.

3. Diseases Based on Virus-Induced Immunological Response

The immune response itself often contribute to the production of disease *e.g.*

1. *Dengue haemorrhagic fever*–this is associated of infection in an individual who had prior infection with a different serotype
2. *RSV*–children immunized with inactivated RSV develop unusually severe disease if subsequently infected with the same virus.

These examples of enhanced viral injury could be due to one or a mixture of the following mechanisms:

1. Increased secondary response to Tc cells
2. Specific ADCC or complement mediated cell lysis
3. Binding of un-neutralized virus-Ab complexes to cell surface Fc receptors, and thus increasing the number of cells infected
4. Immune complex deposition in organs such as the brain or kidney.

The central role of the immune response in the development of some viral diseases is demonstrated in mice infected with LCM virus. In adult mice, severe and often fatal disease follows about a week after intracerebral inoculation, but if the immune response is suppressed (by neonatal thymectomy, chemicals, irradiation, or anti-lymphocytic serum), disease fails to develop although viral multiplication and spread are unrestrained. Moreover, after infection in utero or at birth, specific CMI is not detectable and the mice appear normal for 9 to 12 months in spite of widespread viral multiplication that produces persistent viraemia and viruria, with viral Ags demonstrable in most organs.

In adult mice, viral multiplication and spread are usually restricted in both neural and extraneural tissues, but the CMI response is quick to develop and a critical number of virally infected cells are killed, resulting in lethal disease. In the foetus, neonate, or immunosuppressed adult mouse, infection proceeds unimpeded to eventual involvement of all tissues. Antibodies produced from immune complexes with the virus, and are deposited in the kidney resulting in glomerulonephritis. In addition, necrotic lesions appear in the liver, brain, spleen and other organs, apparently resulting from the reaction between virus-sensitized killer T-lymphocytes and viral Ags present on the surface of many cells. Therefore, in the absence of effective CMI, LCM virus multiplies harmlessly for a long time in mice, producing an inapparent infection.

Non-Specific Systemic Factors

Nonspecific factors that influence resistance to viral infection include various hormones, temperature, NK cells and phagocytes. Hormones have a potential effect on viral infections *e.g.* pregnancy increases the severity of several viral diseases. cortisone enhances the susceptibility of many animals to viral infection and commonly potentiate the severity of the disease. In humans, it aggravates the clinical course of herpetic corneal ulcers and increases the likelihood of VZV pneumonia.

Latent Virus Infections

In latent infections, overt disease is not produced, but the virus is not eradicated. This equilibrium between host and parasite is achieved in various ways by different parasites and hosts. The virus may exist in a truly latent noninfectious occult form, possibly as an integrated genome or an episomal agent, or as an infectious and continuously replicating agent, termed a persistent viral infection.

Infectious agents causing chronic persistent infections have found a way of escaping a cell-mediated immune response. The mechanisms include:

1. Generation of cells that escape a cell-mediated immune response.
2. Down regulation of MHC production in infected cells so that they are not recognized and destroyed by T cells.
3. Infection of cells in immunoprivileged sites such as the brain.

Examples of latent infection include:

1. Chronic Congenital Rubella, CMV, EBV, hepatitis B, HIV
2. Latent HSV, VZV, adenovirus and some retroviral infections
3. SSPE, PML, Kuru, CJD, progressive rubella panencephalitis

1. Chronic Persistent Infections

Enveloped viruses such as paramyxoviruses, some herpesviruses *e.g.* EBV, retroviruses and arenaviruses appear particularly suited to initiate persistent infections. Infection appear to persist because the virus does not disrupt the essential housekeeping functions of the cells. (DNA, RNA and protein synthesis). Some persistently infected cells, such as in measles (SSPE) may be assisted by the capacity of humoral Abs to cap viral Ags on the cell surface. This promotes the shredding of viral Ags from the cell surface, leaving the cell surface free of viral glycoproteins and thus the infected cell is protected from CTLs and K cells.

2. Latent Occult Viral Infections

Some DNA and RNA viruses, may become undetectable following a primary infection only to reappear and produce acute disease. This latency can be accomplished in different ways.

(a) HSV

Primary infection usually occurs between 6 to 18 months of age following which the virus persists and cannot be found except during recurrent acute episodes. The form in which the latent occult virus persists is uncertain. Virus cannot be isolated from tissue homogenates, but by cocultivating cells of sensory ganglia with susceptible cells. Virus has been detected in the trigeminal, thoracic, lumbar and sacral dorsal root ganglia. Hybridization studies have detected the viral genome in normal brains as well as peripheral ganglia. These data suggests that the DNA exist in a linear, unintegrated form, perhaps as episomes.

It may be that, as in virus-carrier cultures, infection is confined to only a small proportion (0.01-0.1 per cent) of the ganglion cells because of Abs, CMI, viral interference or metabolic factors. Because there is humoral Abs present, most of the extracellular virus is neutralized and goes undetected. Acute episodes, in which there is a burst of viral replication, probably depends on a transient change in the local level of immunity or changes in the susceptibility of the uninfected cells induced by a variety of physical and physiological factors such as fever, intense sunlight, fatigue or menstruation. The other herpesviruses that infect humans also commonly produce latent infections: VZV in the sensory ganglia, CMV in lymphocytes and macrophages, and EBV in B-lymphocytes.

(b) Adenovirus

Adenovirus infections in humans are usually self-limiting but the virus frequently establishes a latent, persistent infection of the tonsils and the adenoids. Though these tissues fail to yield infectious virus when homogenized and tested in sensitive cell cultures, cultured fragments of about 85 per cent of these "normal" tonsils and adenoids, after a variable time, show characteristic adenovirus-induced CPE and yield infectious virus. Failure to recover infectious virus initially may be due to the paucity of virions, to their association with either Ab or receptor material, or to the absence of mature virions. The latent infection is probably not the result of lysogeny, since DNA in peripheral lymphocytes appears to be in a linear episomal form.

(c) SSPE

Latency occurs as a result of incomplete viral production. Immature viral measles virus nucleocapsids are produced.

Latent viral infections affect the incidence and pathogenesis of acute viral disease in several ways. A reactivated virus may spread and initiate an epidemic among susceptible contacts *e.g.* VZV. Viral latency can also be seen in the development of several chronic diseases dependent on the immunological response *e.g.* SSPE and PML. Some latent states induce tumourigenesis.

Chapter 7
Viral Genetics

Viruses grow rapidly, there are usually a large number of progeny virions per cell. There is, therefore, more chance of mutations occurring over a short time period. The nature of the viral genome (RNA or DNA; segmented or non-segmented) plays an important role in the genetics of the virus. Viruses may change genetically due to mutation or recombination

Mutants

(a) Origin

Spontaneous Mutations

These arise naturally during viral replication: *e.g.* due to errors by the genome-replicating polymerase or a a result of the incorporation of tautomeric forms of the bases. DNA viruses tend to more genetically stable than RNA viruses. There are error correction mechanisms in the host cell for DNA repair, but probably not for RNA.

Some RNA viruses are remarkably invariant in nature. Probably these viruses have the same high mutation rate as other RNA viruses, but are so precisely adapted for transmission and replication that fairly minor changes result in failure to compete successfully with parental (wild-type, wt) virus.

Mutations that are Induced by Physical or Chemical Means

Chemical: Agents acting directly on bases, *e.g.* nitrous acid Agents acting indirectly, *e.g.* base analogs which mispair more frequently than normal bases thus generating mutations

Physical: Agents such as UV light or X-rays

(b) Types of Mutation

Mutants can be point mutants (one base replaced by another) or insertion/deletion mutants.

(c) Examples of the Kinds of Phenotypic Changes Seen in Virus Mutants

(phenotype = the observed properties of an organism)

Conditional Lethal Mutants

These mutants multiply under some conditions but not others (whereas the wild-type virus grows under both sets of conditions)

e.g. Temperature Sensitive (ts) Mutants

These will grow at low temperature *e.g.* 31 degrees C but not at *e.g.* 39 degrees C, *wild type* grows at 31 and 39 degrees C. It appears that the reason for this is often that the altered protein cannot maintain a functional conformation at the elevated temperature.

e.g. Host Range

These mutants will only grow in a subset of the cell types in which the *wild type* virus will grow—such mutants provide a means to investigate the role of the host cell in viral infection

Plaque Size

Plaques may be larger or smaller than in the *wild type* virus, sometimes such mutants show altered pathogenicity

Drug Resistance

This is important in the development of antiviral agents—the possibility of drug resistant mutants arising must always be considered.

Enzyme-Deficient Mutants

Some viral enzymes are not always essential and so we can isolate viable enzyme-deficient mutants; *e.g.* herpes simplex virus thymidine kinase is usually not required in tissue culture but it is important in infection of neuronal cells.

"Hot" Mutants

These grow better at elevated temperatures than the *wild type* virus. They may be more virulent since host fever may have little effect on the mutants but may slow down the replication of *wild type* virions.

Attenuated Mutants

Many viral mutants cause much milder symptoms (or no symptoms) compared to the parental virus—these are said to be attenuated. These have a potential role in vaccine development and they also are useful tools in determining why the parental virus is harmful.

Exchange of Genetic Material

Recombination

Exchange of genetic information between two genomes.

"Classic" Recombination

This involves breaking of covalent bonds within the nucleic acid, exchange of genetic information, and reforming of covalent bonds.

This kind of break/join recombination is common in DNA viruses or those RNA viruses which have a DNA phase (retroviruses). The host cell has recombination systems for DNA.

Recombination of this type is very rare in RNA viruses (there are probably no host enzymes for RNA recombination). Picornaviruses show a form of very low efficiency recombination. The mechanism is not identical to the standard DNA mechanism, and is probably a "copy choice" kind of mechanism.

Recombination is also common in the coronaviruses–again the mechanism is different from the situation with DNA and probably is a consequence of the unusual way in which RNA is synthesized in this virus.

So far, there is no evidence for recombination in the negative stranded RNA viruses giving rise to viable viruses (In these viruses, the genomic RNA is packaged in nucleocapsids and is not readily available for base pairing).

Various Uses for Recombination Techniques

(a) Mapping genomes (the further apart two genes are, the more likely it is that there will be a recombination event between them).

(b) *Marker rescue*–DNA fragments from *wild type* virus can recombine with mutant virus to generate *wild type* virus–this provides a means to assign a gene function to a particular region of the genome. This also provides a means to insert foreign material into a gene.

Recombination enables a virus to pick up genetic information from viruses of the same type and occasionally from unrelated viruses or even the host genome (as occurs in some retroviruses–see retroviruses).

Reassortment

If a virus has a segmented genome and if two variants of that virus infect a single cell, progeny virions can result with some segments from one parent, some from the other.

This is an efficient process–but is limited to viruses with segmented genomes–so far the only human viruses characterized with segmented genomes are RNA viruses *e.g.* orthomyxoviruses, reoviruses, arenaviruses, bunya viruses.

Reassortment may play an important role in nature in generating novel reassortants and has also been useful in laboratory experiments. It has also been exploited in assigning functions to different segments of the genome. For example, in a reassorted virus if one segment comes from virus A and the rest from virus B, we can see which properties resemble virus A and which virus B.

Reassortment is a non-classical kind of recombination.

Applied Genetics

There is vaccine called Flumist (LAIV, approved June 2003) for influenza virus which involves some of the principles discussed above. The vaccine is trivalent–it contains 3 strains of influenza virus:

The viruses are cold adapted strains which can grow well at 25 degrees C and so grow in the upper respiratory tract where it is cooler. The viruses are temperature-sensitive and grow poorly in the warmer lower respiratory tract. The viruses are attenuated strains and much less pathogenic than wild-type virus. This is due to multiple changes in the various genome segments.

Antibodies to the influenza virus surface proteins (HA–hemagglutinin and NA–neuraminidase) are important in protection against infection. The HA and NA change from year to year. The vaccine technology uses reassortment to generate the viruses which have six gene segments from the attenuated, cold-adapted virus and the HA and NA coding segments from the virus which is likely to be a problem in the up-coming influenza season.

This vaccine is a live vaccine and is given intranasally as a spray and can induce mucosal and systemic immunity.

A live, attenuated reassortant vaccine has recently (2006) been approved for rotaviruses (RotaTeq from Merke). Another attenuated vaccine, Rotarix (Glaxo), is in development.

Complementation

Interaction at a functional level NOT at the nucleic acid level. For example, if we take two mutants with a *ts (temperature-sensitive)* lesion in different genes, neither can grow at a high (non-permissive) temperature. If we infect the same cell with both mutants, each mutant can provide the missing function of the other and therefore they can replicate (nevertheless, the progeny virions will still contain *ts* mutant genomes and be temperature-sensitive).

We can use complementation to group *ts* mutants, since *ts* mutants in the same gene will usually not be able to complement each other. This is a basic tool in genetics to determine if mutations are in the same or a different gene and to determine the minimum number genes affecting a function.

Multiplicity Reactivation

If double stranded DNA viruses are inactivated using ultraviolet irradiation, we often see reactivation if we infect cells with the inactivated virus at a very high multiplicity of infection (*i.e.* a lot of virus particles per cell)–this is because inactivated viruses cooperate in some way. Probably complementation allows viruses to grow initially, as genes inactivated in one virion may still be active in one of the others. As the number of genomes present increases due to replication, recombination can occur, resulting in new genotypes, and sometimes regenerating the *wild type* virus.

Defective Viruses

Defective viruses lack the full complement of genes necessary for a complete infectious cycle (many are deletion mutants)–and so they need another virus to provide the missing functions–this second virus is called a helper virus.

Defective viruses must provide the necessary signals for a polymerase to replicate their genome and for their genome to be packaged but need provide no more. Some defective viruses do more for themselves.

Some Examples of Defective Viruses

Some retroviruses have picked up host cell sequences but have lost some viral functions. These need a closely related virus which retains these functions as a helper.

Some defective viruses can use unrelated viruses as a helper: For example, hepatitis delta virus (an RNA virus) does not code for its own envelope proteins but uses the envelope of hepatitis B virus (a DNA virus).

Defective Interfering Particles

The replication of the helper virus may be less effective than if the defective virus (particle) was not there. This is because the defective particle is competing with the helper for the functions that the helper provides. This phenomenon is known as interference, and defective particles which cause this phenomenon are known as "defective interfering" (DI) particles. Not all defective viruses interfere, but many do. Note that it is possible that defective interfering particles could modulate natural infections.

Phenotypic Mixing

If two different viruses infect a cell, progeny viruses may contain coat components derived from both parents and so they will have coat properties of both parents. This is called phenotypic mixing. IT INVOLVES NO ALTERATION IN GENETIC MATERIAL, the progeny of such virions will be determined by which parental genome is packaged and not by the nature of the envelope.

Phenotypic mixing may occur between related viruses, *e.g.* different members of the Picornavirus family, or between genetically unrelated viruses, *e.g.* Rhabdo- and Paramyxo- viruses. In the latter case the two viruses involved are usually enveloped since it seems there are fewer restraints on packaging nucleocapsids in other viruses' envelopes than on packaging nucleic acids in other viruses' icosahedral capsids.

Chapter 8
Viral Immunology

Viruses are strongly immunogenic and induces 2 types of immune responses; humoral and cellular. The repertoire of specificities of T and B cells are formed by rearrangements and somatic mutations. T and B cells do not generally recognize the same epitopes present on the same virus. B cells see the free unaltered proteins in their native 3-D conformation whereas T cells usually see the Ag in a denatured form in conjunction with MHC molecules. The characteristics of the immune reaction to the same virus may differ in different individuals depending on their genetic constitutions.

Humoral response is responsible for blocking the infectivity of the virus (neutralization). Those of the IgM and IgG class are especially relevant for defense against viral infections accompanied by viraemia, whereas those of the IgA class are important in infections acquired through a mucosa. (the nose, the intestine) In contrast, the cellular response kills the virus-infected cells expressing viral proteins on their surfaces, such as the glycoproteins of enveloped viruses and sometimes core proteins of these viruses.

Humoral Response

Abs are elicited by the surface components of intact virions as well by the internal components of disrupted virions. Also they are elicited by viral products built into the surface of infected cells or released by the cells. Antibodies provide the key to protection against many viral infections. Sometimes, they are also pathogenic *e.g.* immune complexes are thought to be responsible for causing the rash in rubella. Interactions of virions with Abs to different components of their coats have different consequences.

Neutralization

Virus neutralization consists of a decrease in the infectious titre of a viral preparation following its exposure to Abs. The loss of infectivity is bought about by interference by the bound Ab with any one of the steps leading to the release of the viral genome into the host cells. The consequences of the virion-Ab interaction therefore depends on many factors:

1. The structure of the virions
2. The target of the Ab *e.g.* Abs against the HA but not the NA of influenza virus are neutralizing.
3. Mutations affecting surface molecules that may alter the susceptibility to certain Abs
4. The type of Ab, especially its affinity for the components of the virions
5. The number of Ab molecules attached to the virions

Reversible Neutralization

The neutralization process can be reversed by diluting the Ab-Ag mixture within a short time of the formation of the Ag-Ab complexes (30 mins). It is thought that reversible neutralization is due to the interference with attachment of virions to the cellular receptors. The process requires the saturation of the surface of the virus with Abs.

Stable Neutralization

With time, Ag-Ab complexes usually become more stable (several hours) and the process cannot be reversed by dilution. Neither the virions nor the Abs are permanently changed in stable neutralization, for the unchanged components can be recovered. The neutralized virus can be reactivated by proteolytic cleavage. Intact Abs can be recovered by dissociating the Ab- Ag complexes at acid or alkaline pH.

Stable neutralization has a different mechanism to that of reversible neutralization. It had been shown that neutralized virus can attach and that already attached virions can be neutralized. The number of Ab molecules required for stable neutralization is considerably smaller than that of reversible neutralization, Kinetic evidence shows that even a single Ab molecule can neutralize a virion. Such neutralization is generally produced by Ab molecules that establish contact with 2 antigenic sites on different monomers of a virion, greatly increasing the stability of the complexes.

Virion Sites for Neutralization

Only epitopes on molecules involved in the release of the viral genome into the cells are targets of neutralization. In influenza viruses, only the HA and not the NA are targets for neutralization. In polioviruses, all antigenic sites recognizable on the capsid are targets for neutralization, because the capsid is a unit for releasing the nucleic acid. For adenoviruses, the main targets are the hexons rather than the pentons, as the hexons are strongly interconnected and work together for the release of the viral DNA. Occasionally, Abs bound to non-neutralizing epitopes can be detected by neutralization in the presence of complement, whereby the viral enveloped is attacked by the complement cascade.

Protective Role of Neutralizing Antibodies

The neutralizing power of a serum usually reflects the degree of protection in an infected animal. The correlation, however, is not always perfect. Discrepancies may be generated by differences in the neutralizability of a virus in the cells used for assay in vitro compared to those that the virus infects in vivo. *e.g.* the sera of mice protected from yellow fever did not neutralize the virus in vero cells but did so in a mouse neuroblastoma cell line. Another possible reason for discrepancy is that an Ab that does not neutralize in cultures may act in vivo by activating host responses against the virus or virus-infected cells. *e.g.* complement or macrophages. In addition, neutralizing Abs may fail to protect because rapid viral multiplication overcomes the neutralizing power. In the early period of immunization, low affinity Abs act predominantly by activating complement and have low neutralizing power in cultures. The degree of neutralization in cultures is probably best estimated by carrying out neutralization in the presence of complement.

Evolution of Viral Antigens

Viral evolution must tend to select for mutations that change the antigenic determinants involved in neutralization. In contrast, other antigenic sites would tend to remain unchanged because mutations affecting them would not be selected for and could even be detrimental. A virus would thus evolve from an original type to a variety of types, different in neutralization (and sometimes in HI) tests, but retaining some of the original mosaic of antigenic determinants recognizable by CFTs.

These evolutionary arguments are consistent with the observation that the clearest differentiation of types within a family is present in viruses of rather complex architecture, in which the Ags involved in the interaction with the cell vary more than other proteins. Thus enveloped viruses have a strain-specific envelope but a cross-reactive internal capsid; adenoviruses have type-specific fibers and family-specific (and also type-specific) capsomers. Moreover, the C Ag of polioviruses, which appears only after heating, reveals antigenic sites that are normally hidden and hence are not affected by selective pressure. The extent of antigenic variation differs widely among viruses and is most extensive with lentiviruses and influenza viruses.

Types of Virus-Specific Antibodies

Different types of viral preparations elicit the formation of different Abs:

1. Killed virus preparations elicit Abs predominantly directed against the surface of the virions. These Abs have neutralizing and HI activities against the virions as well as CF and precipitating activities against the Ags of the viral coat.

2. Live virus preparations elicit antibodies against all the viral antigens, including both external and internal antigens.

3. Immunization with internal components of the virions produces CF and precipitating Abs active only toward the Ags of these components.

4. Immunization with peptides reproducing segments of virion proteins elicit Abs, the properties of which depend both on the protein and the specific sequences reproduced.

Specificity of Test Methods

The Abs that react in the different tests may overlap though they may not be altogether identical. Neutralization is primarily caused by Ab molecules specific for the sites of the virion that are involved in the release of viral nucleic acid into the cell. CF usually involves additional surface or internal Ags. Neutralization probably requires molecules with a higher affinity for virions than do HI and CF. After viral infection, the titres of Abs to different components rise and fall with quite different time courses.

Because of their high specificity, immunological methods can differentiate not only between viruses of different families but also between closely related viruses of the same family or subfamily. By these means, family Ags may be identified. Usually, antibodies detected by neutralization tend to be less cross-reactive and thus are useful in defining the immunological type. Whereas those detected by CF tend to be more cross-reactive and the useful in defining the family. By proper procedures, however, such as immunization with purified Ags, highly specific CF Abs can be prepared.

The resolving power of Abs is maximized by the use of monoclonal Abs. Whereas all the methods for measuring viral antigens are needed for classifying a new isolate, the method of choice for diagnostic purposes is ELISA, for its high sensitivity and low cost.

Cell-Mediated Immunity

Cytotoxic T Lymphocytes

CMI is very important in localizing viral infections, in recovery, and in the pathogenesis of viral diseases. In experimental animals, primary CTLs reach maximal abundance about 6 days after a viral infection and then disappears as infection subsides. However, memory T cells persists and can be recognized by culturing spleen cells with virus-infected cells where within a few days, secondary CTLs appear in culture with much greater activity than in the initial response.

Formation of CTLs is elicited by cell-associated Ags present at the cell surface, not only for enveloped viruses, but also for other viruses whose core or nonvirion proteins reach the cell surface. As in humoral immunity, type specific and group specific responses can be seen. Even noninfectious or inactivated viruses can elicit a cellular response because their envelopes fuse with the cell plasma membrane in the initial stage of viral penetration. Moreover, the virions themselves may also be able to elicit the response after absorbing to the macrophages. Both internal virion proteins and nonvirion proteins are often recognized by CTLs. An example is the nucleocapsid proteins of enveloped viruses, fragments of which reach the cell surface by an unknown route and are recognized very efficiently, giving rise mainly to cross-reactive CTLs. Often, Abs to viral surface proteins do not block their interaction with CTLs, because the humoral and cellular responses recognize different epitopes.

Antibody-Dependent Cell-Mediated Cytotoxicity

The K cells are the effector cells in ADCC. In vitro, these cells kill virus-infected cells sensitized by IgG from immune donors but not unsensitized targets. ADCC is very efficient in vitro against HSV or VZV infected cells, preventing the usual spread of the virus from infected to neighboring uninfected cells. Therefore, it may play a role in the defense against human infection with these viruses. K cells had been shown to mediate immunity to vaccinia infection rather than Tc cells.

Natural Killer (NK) Cells

In man, the principal NK cell is the large granular lymphocyte (LGL) which comprise 2-5 per cent of peripheral blood lymphocytes. However, not all lytic cells are LGLs and not all LGLs are NK cells. There is overlap of the NK population with K cells. The Fc receptor of the NK cell is however, not involved in the lytic process. There are also mechanistic differences and K cell activity is less consistently augmented by interferon and other immune modulators. NK activity is subject to both positive and negative regulation in vivo and in vitro. Interferon gamma and IL-2 are potent inducers. Besides producing lysis, NK cells can produce alpha-interferon.

The target molecules recognized but NK cells have not been defined but it appears that some determinants are ubiquitous whilst others have a more restricted distribution. An alternative suggestion is that NK cell susceptibility depends on the absence of normal cell surface antigens such as MHC molecules. The importance of NK cells in viral infection is partially understood. It had been shown that mice depleted of NK cells by treatment with Ab against asialo GM1 show an increased susceptibility to CMV.

Chapter 9

Virus Host Interactions

Resistance to and recovery from viral infections will depend on the interactions that occur between virus and host. The defenses mounted by the host may act directly on the virus or indirectly on virus replication by altering or killing the infected cell. The non-specific host defenses function early in the encounter with virus to prevent or limit infection while the specific host defenses function after infection in recovery immunity to subsequent challenges. Although the host defense mechanisms involved in a particular viral infection will vary depending on the virus, dose and portal of entry, some general principals of virus-host interactions are summarized below.

Barriers to Infection

Inherent Barriers

The host has a number of barriers to infection that are inherent to the organism. These represent the first line of defense which function to prevent or limit infection.

Skin

The skin acts a formidable barrier to most viruses and only after this barrier is breached will viruses be able to infect the host.

Lack of Membrane Receptors

Viruses gain entry into host cells by first binding to specific receptors on cells. (Table 1; adapted from: Roitt, Immunology, 5th Ed).

The host range of the virus will depend upon the presence these receptors. Thus, if a host lacks the receptor for a virus or if the host cells lacks some component necessary for the replication of a virus, the host will inherently be resistant to that virus. For example, mice lack receptors for polio viruses and thus are resistant to polio virus. Similarly, humans are inherently resistant to plant and many animal viruses.

Table 1

Virus	Receptor	Cell Type Infected
HIV	CD4	T_H cells
Epstein-Barr virus	CR2 (complement receptor type 2)	B cells
Influenza A	Glycophorin A	Many cell types
Rhino virus	ICAM-1	Many cell types

Mucus

The mucus covering an epithelium acts as a barrier to prevent infection of host cells. In some instances the mucus simply acts as a barrier but in other cases the mucus can prevent infection by competing with virus receptors on cells. For example, orthomyxo- and paramyxovirus families infect the host cells by binding to sialic acid receptors. Sialic acid-containing glycoproteins in mucus can thus compete with the cell receptors and diminish or prevent binding of virus to the cells.

Ciliated Epithelium

The ciliated epithelium which drives the mucociliary elevator can help diminish infectivity of certain viruses. This system has been shown to be important in respiratory infections since, when the activity of this system is inhibited by drugs or infection, there is an increased infection rate with a given inoculum of virus.

Low pH

The low pH of gastric secretions inactivate most viruses. However, enteroviruses are resistant to gastric secretions and thus can survive and replicate in the gut.

Humoral and Cellular Components

Induced Barriers

Changes that occur in the host in response to infection can also help diminish virus infectivity.

Fever

Fever can help to inhibit virus replication by potentiating other immune defenses and by decreasing virus replication. The replication of some viruses is reduced at temperatures above 37degrees C.

Low pH

The pH of inflammatory infiltrates is also low and can help limit viral infections by inactivating viruses.

Humoral Components Involved in Resistance to Viral Infections

Nonspecific

A number of humoral components of the nonspecific immune system function in resistance to viral infection. Some of theses are constitutively present while others are induced by infection.

Interferon (IFN)

IFN was discovered over 40 years ago by Issacs and Lindemann who showed that supernatant fractions from virus-infected cells contained a protein that could confer resistance to infection to other cells. This substance did not act directly on the virus, rather it acted on the cells to make them resistant to infection.

IFN is one of the first lines of defense against viruses because it is induced early after virus infection before any of the other defense mechanisms appear (*e.g.* antibody, Tc cells etc.) (Figure 2). The time after which IFN begins to be made will vary depending on the dose of virus.

There are three types of interferon, IFN-alpha (also known as leukocyte interferon), IFN-beta (also known as fibroblast interferon) and IFN-gamma (also known as immune interferon). IFN-alpha and IFN-beta are also referred to as Type I interferon and IFN-gamma as Type II. There are approximately 20 subtypes of IFN-alpha but only one IFN-beta and IFN-gamma.

The interferons have different characteristics that could be used to distinguish them (*e.g.* pH stability and activity in the presence of SDS) but currently they are identified by using specific antibodies to the interferons.

Inducers of Interferons

Normal cells do not contain preformed IFN nor do they secret interferon constitutively. This is because the interferon genes are not transcribed in normal cells. Transcription of the IFN genes occurs only after exposure of cells to an appropriate inducer. Inducers of IFN-alpha and IFN-beta include virus infection, double stranded RNA (*e.g.* poly inosinic:poly cytidylic acid; [poly I:C]), LPS, and components from some bacteria. Among the viruses, the RNA viruses are the best inducers while DNA viruses are poor IFN inducers, with the exception of poxviruses. Inducers of IFN-gamma include mitogens and antigen (*i.e.* things that activate lymphocytes).

Cellular Events in the Induction of Interferons

The IFN genes are not expressed in normal cells because the cells produce a labile repressor protein that binds to the promoter region upstream of the gene and inhibits transcription. In addition, transcription of the genes require activator proteins to bind to the promoter region and turn on transcription. Inducers of IFN act by either preventing synthesis of the repressor protein or increasing the levels of the activator proteins, thereby turning the IFN gene on. After the inducer is gone, the IFN gene is again turned off by the repressor protein and/or the lack of activator proteins. Once the gene is turned on, it is transcribed, the mRNA is translated and the protein is secreted from the cell. The IFN will bind to IFN-receptors on neighboring cells and induce an antiviral state in the second cell.

Cellular Events in the Action of Interferons

The binding of IFN to its receptor results in the transcription of a group of genes that code for antiviral proteins involved in preventing viral replication in that cell. As a consequence the cell will be protected from infection with a virus until the antiviral proteins are degraded, a process which takes several days. The antiviral state in IFN-treated cells results from the synthesis of two enzymes that result in the inhibition of protein synthesis. One protein indirectly affects protein synthesis by breaking down viral mRNA the other directly affects protein synthesis by inhibiting elongation.

One protein, called 2′5′Oligo A synthetase, is an enzyme that converts ATP into a unique polymer (2′5′ Oligo A) containing 2′- 5′phophodiester bonds. Double stranded RNA is required for the activity of this enzyme. The 2′5′Oligo A in turn activates RNAse L which then breaks down viral mRNA. The second protein is an protein kinase that, in the presence of double stranded RNA, is autophosphorylated and thereby activated. The activated protein kinase in turn phosphorylates elongation factor eIF-2 and inactivates it. By the action of these two IFN-induced enzymes protein synthesis is inhibited. Although the infected cell may die as a consequence of the inhibition of host protein synthesis, the progress of the infection is stopped. Uninfected cells are not killed by IFN treatment since activation of the two enzymes requires double stranded RNA, which is not produced. Some viruses have means of inhibiting the antiviral effects of IFN. For example the adenoviruses produce an RNA which prevents the activation of the protein kinase by double stranded RNA thereby reducing the antiviral effects of IFN.

Other Biological Activities of Interferons

IFN not only induces the production of antiviral proteins, it also has other effects on cells, some of which indirectly contribute to the ability of the host to resist or recover from a viral infection (Figure 5). IFN can help modulate immune responses by its effects on Class I and Class II MHC molecules. IFN-alpha, IFN-beta and IFN-gamma increase expression of Class I molecules on all cells thereby promoting recognition by Tc cells which can destroy virus infected cells. IFN-gamma can also increase expression of Class II MHC molecules on antigen presenting cells resulting in better presentation of viral antigens to CD4$^+$ T helper cells. Furthermore, IFN-gamma can activate NK cells which can kill virus infected cells. IFNs also activate the intrinsic and extrinsic antiviral activities of macrophages. Intrinsic antiviral activity is the ability of macrophages to resist infection with a virus and extrinsic antiviral activity is the ability of macrophages to kill other cells infected with virus. The IFNs also have anti-proliferative activity making them useful in the treatment of some malignancies.

Clinical Uses of Interferons

IFNs have been used in the treatment of a number of viral and other diseases (Tables 2, 3 and 4)

Table 2: Clinical Uses of Interferons

Interferon	Therapeutic Use
IFN-alpha, IFN-beta	Hepatitis B (chronic)
	Hepatitis C
	Herpes zoster
	Papilloma virus
	Rhino virus (prophylactic only)
	Warts

Contd...

Table 2–Contd...

Interferon	Therapeutic Use
IFN-gamma	Lepromatous leprosy
	Leshmaniasis
	Toxoplasmosis
	Chronic granulomatous disease (CGD)

Table 3: Use of Interferons on Cancer Treatment

Tumor	Percent Complete or Partial Remissions
Hairy cell leukemia	90
Chronic myelocytic leukemia	90
T cell lymphoma	53
Kaposi's sarcoma	42
Endocrine pancreatic neoplasms	30
Non-Hodgkin's lymphomas	25–35

Table 4: Common Side Effects of Interferons

Interferons	Fever
	Malaise
	Fatigue
	Muscle pains
	Toxicity to: kidney, liver
	Bone marrow
	Heart

Complement

Most viruses do not fix complement by the alternative route. However, the interaction of a complement-fixing antibody with a virus infected cell or with an enveloped virus can result in the lysis of the cell or virus. Thus, by interfacing with the specific immune system, complement also plays a role in resistance to viral infections.

Cytokines

Cytokines other than IFN also may play a role in resistance to virus infection. Tumor necrosis factor alpha (TNF-α), interleukin-1 (IL-1) and IL-6 have been shown to have antiviral activities *in vitro*. These cytokines are produced by activated macrophages but their contribution to resistance *in vivo* has not been fully elucidated.

Specific

Antibody produce by the specific immune system is involved primarily in the recovery from viral infection and in resistance to subsequent challenge with the virus. IgG, IgM and IgA antibodies can all play a role in immunity to virus infection but the relative contributions of the different classes depends on the virus and the portal of entry. For example, IgA will be more important in viruses that infect the mucosa while IgG antibodies will be more important in infections in which viremia is a prominent feature. Antibodies can have both beneficial and harmful effects for the host.

Beneficial Effects

Antibody can directly neutralize virus infectivity by preventing the attachment of virus to receptors on host cells or entry of the virus into the cell. Antibodies can also prevent uncoating of virus by interfering with the interaction of viral proteins involved in uncoating. Complement fixing antibodies can assist in the lysis of viral infected cells or enveloped viruses. Antibodies can also act as an opsonins and augment phagocytosis of viruses either by promoting their uptake via Fc or C3b receptors or by agglutinating viruses to make them more easily phagocytosed. Antibody coated virus infected cells can be killed by K cells thereby preventing the spread of the infection.

Harmful Effects

(a) Immunopathological Damage

Fixation of complement by immune complexes can result in the release of vasoactive amines, recruitment of inflammatory cells and subsequent damage to host tissue. Some viruses such a lymphocytic choriomeningitis virus produce large amounts of immune complexes in the circulation which lodge in the vascular beds and in the kidneys where they fix complement and result in tissue damage. Other examples of viruses that cause these effects are: measles, respiratory syncytial virus, dengue and serum hepatitis virus.

(b) Immune Adherence

Opsonization of viruses with antibody can enhance their uptake by phagocytic cells. If the virus is able to survive in the phagocyte, this allows for the spread of virus infection. Dengue and HIV are examples of viruses that can survive in macrophages.

Serology

Since the isolation and identification of viruses is not commonly done in the clinical laboratory, the clinical picture and serology plays a greater role in the diagnosis of viral disease. The major types of antibodies that are assayed for are neutralizing, hemagglutination inhibiting and complement fixing antibodies. Complement fixing antibodies follow the kinetics of IgM and are most useful in indicating a current or recent infection. In contrast the neutralizing and hemagglutinating antibodies follow the kinetics of IgG, persist for a long time and are used to assess immunity. The development of antibodies to different components of the virus is used in staging the disease. For example in hepatitis B and HIV infections this approach is used.

Cellular Components

In addition to the barriers and humoral components involved in resistance to and recovery from viral infections, there are several different cells that play a role in our antiviral defenses.

Nonspecific

Macrophages

By virtue of the location at various sites in the body, macrophages are one of the first cells to encounter viruses. Experimental evidence suggest that these cells play an important role in resistance to viral infection. For example, newborn mice are susceptible to infection with herpes virus type 1 due to a defect in the ability of macrophages to prevent replication of the virus. Macrophages from adult mice however, are able to prevent replication of the virus and these mice are resistant to infection with this virus. Also, animals in which macrophages have been depleted are more susceptible to infection with a variety of viruses. Macrophages contribute to antiviral defenses in a number of ways.

(a) Intrinsic antiviral activity–Macrophages can be infected with viruses but many viruses are incapable of replicating in macrophages. Macrophages that are activated (*e.g.* by IFN-γ) are even more capable of resisting viral replication. Thus, macrophages help limit viral infections by virtue of their intrinsic ability to prevent replication of viruses. However some viruses are able to replicate or at least survive in macrophages and thus can be spread by macrophages (see above).

(b) Extrinsic antiviral activity–Macrophages are also able to recognize virus infected cells and to kill them. Thus, macrophages also contribute to antiviral defenses by virtue of their cytotoxic activity.

(c) ADCC–Virus infected cells that are coated with IgG antibodies can be killed by macrophages by ADCC

(d) IFN production–Macrophages are a source of IFN.

NK Cells

Experimental evidence also suggests that NK cells also play a role in resistance to viral infection. Mice that are depleted of NK cells are more susceptible to infection with certain viruses. Also, patients with low NK cell activity are more susceptible to reoccurrences with herpes simplex type 1 virus. NK cells act by recognizing and killing virus infected cells. The recognition of virus infected cells is not MHC-restricted or antigen specific. Thus, NK cells will kill cells infected with many different viruses. NK cells can also mediate ADCC and can kill virus infected cells by this mechanism. The activities of NK cells can be enhanced by IFN-γ and Il-2.

Specific

T Cells

T cells play a major role in recovery from viral infections. Cytotoxic T cells (CTLs) generated in response to viral antigens on infected cells can kill the infected cells thereby preventing the spread of infection. Helper T cells are involved in generation of CTLs and in assisting B cells to make antibody. In addition, lymphokines secreted by T cells can recruit and activate macrophages and NK cells thereby mobilizing a concerted attack in the virus.

Summary of Defenses

Table 5 (Adapted from: Baron, Medical Microbiology, 2nd Ed., Table 69-2) summarizes the host defenses against viral infections and it indicates the targets for each of these defenses.

Table 5: Host Effector Functions in Viral Infections

Host Defense	Effector	Target of Effector
Early nonspecific	Fever	Virus replication
responses	Phagocytosis	Virus
	Inflammation	Virus replication
	NK cell activity	Virus-infected cell
	Interferon	Virus replication, immunomodulation
Immune responses	Cytotoxic T lymphocytes	Virus infected cell
mediated by cells	Activated macrophages	Virus, virus-infected cell
	Lymphokines	Virus-infected cells, immunomodulation
	ADCC	Virus-infected cell
Humoral immune	Antibody	Virus, Virus-infected cell
responses	Antibody + complement	Virus, virus-infected cell

Relative Contributions of the Host Defense Mechanisms

The relative contribution of the various host defense mechanisms will depend on the nature of the virus and the portal of entry. Antibodies will be more important in infections in which viremia is a prominent feature. However, antibodies may not be helpful in infections with herpes or paramyxoviruses in which the virus can be passed from cell to cell by cell fusion. In this instance cell mediated immunity is more important. If a virus only infects cells in the mucosal surface, IgA antibodies may be important.

An understanding of the host defense mechanisms is important for vaccine development and for proper administration of vaccines. If IgA antibodies are important for protection against a particular virus, then any vaccine must be able to stimulate production of IgA antibodies in the appropriate mucosal surface. Alternatively if CTLs are important then the vaccine must be able to stimulate CTL production. That is why live vaccines are often preferable to a killed vaccine because live vaccines usually lead to the generation of CTLs while killed vaccines do not.

Virus-Induced Immunopathology

Although the host has a variety of defenses to protect against viral infections, sometimes it is the immune response to the infection that is the direct cause of tissue injury. For example, infants infected with cytomegalovirus have circulating immune complexes that are deposited in the kidneys and joints resulting in pathology such as arthritis and glomerular nephritis. Another example is fatal hemorrhagic shock syndrome associated with dengue virus infection. In this instance fixation of complement by circulating immune complexes results in release of products of the complement cascade leading to sudden increased vascular permeability, shock and death.

Immunosuppression

Many virus are able to suppress immune responses and thereby overcome or minimize host defenses. The best example is HIV which infects the CD4$^+$ cells thereby destroying the specific immune system. Other viruses (*e.g.* measles virus) can also infect lymphocytes and affect their replication and differentiation. Virus-induced immunosuppression is major concern in vaccine development. Some of the mechanisms by which viruses can evade host defenses.

Viral Products that Interfere with Host Defenses

Host Defense Affected	Virus	Virus Product	Mechanism
Interferon	EBV	EBERS (small RNAs)	Blocks protein kinase activation
	Vaccinia	eIF-2alpha homolog	Prevents phosphorylation of eIF-2alpha by protein kinase
Complement	Vaccinia	Homologues of complement control proteins	Blocks complement activation
Antibody	HSV-1	gE/gI	Binds Fc-gamma and blocks function
Cytokines	Myxoma	IFN-gamma receptor homolog	Competes for IFN-gamma and blocks function
	Shope fibroma virus	TNF receptor	Competes for TNF and blocks function
	EBV	IL-10 homolog	Reduces IFN-gamma function
MHC Class I	CMV	Early protein	Prevents transport of peptide-loaded MHC
	Adenovirus	E3	Blocks transport of MHC to surface
Apoptosis	Adenovirus	14.7K	Inhibits capsases
	EBV	Bcl-2 homolog	Anti-apoptotic
NK cells	HCMV	UL-18	MHC homolog inhibits NK cells

Chapter 10
Bacteriophage

Bacteriophage (phage) are obligate intracellular parasites that multiply inside bacteria by making use of some or all of the host biosynthetic machinery (*i.e.*, viruses that infect bacteria).

There are many similarities between bacteriophages and animal cell viruses. Thus, bacteriophage can be viewed as model systems for animal cell viruses. In addition a knowledge of the life cycle of bacteriophage is necessary to understand one of the mechanisms by which bacterial genes can be transferred from one bacterium to another.

At one time it was thought that the use of bacteriophage might be an effective way to treat bacterial infections, but it soon became apparent that phage are quickly removed from the body and thus, were of little clinical value. However, bacteriophage are used in the diagnostic laboratory for the identification of pathogenic bacteria (phage typing). Although phage typing is not used in the routine clinical laboratory, it is used in reference laboratories for epidemiological purposes. Recently, new interest has developed in the possible use of bacteriophage for treatment of bacterial infections and in prophylaxis. Whether bacteriophage will be used in clinical medicine remains to be determined.

Classification

The dsDNA tailed phages, or *Caudovirales*, account for 95 per cent of all the phages reported in the scientific literature, and possibly make up the majority of phages on the planet.[1] However, there are other phages that occur abundantly in the biosphere, phages with different virions, genomes and lifestyles. Phages are classified by the International Committee on Taxonomy of Viruses (ICTV) according to morphology and nucleic acid.

ICTV classification of phages

Order	Family	Morphology	Nucleic acid
Caudovirales	Myoviridae	Non-enveloped, contractile tail	Linear dsDNA
	Siphoviridae	Non-enveloped, long non-contractile tail	Linear dsDNA
	Podoviridae	Non-enveloped, short non contractile tail	Linear dsDNA
	Tectiviridae	Non-enveloped, isometric	Linear dsDNA
	Corticoviridae	Non-enveloped, isometric	Circular dsDNA
	Lipothrixviridae	Enveloped,rod-shaped	Linear dsDNA
	Plasmaviridae	Enveloped, pleomorphic	Circular dsDNA
	Rudiviridae	Non-enveloped, rod-shaped	Linear dsDNA
	Fuselloviridae	Non-enveloped, lemon-shaped	Circular dsDNA
	Inoviridae	Non-enveloped, filamentous	Circular ssDNA
	Microviridae	Non-enveloped, isometric	Circular ssDNA
	Leviviridae	Non-enveloped, isometric	Linear ssRNA
	Cystoviridae	Enveloped, spherical	Segmented dsRNA

Composition and Structure

A. Composition

Although different bacteriophages may contain different materials they *all* contain nucleic acid and protein.

Depending upon the phage, the nucleic acid can be either DNA or RNA but not both and it can exist in various forms. The nucleic acids of phages often contain unusual or modified bases. These modified bases protect phage nucleic acid from nucleases that break down host nucleic acids during phage infection. The size of the nucleic acid varies depending upon the phage. The simplest phages only have enough nucleic acid to code for 3-5 average size gene products while the more complex phages may code for over 100 gene products.

The number of different kinds of protein and the amount of each kind of protein in the phage particle will vary depending upon the phage. The simplest phage have many copies of only one or two different proteins while more complex phages may have many different kinds. The proteins function in infection and to protect the nucleic acid from nucleases in the environment.

B. Structure

Bacteriophage come in many different sizes and shapes. The basic structural features of bacteriophages are illustrated in Figure, which depicts the phage called T4 (Figures 8a and b)

1. Size

T4 is among the largest phages; it is approximately 200 nm long and 80-100 nm wide. Other phages are smaller. Most phages range in size from 24-200 nm in length.

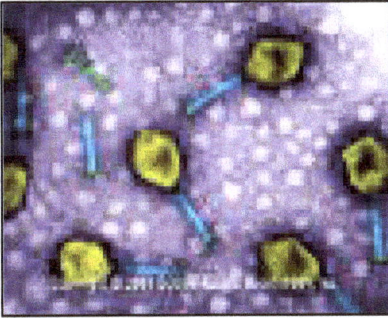

Figure 8a: T4 Bacteriophage (TEM x390,000)

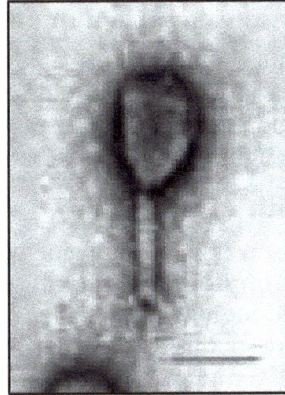

Figure 8b: T4 Bacteriophage Negative Stain Electronmicrograph

2. Head or Capsid

All phages contain a head structure which can vary in size and shape. Some are icosahedral (20 sides) others are filamentous. The head or capsid is composed of many copies of one or more different proteins. Inside the head is found the nucleic acid. The head acts as the protective covering for the nucleic acid.

3. Tail

Many but not all phages have tails attached to the phage head. The tail is a hollow tube through which the nucleic acid passes during infection. The size of the tail can vary and some phages do not even have a tail structure. In the more complex phages like T4 the tail is surrounded by a contractile sheath which contracts during infection of the bacterium. At the end of the tail the more complex phages like T4 have a base plate and one or more tail fibers attached to it. The base plate and tail fibers are involved in the binding of the phage to the bacterial cell. Not all phages have base plates and tail fibers. In these instances other structures are involved in binding of the phage particle to the bacterium.

Infection of Host Cells

A. Adsorption

The first step in the infection process is the adsorption of the phage to the bacterial cell. This step is mediated by the tail fibers or by some analogous structure on those phages that lack tail fibers and it is reversible. The tail fibers attach to specific receptors on the bacterial cell and the host specificity of the phage (*i.e.* the bacteria that it is able to infect) is usually determined by the type of tail fibers that a phage has. The nature of the bacterial receptor varies for different bacteria. Examples include proteins on the outer surface of the bacterium, LPS, pili, and lipoprotein. These receptors are on the bacteria for other purposes and phage have evolved to use these receptors for infection.

B. Irreversible Attachment

The attachment of the phage to the bacterium via the tail fibers is a weak one and is reversible. Irreversible binding of phage to a bacterium is mediated by one or more of the components of the base plate. Phages lacking base plates have other ways of becoming tightly bound to the bacterial cell.

C. Sheath Contraction

The irreversible binding of the phage to the bacterium results in the contraction of the sheath (for those phages which have a sheath) and the hollow tail fiber is pushed through the bacterial envelope. Phages that don't have contractile sheaths use other mechanisms to get the phage particle through the bacterial envelope. Some phages have enzymes that digest various components of the bacterial envelope.

D. Nucleic Acid Injection

When the phage has gotten through the bacterial envelope the nucleic acid from the head passes through the hollow tail and enters the bacterial cell. Usually, the only phage component that actually enters the cell is the nucleic acid. The remainder of the phage remains on the outside of the bacterium. There are some exceptions to this rule. This is different from animal cell viruses in which most of the virus particle usually gets into the cell. This difference is probably due to the inability of bacteria to engulf materials.

Phage Multiplication Cycle (Figure 9)

A. Lytic or Virulent Phages

1. Definition

Lytic or virulent phages are phages which can only multiply on bacteria and kill the cell by lysis at the end of the life cycle.

2. Life Cycle

The life cycle of a lytic phage is illustrated in Figure 9.

(a) Eclipse Period

During the eclipse phase, no infectious phage particles can be found either inside or outside the bacterial cell. The phage nucleic acid takes over the host biosynthetic machinery and phage specified m-RNA's and proteins are made. There is an orderly expression of phage directed macromolecular synthesis, just as one sees in animal virus infections. Early m-RNA's code for early proteins which are needed for phage DNA synthesis and for shutting off host DNA, RNA and protein biosynthesis. In some cases the early proteins actually degrade the host chromosome. After phage DNA is made late m-RNA's and late proteins are made. The late proteins are the structural proteins that comprise the phage as well as the proteins needed for lysis of the bacterial cell.

(b) Intracellular Accumulation Phase

In this phase the nucleic acid and structural proteins that have been made are assembled and infectious phage particles accumulate within the cell.

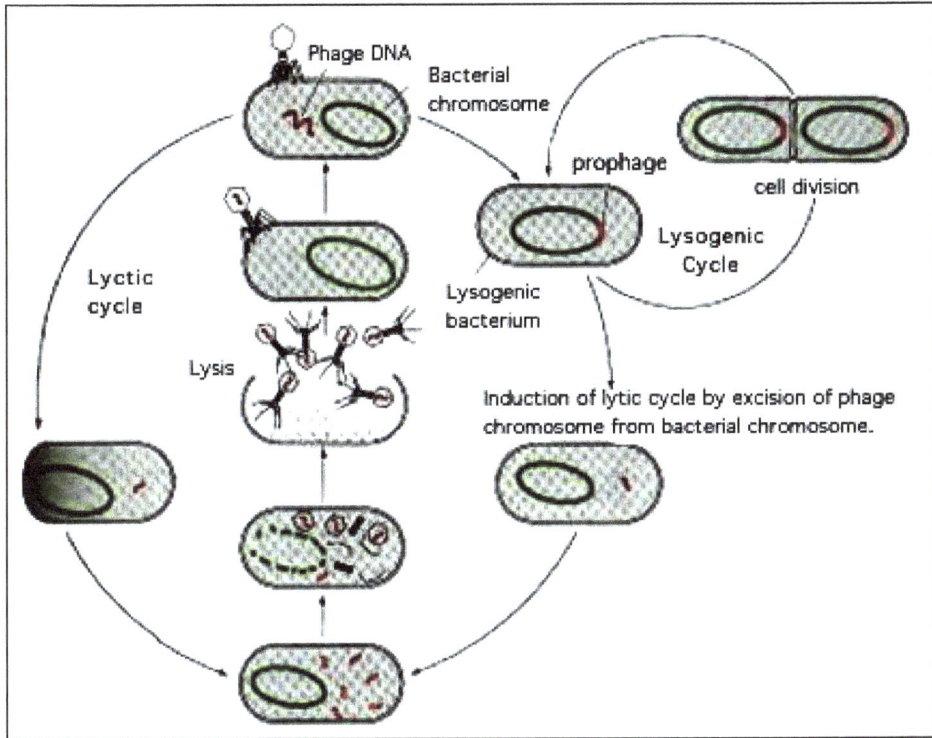

Figure 9: Lytic and Lysogenic Cycles

(c) Lysis and Release Phase

After a while the bacteria begin to lyse due to the accumulation of the phage lysis protein and intracellular phage are released into the medium. The number of particles released per infected bacteria may be as high as 1000.

3. Assay for Lytic Phage

(a) Plaque Assay

Lytic phage are enumerated by a plaque assay. A plaque is a clear area which results from the lysis of bacteria. Each plaque arises from a single infectious phage. The infectious particle that gives rise to a plaque is called a pfu (plaque forming unit).

(b) Intracellular Accumulation Phase

In this phase the nucleic acid and structural proteins that have been made are assembled and infectious phage particles accumulate within the cell.

(c) Lysis and Release Phase

After a while the bacteria begin to lyse due to the accumulation of the phage lysis protein and intracellular phage are released into the medium. The number of particles released per infected bacteria may be as high as 1000.

B. Lysogenic or Temperate Phage

1. Definition

Lysogenic or temperate phages are those that can either multiply via the lytic cycle or enter a quiescent state in the cell. In this quiescent state most of the phage genes are not transcribed; the phage genome exists in a repressed state. The phage DNA in this repressed state is called a *prophage* because it is not a phage but it has the potential to produce phage. In most cases the phage DNA actually integrates into the host chromosome and is replicated along with the host chromosome and passed on to the daughter cells. The cell harboring a prophage is not adversely affected by the presence of the prophage and the lysogenic state may persist indefinitely. The cell harboring a prophage is termed a *lysogen*.

2. Events Leading to Lysogeny

The Prototype Phage: Lambda

(a) Circularization of the Phage Chromosome

Lambda DNA is a double stranded linear molecule with small single stranded regions at the 5' ends. These single stranded ends are complementary (*cohesive ends*) so that they can base pair and produce a circular molecule. In the cell the free ends of the circle can be ligated to form a covalently closed circle.

(b) Site-specific Recombination

A recombination event, catalyzed by a phage coded enzyme, occurs between a particular site on the circularized phage DNA and a particular site on the host chromosome. The result is the integration of the phage DNA into the host chromosome.

(c) Repression of the Phage Genome

A phage coded protein, called a *repressor*, is made which binds to a particular site on the phage DNA, called the *operator*, and shuts off transcription of most phage genes EXCEPT the repressor gene. The result is a stable repressed phage genome which is integrated into the host chromosome. Each temperate phage will only repress its own DNA and not that from other phage, so that repression is very specific (immunity to superinfection with the same phage).

3. Events Leading to Termination of Lysogeny

Anytime a lysogenic bacterium is exposed to adverse conditions, the lysogenic state can be terminated. This process is called *induction*. Conditions which favor the termination of the lysogenic state include: desiccation, exposure to UV or ionizing radiation, exposure to mutagenic chemicals, etc. Adverse conditions lead to the production of proteases (rec A protein) which destroy the repressor protein. This in turn leads to the expression of the phage genes, reversal of the integration process and lytic multiplication.

4. Lytic vs Lysogenic Cycle

The decision for lambda to enter the lytic or lysogenic cycle when it first enters a cell is determined by the concentration of the repressor and another phage protein called *cro* in the cell. The cro protein turns off the synthesis of the repressor and thus prevents the establishment of lysogeny. Environmental conditions that favor the production of cro will lead to the lytic cycle while those that favor the production of the repressor will favor lysogeny.

5. Significance of Lysogeny

(a) Model for Animal Virus Transformation

Lysogeny is a model system for virus transformation of animal cells

(b) Lysogenic Conversion

When a cell becomes lysogenized, occasionally extra genes carried by the phage get expressed in the cell. These genes can change the properties of the bacterial cell. This process is called lysogenic or phage conversion. This can be of significance clinically. *e.g.* Lysogenic phages have been shown to carry genes that can modify the Salmonella O antigen, which is one of the major antigens to which the immune response is directed. Toxin production by Corynebacterium diphtheriae is mediated by a gene carried by a phage. Only those strain that have been converted by lysogeny are pathogenic.

Phage Therapy

Phages were discovered to be anti-bacterial agents and put to use as such soon after they were discovered, with varying success. However, antibiotics were discovered some years later and marketed widely, popular because of their broad spectrum; also easier to manufacture in bulk, store and prescribe. Hence development of phage therapy was largely abandoned in the West, but continued throughout 1940s in the former Soviet Union for treating bacterial infections, with widespread use including the soldiers in the Red Army–much of the literature being in Russian or Georgian, and unavailable for many years in the West. This has continued after the war, with widespread use continuing in Georgia and elsewhere in Eastern Europe. There is anecdotal evidence there, but no completed clinical trials in the US or Western Europe.

Bacteriophages in the Environment

Some time ago it was detected that phages are much more abundant in the water column of freshwater and marine habitats than previously thought and that they can cause significant mortality of bacterioplankton. Methods in phage community ecology have been developed to assess phage-induced mortality of bacterioplankton and its role for food web process and biogeochemical cycles, to genetically fingerprint phage communities or populations and estimate viral biodiversity by metagenomics. The release of lysis products by phages converts organic carbon from particulate (cells) to dissolved forms (lysis products), which makes organic carbon more bio-available and thus acts as a catalyst of geochemical nutrient cycles. Phages are not only the most abundant biological entities but probably also the most diverse ones. The majority of the sequence data obtained from phage communities has no equivalent in data bases. These data and other detailed analyses indicate that phage-specific genes and ecological traits are much more frequent than previously thought. In order to reveal the meaning of this genetic and ecological versatility, studies have to be performed with communities and at spatiotemporal scales relevant for microorganisms. Bacteriophages have also been used in hydrological tracing and modelling in river systems especially where surface water and groundwater interactions occur. The use of phages is preferred to the more conventional dye marker because they are significantly less adsorbed when passing through ground-waters and they are readily detected at very low concentrations.

Bacteriophages and Food Fermentation

A broad number of food products, commodity chemicals, and biotechnology products are manufactured industrially by large-scale bacterial fermentation of various organic substrates. Because enormous amounts of bacteria are being cultivated each day in large fermentation vats, the risk that bacteriophage contamination rapidly brings fermentations to a halt and cause economical setbacks is a serious threat in these industries. The relationship between bacteriophages and their bacterial hosts is very important in the context of the food fermentation industry. Sources of phage contamination, measures to control their propagation and dissemination, and biotechnological defense strategies developed to restrain phages are of interest. The dairy fermentation industry has openly acknowledged the problem of phage and has been working with academia and starter culture companies to develop defense strategies and systems to curtail the propagation and evolution of phages for decades.

Bacteriophage Typing

Strains within a particular salmonella serotype may be differentiated into a number of phage-types by their patterns of susceptibility to lysis by a series of phages with different specificities. The determination of the phage-type of strains isolated from different patients, carriers or other sources is valuable in the epidemiological study of infections as it helps to define groups of infections because it helps to define groups of persons who have been infected with the same strain from the same source. There is a high degree of correlation between the phage-type and the epidemic source. The phage-typing method has thus become a well-established procedure in the routine epidemiological investigation of typhoid fever.

S.typhi, S.paratyphi A, S.paratyphi B, S.typhimurium and *S.enteritidis* serotypes can be subdivided by phage typing. Around 106 different phage-types of *S.typhi* and 232 different phage-types of *S.typhimurium* have been distinguished.

Bacteriophage Typing of *S. typhi*

This is done by determining the sensitivity of the culture to a series of variants of a single phage, Vi-phage II, which have been adapted to the different types of typhoid bacillus. The Vi-phage typing method for Typhi (Craigie and Yen 1938) was the first phage-typing method to be developed and is still in many ways unique. The special contribution of Craigie and his co-workers was their observation that the Vi-II phage was adaptable and showed a high degree of specificity for the last strain on which it had been propagated. This is due to the selection of spontaneously occurring host-range mutants of the phage by the bacterium and due to a non-mutational phenotypic modification of the phage by the host strain. With the help of a series of these adapted Vi-II phages they classified most of the strains into 11 phage types and so far about 106 different phage-types of *S.typhi* have been distinguished and are designated by letter or number.

The method however has its own limitations.(*e.g.* about one-quarter of the strains isolated in the United Kingdom are untypable). Some strains are untypable because they have no Vi-antigen and others because they are degraded Vi-strains(DVS) which are sensitive to many of the battery of Vi-typing phages and hence do not conform to a specific typing pattern. Some strains are resistant to all the specific adaptations of Vi-II but are sensitive to the Craigie and Yen phages Vi-I and Vi-IV.

A more serious difficulty is that one type (usually A or E1) may be so common in a country as to limit the usefulness of the epidemiological information. The common phage types have therefore been further subdivided by biochemical tests and by the use of a battery of unadapted O-phages.

The techniques of phage-typing are complicated and it is usual for cultures for typing to be sent to a reference laboratory:- Lady Hardinge Medical College, New Delhi; Central Research Institute, Kasauli (for India)

Bacteriophage Typing of *S. paratyphi* B

Felix and Callow (1943) originally recognised only 4 phage-types of *S. paratyphi* B. This scheme was then modified/expanded by the introduction of diverse O-phages to the system. Currently 53 phage types are recognised. The Scholtens modification of the scheme has further subdivided these existing phage-types.

Bacteriophage Typing of *S. typhimurium*

Felix (1956) and Callow (1959) developed a system for phage typing strains of Typhimurium that has become established internationally as the method of choice for the epidemiological study of this most frequently isolated serotype. This system originally distinguished 34 phage types, and now has been progressively expanded. With a battery of 36 phages this system provides a very fine degree of discrimination for the 232 definitive types which are currently recognised.

The findings that certain phage types are commonly associated with particular host animals and that there is a close approximation between the principal types responsible for human infections and those found in animal incidents have proved invaluable in the rapid identification of the likely sources of human infection.

Bacteriophage typing of other Salmonellae: Many a times some other serotypes of Salmonella are implicated in human disease to an extent to cause concern and phage typing schemes have also been developed for such serotypes.

Biotyping

Strains in a particular serotype may be differentiated into biotypes by their different fermentation reaction with selected substrates. In many serotypes there are few biochemical tests in which significant numbers of strains behave differently and so the number of identifiable biotypes is small; but it is a useful supplement to phage typing for it can subdivide a large group of untypable strains or members of common phage types.

Dudguid *et al.* in 1975 developed a scheme for biotyping to study the epidemiology of infections with Typhimurium. This scheme was based on the use of 15 biochemical characters. Thirty-two potential primary biotypes were defined by the combinations of positive and negative reactions shown in the 5 tests (D-xylose, m-inositol, L-rhamnose, d-tartarate and m-tartarate) most discriminating in this type(S.typhimurium).These primary biotypes were designated by numbers (1-32) and the full biotypes by appending to these numbers letters which indicated results in 10 secondary tests. By now 24 primary and 184 full biotypes have been identified.

Plasmid Typing

The wild-type strains of salmonellae mostly carry plasmids differing in size and number. They can be separated by electrophoresis in agarose gels on the basis of molecular size, to give a plasmid profile. Plasmid typing is gaining increasing importance as an alternative means for typing salmonellae provided that the strain contains plasmids.

The technique is inexpensive, rapid, can be performed independently of the reference laboratories and now is compared favourably with phage typing for the definition of types within a serotype in epidemic conditions.

M13 Bacteriophage

M13 is a filamentous bacteriophage composed of circular single stranded DNA (ssDNA) which is 6407 nucleotides long encapsulated in approximately 2700 copies of the major coat protein P8, and capped with 5 copies of two different minor coat proteins (P9, P6, P3) on the ends. The minor coat protein P3 attaches to the receptor at the tip of the F pilus of the host *Escherichia coli*. Infection with filamentous phages is not lethal, however the infection causes turbid plaques in *E. coli*. It is a non-lytic virus. However a decrease in the rate of cell growth is seen in the infected cells. M13 plasmids are used for many recombinant DNA processes, and the virus has also been studied for its uses in nanostructures and nanotechnology.

Structure of M13 Phage

M13 phage is a filamentous virus that infect F+ strains of *E.coli*. It looks like a long straight or flexuous rod. M13 phage is about 760-1950 nm long and 6-9 nm wide. The virion consists of an elongated capsid and a circular single stranded DNA (genome).

Capsid

The capsid is composed of helically arranged capsomeres The helix is left handed and composed ofa major protein called VIIIprotein or B-protein. The distal end of a capsomere overlaps the base of the proceeding capsomere. The major coat protein is the product of phage gene VIII (g8p). There are 2700 to 3000 copies of B-protein in a capsid. The major proteins are held togetherbypeptide bonds The ends of the virus capsid has an A-protein which is a multimeric protein composed of four minor capsid proteins. Four minor capsid proteins are the product of the genes gIII, gVI, gVII and gIX. Gp^8 proteins consist of approximately 50 amino acid residues. It assumes the form of a helix and appears as a shot rod. The A protein attaches the phage to the tip of sex filus of *E.coli*.

Genome

M 13 has circular single stranded DNA (ssDNA) as its genome. It is 6,407 base pair long and non-segmented. The genome constitutes about 21 per cent of virion by weight. It forms a double stranded replicative form during multiplication.

Genetic Map of M13

The genome of M13 contains 10 genes which are named Using Roman numerals I through X. It has a unique origin of replication. It is 6407 bp in size. The origin of replication (ori) occurs in between the genes IV, and II. There is a short intergenic sequence (IS) between the *Ori* and gene II. The IS is not useful for phage multiplication (Figure 10).

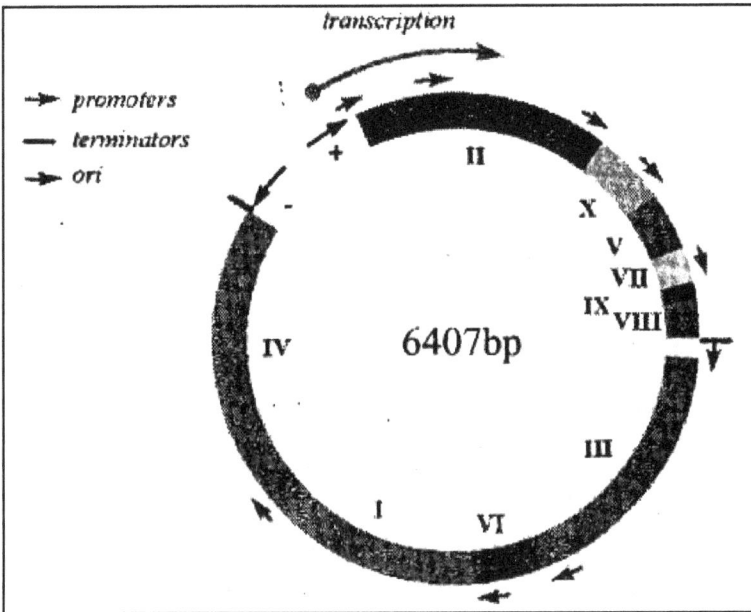

Figure 10: Circular Single Stranded ssDNA of M13 Phage

The functions of various genes in Ml3 genome are given in below.

Gene	Proteins	Functions
I	Gp^1	–
II	Gp^2	Nick formation and initiation of rolling circle replication
III	Gp^3	Minor coat protein. Binds to receptor on F pilus of host cell
IV	Gp^4	Forms a channal for movement of M13 phage from the cytoplasm to the exterior
V	Gp^5	Binding protein (Binds strongly to new (t) strands
VI	Gp^6	Minor coat protein at the proximal,it is associated with PIII

Gene	Proteins	Functions
VII	Gp^7	Protein required for virus assembly
VIII	Gp^8	Major coat protein which binds at the inner surface of plasma membrane
IX	Gp^9	Five copies at the end of bacteriophage needed for assembly
X	Gp^{10}	Repressor to control RF formation

Life Cycle

The Ml3 phage infects only F$^+$ cells of *Escherichia coli*. It does not lyse the host cell dining the multiplication (Figure 11)

Life Cycle of Ml3 Phage Involves the Following Steps

1. Attachment, 2. Entry of phage DNA 3. Formation of Replicative forms 4. Replication of RF 5. Synthesis of ssDNA 6. Transcription 7. Assembly of phages 8. Release of progeny phages

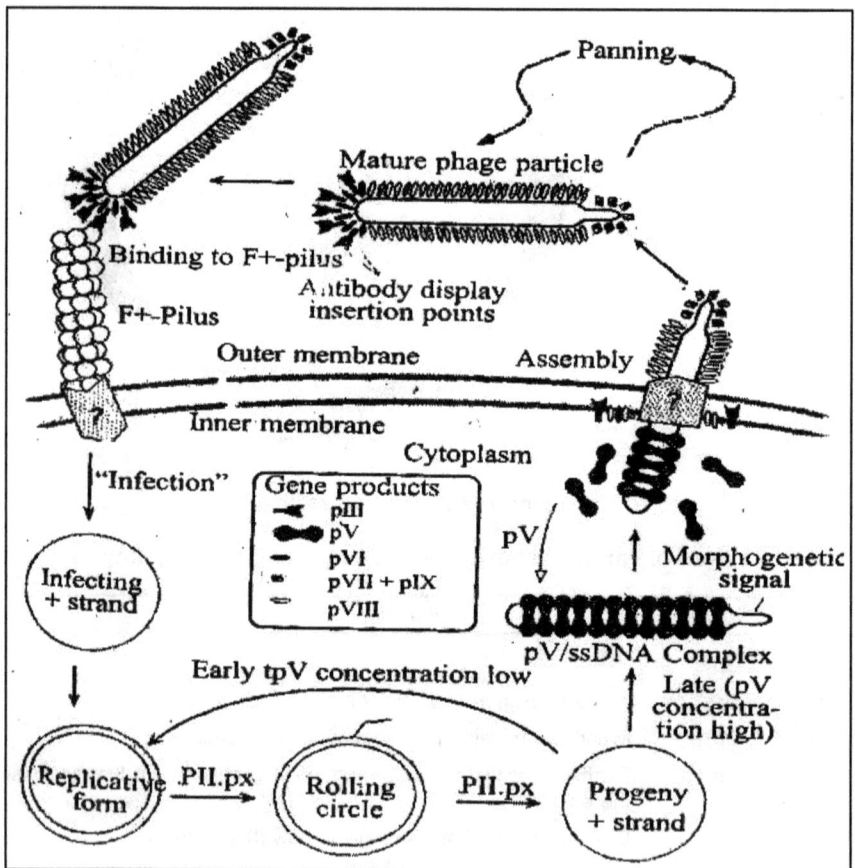

Figure 11. Life Cycle of M13 Phage

1. Attachment

The phage attaches with tip of F pilus of *E. coli* by A- protein. Tip of pilus or base of it has a receptor to accept the phage.

2. Entry of phage DNA

Phage DNA and its A-protein (protein III) enters the cell through pilus. The capsid protein remains attached with the cell membrane and the DNA is released into the cytoplasm. The capsid attached to the membrane may be reused for virus assembly of bacteriophage.

3. Formation of Replicative Forms (RF)

This is the first stage of replication of ssDNA genome. Positive ssDNA is converted into double stranded circular form called Replicative Form (RF) with the help of *E.coli* polymerase. It occurs on the inner surface of cell membrane.The +strand of Ml 3 is used as a template for the replication. The primer is synthesized by RNA polymerase III 8Yflthesis of daughter strand is done by DNA polymerase III. Replicative form (RF) formed during stage I undergoes several steps to synthesize many RF molecules.

4. Replication of RF

This is the second stage of DNA replication in the life cycle of Ml3 phage. It produces 100-200 copies of RF molecules. It does not give + strands directly.

☆ Ml 3 gene II protein nicks the outer (+) strand at the origin of replication.

☆ Using *E.coli* SSB (single strand binding) protein and DNA Pol III, the (+) strand is extended from 3'OH end.

☆ Replication takes place by rolling circle model.

☆ When the new (+) strand reaches the origin, cleaved again by gene II protein.

☆ The old (+) strand is freed and its 3' OH and 5'-P ends are joined by gene II protein.

☆ DNA ends are joined by DNA ligase.

☆ RF molecule replication is controlled by gene X acts a repressor.

☆ The (-) strand is synthesized on (+)strand during this stage.

☆ Host genes dnaC, dna E and polA are useful for lication of RF of M13 phage.

5. Synthesis of ssDNA

This is the third stage of DNA replication in M13 phage life cycle. The (+) strand is used as (+) strand genome of virus.

The (+) strand synthesized in RF replication is used as single stranded DNA genome of phage.

☆ Gene V protein binds to new (+) strand and thereby prevents (-) strand synthesis.

6. Transcription

The phage genes III, IV and V are transcribed into mRNA which are in turn translated into proteins III, IV and V. These proteins are used in the assembly of progeny phages. A few parental capsids attached to the cell membrane are also used to assemble new phage particles.

7. Assembly of Phages

Major capsid protein is added to the inner surface of cell membrane to form a helical capsid: It is composed of B- protein (or IV protein). The (+) strand of phage DNA is inserted while the length of the capsid is increasing. Protein V helps for packaging of the DNA in the capsid. Finally, A- protein (protein III) is added to the ends of the viral coat. As a result, mature Ml 3 phages are formed on the surface of cell membrane.

8. Release of Progeny phages

The complete M13 phage is extruded out through F-pilus. M 13 phages establish permanent infection without 1ysis and produce 300 particles / infected cell.

Application of M13 Phage

☆ It acts as a cloning vector. Replicative form (RFJf M13 is the source for the vectors like $M13MP_1$, M13 MP_2, MI3 MP_7 and $M13MP_8$.

☆ Insertion of Lac Z gene in M13RF leads to $M13MP_1$

☆ Site specific mutation of $M13MP_1$ leads to $M13MP_2$.

☆ Other vectors are constructed from Ml $3MP_2$ by using restriction sites and polylinkers.

☆ Large sized DNA segments can be packed within MI3 phage due to its filamentous nature.

☆ Host cells are not lysed while releasing mature ages because M13 is a leaky phage.

☆ M13 DNA is used for constructing other plasmids like pUC and phagmids. *e.g. pEMBL8.*

Enterobacteria Phage T4

Group	–	Group I (dsDNA)
Order	–	*Caudovirales*
Family	–	*Myoviridae*
Genus	–	*T4-like viruses*
Species	–	*T4 Phage*

Enterobacteria phage T4 is a phage that infects *E. coli* bacteria. Its DNA is 169-170 kbp long; one of the longest DNAs in phages, and is held in an icosahedral head. T4 is also one of the largest phages, at approximately 90 nm wide and 200 nm long (most phages range from 25 to 200 nm in length). Its tail fibres allow attachment to a host cell, and the T4's tail is hollow so that it can pass its nucleic acid to the cell it is infecting during attachment. T4 is only capable of undergoing a lytic lifecycle and not the lysogenic life cycle.

Life Cycle

The lytic lifecycle (from entering a bacterium to its destruction) takes approximately 30 minutes (at 37 °C) and consists of:

- ✰ Adsorption and penetration (starting immediately)
- ✰ Arrest of host gene expression (starting immediately)
- ✰ Enzyme synthesis (starting after 5 minutes)
- ✰ DNA replication (starting after 10 minutes)
- ✰ Formation of new virus particles (starting after 12 minutes)

After the lifecycle is complete the host cell bursts open and ejects the newly built viruses into the environment, at which point the host cell is destroyed.

Infection Process

The T4 Phage initiates infection of an *E. coli* bacterium by recognizing cell surface receptors of the host with its long tail fibers (LTF). A recognition signal is sent through the LTFs to the baseplate. This unravels the short tail fibers (STF) that bind irreversibly to the *E. coli* cell surface. The baseplate changes conformation and the tail sheath contracts causing GP5 at the end of the tail tube to puncture the outer membrane of the cell. The lysozyme domain of GP5 is activated and degrades the periplasmic peptidogliycan layer. The remaining part of the membrane is degraded and DNA from the head of the Phage can travel through the tail tube and enter the *E. coli*.

Interesting Features

The T4 phage has some unique features, such as:

Eukaryote-like introns

- ✰ High speed DNA copying mechanism, with only 1 error in 300 copies
- ✰ Special DNA repair mechanisms
- ✰ It infects *E. coli* O157:H7
- ✰ Genome terminally redundant
- ✰ Genome first replicated as a unit, and then several genomic units are recombined end-to-end to form a concatemer. When packaged, the concatemer is cut at unspecific positions but of same length, leading to several genomes that represent Circular permutations of the original.

In addition, a number of Nobel Prize winners worked with phage T4 or T4-like phages including Max Delbrück, Salvador Luria, Alfred Hershey, James D. Watson, and Francis Crick. Other important scientists who worked with phage T4 include Michael Rossmann, Seymour Benzer, Bruce Alberts, Gisela Mosig, Richard Lenski, and James Bull.

Phi X 174

| Group | – | *Group II (ssDNA)* |
| Family | – | *Microviridae* |

Genus	–	*Microvirus*
Species	–	*phi X 174 phage*

PhiX174 was the first DNA virus discovered to have a single-stranded, circular genome. PhiX174 DNA is 5386 nt in length. The DNA strand packaged into the virion is termed the "plus" strand. After entering the cell, PhiX174 DNA is used as a template for minus-strand synthesis, producing double-stranded DNA. The conversion of plus DNA strands to double-strands does not require any of the phage genes to function. The double-stranded DNA can then be transcribed, resulting in synthesis of phage-encoded proteins. Synthesis of single-stranded (plus) DNA requires the phage-encoded gene A protein. DNA synthesis is initiated at ori (+) and proceeds in the direction indicated. Late in infection, the single-stranded circles are encapsidated into new virions. The cycle terminates by cellular lysis, mediated by phage gene E encoded protein.

The genes identified in phage PhiX174 are shown on the map. All genes are transcribed clockwise. Enumeration of phage DNA begins with the unique PstI site and continues clockwise around the viral (+) strand in the 5'=>3' direction. The map shows enzymes that cut PhiX174 DNA once. Enzymes produced by Fermentas are shown in blue. The coordinates refer to the position of the first nucleotide in each recognition sequence.

Enzymes which Cut PhiX174 DNA Once

AasI 5171, *AatII* 2782, *AdeI* 5183, *Alw44I* 4779, *BaeI* 2581, *BcnI* 2800, *BoxI* 1694, *BseSI* 4779, *Bsh1285I* 4601, *BtsI* 2203, *Cfr42I* 2859, *Eam1105I* 1760, *Eco47I* 5042, *Eco88I* 162, *Eco147I* 4486, *FalI* 3172, *MbiI* 530, *MunI* 3939, *NsbI* 155, *OliI* 2912, *PauI* 5348, *PsiI* 2304, *PstI* 5382, *SapI* 3745, *SexAI* 3499, *SspI* 1007, *XhoI* 162. (Figure 12).

Coordinates of PhiX174 Genes (Termination Codons Included)

A 3981-136, *A** 4497-136, *B* 5075-51,*C* 133-393, *D* 390-848,*E* 568-843,*F* 1001-2284,*G* 2395-2922, *H* 2931-3917,*J* 848-964,*K* 51-221.

Lambda Phage

Group	–	*Group I (dsDNA)*
Order	–	*Caudovirales*
Family	–	*Siphoviridae*
Genus	–	*λ-like viruses*
Species	–	*λ Phage*

Enterobacteria phage λ (lambda phage) is a temperate bacteriophage that infects *Escherichia coli*.

Structure

Lambda phage is a virus particle consisting of a head, containing double-stranded linear DNA as its genetic material, and a tail that can have tail fibers. The

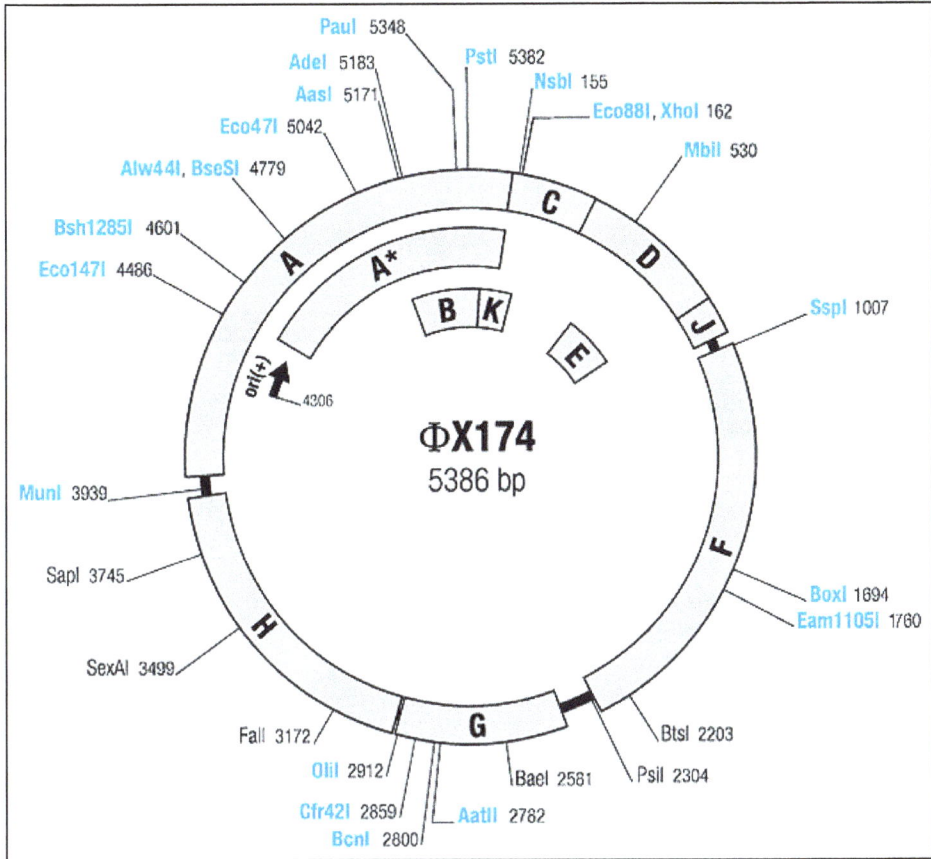

Figure 12. Structure of Phi X 174

phage particle injects its DNA into its host through the tail, and the phage will then usually enter the lytic pathway where it replicates its DNA, degrades the host DNA and hijacks the cell's replication, transcription and translation mechanisms to produce as many phage particles as cell resources allow. When cell resources are depleted, the phage will lyse (break open) the host cell, releasing the new phage particles. However, under certain conditions, the phage DNA may integrate itself into the host cell chromosome in the lysogenic pathway. In this state, the λ DNA is called a prophage and stays resident within the host's genome without apparent harm to the host, which can be termed a lysogen when a prophage is present. The prophage is duplicated with every subsequent cell division of the host. The phage genes expressed in this dormant state code for proteins that repress expression of other phage genes. These proteins are broken down when the host cell is under stress, resulting in the expression of the repressed phage genes. Stress can be from starvation, poisons (like antibiotics), or other factors that can damage or destroy the host. In response to stress, the activated prophage is excised from the DNA of the host cell by one of the newly expressed gene products and enters its lytic pathway.

Lambda phage was discovered by Esther Lederberg in 1950 It has been used heavily as a model organism, and has been a rich source for useful tools in molecular biology. Uses include its application as a vector for the cloning of recombinant DNA, the use of its site specific recombinase, int, for the shuffling of cloned DNAs by the 'Gateway' method, and the application of its Red operon, including the proteins Red alpha (also called 'exo'), beta and gamma in the DNA engineering method called recombineering.

In the following page, we will write genes in italics and their associated proteins in Roman. For instance, *cI* refers to the gene, while cI is the resulting protein encoded by that gene.

Anatomy

The virus particle consists of a head and a tail that can have tail fibres. The head contains 48,490 base pairs of double-stranded, linear DNA flanked by 12-base-pair, single-stranded segments that make up the two strands of the *cos* site. In its circular form in the host cytoplasm, the phage genome therefore is 48,502 base pairs in length. The prophage exists as a linear section of DNA inserted into the host chromosome (Figure 13).

Genetic Map of Lambda Phage

Phage lambda genome is a linear dsDNA consisting of 48,502 basepairs. The genetic map has about 60 genes, of which most of them doing related functions are grouped in clusters and transcribed as polycistronic mRNAs. The other genes are found as individual genes along the genetic map. Genes associated with following functions are clustered together in the genetic map:

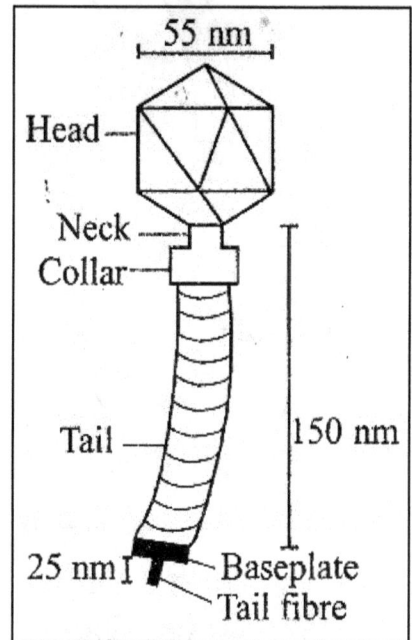

Figure 13. Structure of Lambda Phage

Genes for head synthesis, Genes for tail synthesis, Genes for expression and integration, Genes for recombination, Genes for repression of expression, Genes for DNA synthesis, Genes for cell lysis.

The genes for control of early genes, control of late genes, phage maturation and integration occur as individual genes. Genetic map of lambda phage is illustrated in Figure 14.

cro (Control of Repressor's Operator)

Transcription inhibitor, binds OR3, OR2 and OR1 (affinity OR3 > OR2 = OR1, *i.e.* preferentially binds OR3). At low concentrations blocks the pRM promoter (preventing cI production). At high concentrations downregulates its own production through OR2 and OR1 binding. No cooperative binding (c.f. below for cI binding)

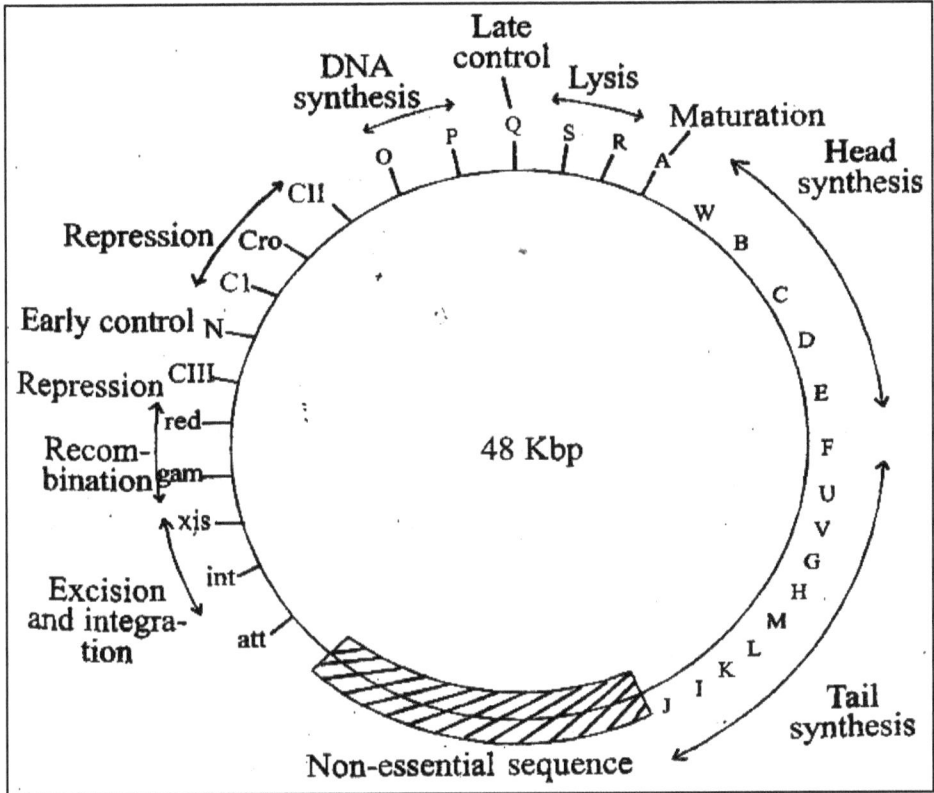

Figure 14. Genetic Map of Lambda Phage

cI; (Clear 1)

Transcription inhibitor, binds OR1, OR2 and OR3 (affinity OR1 > OR2 = OR3, *i.e.* preferentially binds OR1). At low concentrations blocks the pR promoter (preventing cro production). At high concentrations downregulates its own production through OR3 binding. Binding of cI at OR1 stimulates an almost simultaneous cI binding to OR2 via cooperative binding (via cI C terminal domain interactions) N terminal domain of cI on OR2 tightens the binding of RNA polymerase to pRM and hence stimulate its own transcription. Repressor also inhibits transcription from the pL promoter. Susceptible to cleavage by RecA* in cells undergoing the SOS response.

cII; (Clear 2) Transcription Activator

Activates transcription from the pAQ, pRE and pI promoters. Low stability due to susceptibility to cellular HflB (FtsH) proteases (especially in healthy cells and cells undergoing the SOS response).

cIII; (Clear 3)

HflB (FtsH) binding protein, protects cII from degradation by proteases.

N; (aNtiterminator)

RNA binding protein and RNA polymerase cofactor, binds RNA (at Nut sites) and transfers onto the nascent RNApol that just transcribed the nut site. This RNApol modification prevents its recognition of termination sites, so normal RNA polymerase termination signals are ignored and RNA synthesis continues into distal phage genes.

Q

DNA binding protein and RNApol cofactor, binds DNA (at Qut sites) and transfers onto the initiating RNApol. This RNApol modification alters its recognition of termination sequences, so normal ones are ignored; special Q termination sequences some 20,000 bp away are effective.

xis; (eXclSion)

Excisionase and Int protein regulator, manages excision and insertion of phage genome into the host's genome.

int; (INTegration)

Int protein, manages insertion of phage genome into the host's genome. In Conditions of low int concentration there is no effect. If xis is low in concentration and int high then this leads to the insertion of the phage genome. If xis and int have high (and approximately equal) concentrations this leads to the excision of phage genomes from the host's genome.

Repressor

The repressor found in the phage lambda is a notable example of the level of control possible over gene expression by a very simple system. It forms a 'binary switch' with two genes under mutually exclusive expression, as discovered by Barbara J. Meyer.

The lambda repressor gene system consists of (from left to right on the chromosome):

cI gene,OR3,OR2,OR1 and *cro* gene

The lambda repressor is a dimer also known as the cI protein. It regulates the transcription of the cI protein and the Cro protein.

The life cycle of lambda phages is controlled by cI and Cro proteins. The lambda phage will remain in the lysogenic state if cI proteins predominate, but will be transformed into the lytic cycle if cro proteins predominate.

The cI dimer may bind to any of three operators, OR1, OR2, and OR3, in the order OR1 = OR2 > OR3. Binding of a cI dimer to OR1 enhances binding of a second cI dimer to OR2, an effect called cooperativity. Thus, OR1 and OR2 are almost always simultaneously occupied by cI. However, this does not increase the affinity between cI and OR3, which will be occupied only when the cI concentration is high. This cooperative action is shown by the relative affinity of the repressor for the native sequences individually, which is OR1 > OR2 = OR3; different from the actual order of binding.

In the absence of cI proteins, the *cro* gene may be transcribed.

In the presence of cI proteins, only the *cI* gene may be transcribed.

At high concentration of cI, transcriptions of both genes are repressed.

Lifecycle

Infection

1. Bacteriophage Lambda binds to the target *E. coli* cell, the J protein in the tail tip interacting with the lamB gene product of *E. coli*, a porin molecule which is part of the maltose operon.

2. The linear phage genome is injected past the cell outer membrane.

3. The DNA passes through a separate sugar transport protein (ptsG) in the inner membrane, and immediately circularises using the *cos* sites, 12-base G-C rich cohesive "sticky ends". The single-stranded nicks are ligated by host DNA ligase.

4. Host DNA gyrase puts negative supercoils in the circular chromosome, causing A-T rich regions to unwind and drive transcription.

5. Transcription starts from the constitutive P_L, P_R and $P_{R'}$ promoters producing the 'immediate early' transcripts. Initially these express the *N* and *cro* genes, producing N, Cro and a short inactive protein.

6. Cro binds to *OR3* preventing access to the P_{RM} promoter preventing expression of the *cI* gene. N binds to the two *Nut* (N utilisation) sites, one in the *N* gene in the P_L reading frame, and one in the *cro* gene in the P_R reading frame.

7. The N protein is an antiterminator, and functions to extend the reading frames that it is bound to. When RNA polymerase transcribes these regions, it recruits the N and forms a complex with several host Nus proteins. This complex skips through most termination sequences. The extended transcripts (the 'late early' transcripts) include the *N* and *cro* genes along with *cII* and *cIII* genes, and *xis*, *int*, *OP* and *Q* genes discussed later.

8. The cIII protein acts to protect the cII protein from proteolysis by FtsH (a membrane-bound essential E. coli protease) by acting as a competitive inhibitor. This inhibition can induce a bacteriostatic state, which favours lysogeny. cIII also directly stabilises the cII protein On initial infection, the stability of cII determines the lifestyle of the phage; stable cII will lead to the lysogenic pathway, whereas if cII is degraded the phage will go into the lytic pathway. Low temperature, starvation of the cells and high multiplicity of infection (MOI) are known to favor lysogeny.

N Antitermination

This occurs without the N protein interacting with the DNA; the protein instead binds to the freshly transcribed mRNA. Nut sites contain 3 conserved "box's", of which only BoxB is essential.

1. The boxB RNA sequences are located close to the 5′ end of the pL and pR transcripts. When transcribed, each sequence forms a hairpin loop structure that the N protein can bind to.

2. N protein binds to boxB in each transcript, and contacts the transcribing RNA polymerase via RNA looping. The N-RNAP complex is stabilized by subsequent binding of several host Nus (N utilisation substance) proteins (which include transcription termination/antitermination factors and, bizarrely, a ribosome subunit).

3. The entire complex (including the bound *Nut* site on the mRNA) continues transcription, and can skip through termination sequences.

Lytic Lifestyle

This is the lifecycle that the phage follows following most infections, where the cII protein does not reach a high enough concentration due to degradation, so does not activate its promoters.

1. The 'late early' transcripts continue being written, including *xis*, *int*, Q and genes for replication of the lambda genome (*OP*). Cro dominates the repressor site (see "Repressor"), repressing synthesis from the P_{RM} promoter.

2. The O and P proteins initiate replication of the phage chromosome.

3. Q, another antiterminator, binds to *Qut* sites.

4. Transcription from the $P_{R'}$ promoter can now extend to produce mRNA for the lysis and the head and tail proteins.

5. Structural proteins and phage genomes self assemble into new phage particles.

6. Products of the lysis genes *R* and *S*, cause cell lysis at high enough concentrations. S is a holin which makes holes in the membrane. R is an endolysin which cleaves the cell wall. Around 100 new phage are released.

Rightward Transcription

Rightward transcription expresses the *O*, *P* and *Q* genes. O and P are responsible for initiating replication, and Q is another antiterminator which allows the expression of head, tail and lysis genes from $P_{R'}$.

Lytic Replication

1. For the first few replication cycles, the lambda genome undergoes θ replication (circle-to-circle).

2. This is initiated at the *ori* site located in the *O* gene. O protein binds the *ori* site, and P protein binds the DnaB subunit of the host replication machinery as well as binding O. This effectively commandeers the host DNA polymerase.

3. Soon, the phage switches to a rolling-circle type of replication similar to that used by phage M13. The DNA is nicked and the 3′ end serves as a

primer. Notably, this doesn't release single copies of the phage genome, but rather one long molecule with many copies of the genome: a concatemer.

4. These concatemers are cleaved at their *cos* sites as they are packaged. Packaging cannot occur from circular phage DNA, only from concatomeric DNA.

Q Antitermination

Q is similar to N in its effect: Q binds to RNA polymerase in *Qut* sites and the resulting complex can ignore terminators, however the mechanism is very different; the Q protein first associates with a DNA sequence rather than an mRNA sequence.

1. The *Qut* site is very close to the $P_{R'}$ promoter, close enough that the σ factor has not been released from the RNA polymerase holoenzyme. Part of the *Qut* site resembles the -10 Pribnow box, causing the holoenzyme to pause.

2. Q protein then binds and displaces part of the σ factor and transcription re-initiates.

3. The head and tail proteins are transcribed and self-assemble.

Leftward Transcription

Leftward transcription expresses the *gam*, *red*, *xis* and *int* genes. Gam and red proteins are involved in recombination. Gam is also important in that it inhibits the host RecBCD nuclease from degrading the 3' ends in rolling circle replication. Int and xis are integration and excision proteins which are vital to lysogeny.

xis and *int* Regulation of Insertion and Excision

1. *xis* and *int* are found on the same piece of mRNA, so approximately equal concentrations of xis and int proteins are produced. This results (initially) in the excision of any inserted genomes from the host genome.

2. The mRNA from the P_L promoter forms a stable secondary structure with a bobby pin loop in the *sib* section of the mRNA. This targets the 3' (*sib*) end of the mRNA for RNAaseIII degradation, which results in a lower effective concentration of *int* mRNA than *xis* mRNA (as the *int* cistron is nearer to the *sib* sequence than the *xis* cistron is to the *sib* sequence), so a higher concentrations of xis than int is observed.

3. Higher concentrations of xis than int result in no insertion or excision of phage genomes, the evolutionarily favoured action–leaving any pre-inserted phage genomes inserted (so reducing competition) and preventing the insertion of the phage genome into the genome of a doomed host.

Lysogenic (or Lysenogenic) Lifestyle

This is the lifecycle that the phage follows after a small number of infections in specific conditions, where the cII protein reaches a high enough concentration due to stabilisation and lack of degradation, and so activates its promoters.

1. The 'late early' transcripts continue being written, including *xis*, *int*, Q and genes for replication of the lambda genome.

2. The stabilized cII acts to promote transcription from the P_{RE}, P_I and P_{antiq} promoters.

3. The P_{antiq} promoter produces antisense mRNA to the Q gene message of the P_R promoter transcript, thereby switching off Q production. The P_{RE} promoter produces antisense mRNA to the cro section of the P_R promoter transcript, turning down cro production, and has a transcript of the *cI* gene. This is expressed, turning on cI repressor production. The P_I promoter expresses the *int* gene, resulting in high concentrations of int protein. This int protein integrates the phage DNA into the host chromosome (see "Prophage Integration").

4. No Q results in no extension of the $P_{R'}$ promoter's reading frame, so no lytic or structural proteins are made. Elevated levels of int (much higher than that of xis) result in the insertion of the lambda genome into the hosts genome (see diagram). Production of cI leads to the binding of cI to the *OR1* and *OR2* sites in the P_R promoter, turning off *cro* and other early gene expression. cI also binds to the P_L promoter, turning off transcription there too.

5. Lack of cro leaves the *OR3* site unbound, so transcription from the P_{RM} promoter may occur, maintaining levels of cI.

6. Lack of transcription from the P_L and P_R promoters leads to no further production of cII and cIII.

7. As cII and cIII concentrations decrease, transcription from the P_{antiq}, P_{RE} and P_I stop being promoted since they are no longer needed.

8. Only the P_{RM} and $P_{R'}$ promoters are left active, the former producing cI protein and the latter a short inactive transcript. The genome remains inserted into the host genome in a dormant state.

Prophage Integration

The *integration* of phage λ takes place at a special attachment site in the bacterial and phage genomes, called *att*[1]. The sequence of the bacterial att site is called *attB*, between the *gal* and *bio* operons, and consists of the parts B-O-B', whereas the complementary sequence in the circular phage genome is called *attP* and consists of the parts P-O-P'. The integration itself is a sequential exchange (see genetic recombination) via a Holiday junction and requires both the phage protein Int and the bacterial protein IHF (*integration host factor*). Both Int and IHF bind to *attP* and form an intasome, a DNA-protein-complex designed for site-specific recombination of the phage and host DNA. The original B-O-B' sequence is changed by the integration to B-O-P'-phage DNA-P-O-B'. The phage DNA is now part of the host's genome.

Maintenance of Lysogeny

☆ Lysogeny is maintained solely by cI. cI represses transcription from P_L and P_R while upregulating and controlling its own expression from P_{RM}. It is therefore the only protein expressed by lysogenic phage.

☆ This is coordinated by the P_L and P_R operators. Both operators have three binding sites for cI: *OL1*, *OL2*, and *OL3* for P_L, and *OR1*, *OR2* and *OR3* for P_R.

☆ cI binds most favorably to *OR1*; binding here inhibits transcription from P_R. As cI easily dimerises, the binding of cI to *OR1* greatly increases the affinity of the binding of cI to *OR2*, and this happens almost immediately after *OR1* binding. This activates transcription in the other direction from P_{RM}, as the N terminal domain of cI on *OR2* tightens the binding of RNA polymerase to P_{RM} and hence cI stimulates its own transcription. When it is present at a much higher concentration, it also binds to *OR3*, inhibiting transcription from P_{RM}, thus regulating its own levels in a negative feedback loop.

☆ cI binding to the P_L operator is very similar, except that it has no direct effect on cI transcription. As an additional repression of its own expression, however, cI dimers bound to *OR3* and *OL3* bend the DNA between them to tetramerise.

☆ The presence of cI causes immunity to superinfection by other lambda phages, as it will inhibit their P_L and P_R promoters.

Induction

The classic induction of a lysogen involved irradiating the infected cells with UV light. Any situation where a lysogen undergoes DNA damage or the SOS response of the host is otherwise stimulated leads to induction.

1. The host cell, containing a dormant phage genome, experiences DNA damage due to a high stress environment, and starts to undergo the SOS response.

2. RecA (a cellular protein) detects DNA damage and becomes activated. It is now RecA*, a highly specific co-protease.

3. Normally RecA* binds LexA (a transcription repressor), activating LexA auto-protease activity, which destroys LexA repressor allowing production of DNA repair proteins. In lysogenic cells this response is hijacked, and RecA* stimulates cI autocleavage. This is because cI mimics the structure of LexA at the autocleavage site.

4. Cleaved cI can no longer dimerise, and loses its affinity for DNA binding.

5. The P_R and P_L promoters are no longer repressed and switch on, and the cell returns to the lytic sequence of expression events (note that cII is not stable in cells undergoing the SOS response). There is however one notable difference.

Control of Phage Genome Excision in Induction

Schematic representation of the insertion of the bacteriophage lambda. Note how *sib* is displaced by the recombination event from the N extended P_L promoter open reading frame.

1. The phage genome is still inserted in the host genome and needs excision for DNA replication to occur. The *sib* section beyond the normal P_L promoter

transcript is, however, no longer included in this reading frame (see diagram).

2. No *sib* domain on the P_L promoter mRNA results in no hairpin loop on the 3′ end, and the transcript is no longer targeted for RNAaseIII degradation.

3. The new intact transcript has one copy of both *xis* and *int*, so approximately equal concentrations of xis and int proteins are produced.

4. Equal concentrations of xis and int result in the excision of the inserted genome from the host genome for replication and later phage production.

Limitations of Lambda Vectors

1. For successful packaging of lambda DNA in capsid, the two cos-sites should be separated by distance ranging from 37,000 to 54,000 baepairs. If the distance is lower than 37,000 bp or higher than 54,000 bp, the packaging of DNA does not take place.

2. Sometimes, many lambda DNA join together by base pairing between their cos-sites and form relatively long DNA. Here packaging of DNA does not take place successfully.

3. The lambda DNA has no easily selectable marker gene. So transformants should be identified by DNA hybridization method.

4. The recombinant lambda DNA may enter the lytic life cycle.

5. The lambda phage has narrow host-range.

Advantages

1. The efficiency of gene transfer through phage is high.

2. Foreign DNAs upto 23 kbp can be packed in virus Capsid and transduced to *E. coli*.

P1 Phage

P1 phage is a temperate phage belonging to the Family *Myovridae*. It infects some strains of *Escherichia coli* and *Shigella* sp. P1 phage performs non-integrative lysogenic cycle in its host. In non-integrative lysogenic process, DNA of phage is not integrated into the host chromosome. Viral DNA is maintained within a host cell as like a plasmid. Non-integrative lysogenic life cycle was discovered by K.Ieda and J.Tomizawa.

Structure

Structure of P1 phage is similar to that of T4 phage. It consists of a head and a tail. It is tadpole shaped. The head consists of an outer protein coat called capsid and inner DNA that represents the viral genome. The capsid hexagonal in outline and shows icosahedral symmetry. It is 83-87nm in diameter. The capsid is composed of 152 capsomeres including pentamers and hexamers. The capsid triangulation number is 13 (T= 13).

The tail consists of a neck, collar, baseplate and six tail fibres. The tail is attached to the head by a short tube called neck. Tail is 16-20nm wide and 216 nm long. It is tubular, rigid and thick. Tubular tail is bounded by a sheath of stacked rings. These rings slide over one another during contract so that tail sheath is called contractile sheath. The tail shows helical symmetry. The distal end of tail has a baseplate, which is hexagonal in outline; the base plate is smaller than that of T4 phage, Six tail fibres are found attached to the baseplate. Th tail fibres are long and terminal in position (Figure 15)

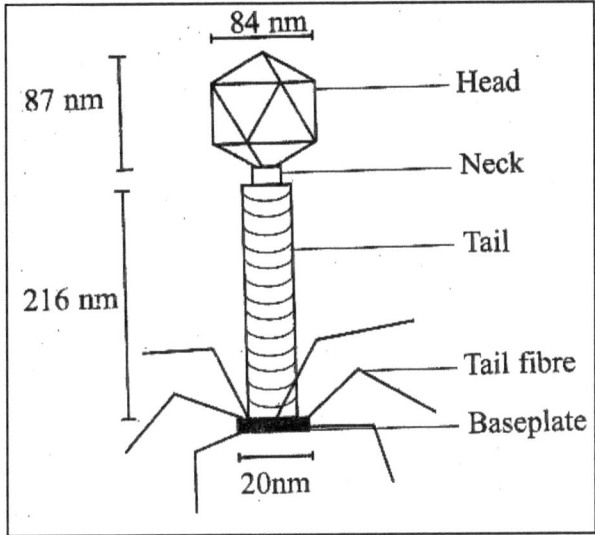

Figure 15. Structure of P1 Phage

Life Cycle

Phage P1 exhibits two different types of life cycle. They are

1. Lytic cycle or Virulent cycle
2. Lysogenic life cycle

Whether lytic cycle or lysogenic cycle has to be followed is decided by a switching on/off mechanism.

1. Lytic Cycle or Virulent Cycle

In the lytic cycle, intracellular multiplication of the phage ends in the lysis of the host bacterium and release of progeny virions. The lytic cycle involves the following steps:

1. Adsorption of p1 phage on the surface of *E.coli* cells
2. Penetration of phage DNA into the host cell.
3. Transcription and protein synthesis
4. Breakdown of host chromosome by nuclease enzymes by phage genome.
5. Replication of phage DNA
6. Assembly of phages from viral components. DNA packaging occurs by a "headful mechanism" starting at Pac site.
7. Release of progeny virions by lysis of host cell.

2. Lysogenic Life Cycle

If lysogenic route is selected, P1 DNA follows the folllowing steps:

☆ The phage nucleic acid is inserted into the host cell following adsorption and penetration.

☆ P1 phage genome does not normally integrate into the host chromosome. The phage DNA is maintained in an autonomous self-replicating state in the form of a plasmid.

☆ There occurs one or two copies of P1 plasmid per host cell (Low copy number).

☆ Replication control system of P1 plasmid involves a plasmid-encoded protein Rep A.

☆ Rep A is essential for replication and is an autoregulator.

☆ Rep A gene is flanked on each side by 19 bp repeat sequences.

☆ One of these repeats is *IncA* which is involved in controlling the copy number of P1 plasmid.

☆ The P1 phage genome is inherited to daughter cells by means of cell division. P1 phage encodes a site-specific recombination (SSR) system. It helps in accurate partition by resolving plasmid dimmers into monomers. SSR system operates at lox p site on the plasmid (P 1).

☆ The insertion of P1 DNA into the vector involves the process of recombination.

☆ Lox P site of P1 is involved in recombination and the process is catalyzed by *Cyclization Recombination Protein* (Crp).

Under certain conditions the plasmid enters in lytic cycle ci produces progeny P1 phages. They are released by lysis of host cell.

Applications of P1 phage

☆ P1 phage vectors are developed at *Natstrenberg laboratory*

☆ P1 phage has been widely used as a vector in molecular cloning.

☆ P1 phage can be used to transfer larger desired foregin DNA to bacteria.

☆ About 100 kb of foreign DNA can be inserted into p1 and can be efficiently packaged into it.

☆ P1 phage infects *E. coli* and p1 DNA along with insert DNA behaves as a plasmid within the host cell

Chapter 11
Plant Viruses

Viruses are very small (submicroscopic) infectious particles (virions) composed of a protein coat and a nucleic acid core. They carry genetic information encoded in their nucleic acid, which typically specifies two or more proteins. Translation of the genome (to produce proteins) or transcription and replication (to produce more nucleic acid) takes place within the host cell and uses some of the host's biochemical "machinery". Viruses do not capture or store free energy and are not functionally active outside their host. They are therefore parasites (and usually pathogens) but are not usually regarded as genuine microorganisms.

Most viruses are restricted to a particular type of host. Some infect bacteria, and are known as bacteriophages, whereas others are known that infect algae, protozoa, fungi (mycoviruses), invertebrates, vertebrates or vascular plants. However, some viruses that are transmitted between vertebrate or plant hosts by feeding insects (vectors) can replicate within both their host and their vector. This web site is mostly concerned with those viruses that infect plants but we also provide some taxonomic and genome information about viruses of fungi, protozoa, vertebrates and invertebrates where these are related to plant viruses.

Virus Classification

The highest level of virus classification recognises six major groups, based on the nature of the genome:

Double-stranded DNA (dsDNA)

There are no plant viruses in this group, which is defined to include only those viruses that replicate without an RNA intermediate. It includes those viruses with the largest known genomes (up to about 400,000 base pairs) and there is only one genome component, which may be linear or circular. Well-known viruses in this group include the herpes and pox viruses.

Single-stranded DNA (ssDNA)

There are two families of plant viruses in this group and both of these have small circular genome components, often with two or more segments.

Reverse-transcribing Viruses

These have dsDNA or ssRNA genomes and their replication includes the synthesis of DNA from RNA by the enzyme reverse transcriptase; many integrate into their host genomes. The group includes the retroviruses, of which Human immunodeficiency virus (HIV), the cause of AIDS, is a member. There is a single family of plant viruses in this group and this is characterised by a single component of circular dsDNA, the replication of which is *via* an RNA intermediate.

Double-stranded RNA (dsRNA)

Some plant viruses and many of the mycoviruses are included in this group.

Negative Sense Single-stranded RNA (ssRNA-)

In this group, some or all of the genes are translated into protein from an RNA strand complementary to that of the genome (as packaged in the virus particle). There are some plant viruses in this group and it also includes the viruses that cause measles, influenza and rabies.

Positive Sense Single-stranded RNA (ssRNA+)

The majority of plant viruses are included in this group. It also includes the SARS coronavirus and many other viruses that cause respiratory diseases (including the "common cold"), and the causal agents of polio and foot-and-mouth disease.

Within each of these groups, many different characteristics are used to classify the viruses into families, genera and species. Typically, a combination of characters are used and some of the most important are:

Particle Morphology

The shape and size of particles as seen under the electron microscope.

Genome Properties

This includes the number of genome components and the translation strategy. Where genome sequences have been determined, the relatedness of different sequences is often an important factor in discriminating between species.

Biological Properties

This may include the type of host and also the mode of transmission.

Serological Properties

The relatedness (or otherwise) of the virion protein(s).

Particle Morphology

Amongst plant viruses, the most frequently encountered shapes are: (Figures 16, 17, 18, 19 and 20)

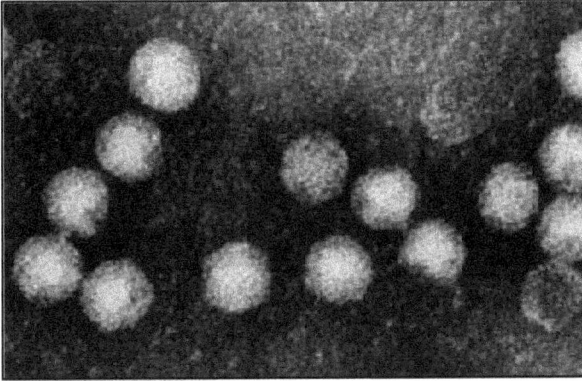

Figure 16. Isometric: Apparently Spherical and (Depending on the species) from About 18nm in Diameter Upwards. The example here shows *Tobacco necrosis virus*, genus *Necrovirus* with particles 26nm in diameter.

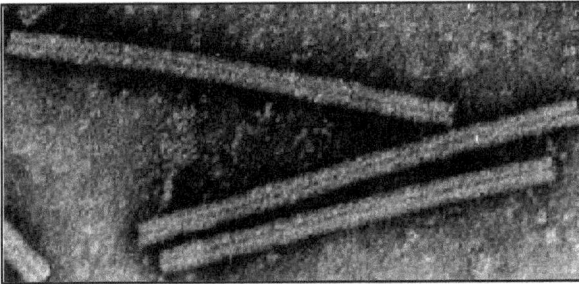

Figure 17. Rod-Shaped: About 20–25 nm in Diameter and from About 100 to 300 nm Long. These appear rigid and often have a clear central canal (depending on the staining method used). Some viruses have two or more different lengths of particle and these contain different genome components. The example here shows *Tobacco mosaic virus*, genus *Tobamovirus* with particles 300 nm long.

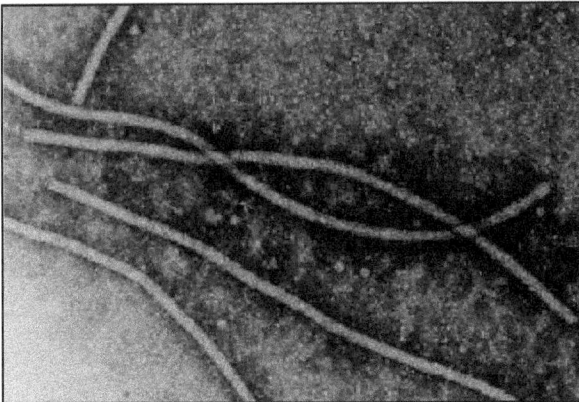

Figure 18. Filamentous: Usually About 12 nm in Diameter and More Flexuous than the Rod-shaped Particles. They can be up to 1000 nm long, or even longer in some instances. Some viruses have two or more different lengths of particle and these contain different genome components. The example here shows *Potato virus Y*, genus *Potyvirus* with particles 740 nm long.

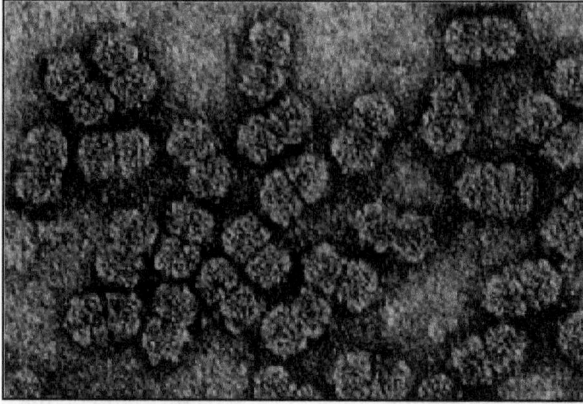

**Figure 19. Geminate: Twinned Isometric Particles about 30 x 18 nm.
These particles are diagnostic for viruses in the family *Geminiviridae*
which are widespread in many crops especially in tropical regions.
The example here shows *Maize streak virus*, genus *Mastrevirus*.**

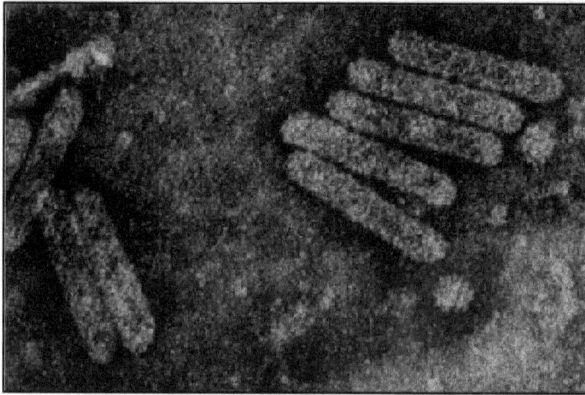

**Figure 20. Bacilliform: Short Round-ended Rods. These come in
various forms up to about 30 nm wide and 300 nm long. The example here shows
Cocoa swollen shoot virus, genus *Badnavirus* with particles 28 x 130 nm.**

Genome Properties

Important features include:

(a) Nature of the Genome

Circular (as in all known plant DNA viruses) or linear.

(b) Number of Genome Components

This varies from a single component (*e.g.* in the genera *Potyvirus* and *Tobamovirus*) to 11 (in some members of the genus *Nanovirus*). Individual components vary in size from about 1kb (*Nanovirus* components) to about 20 kb (in the genus *Closterovirus*).

(c) Number of Genes

These vary considerably. Most plant viruses have at least 3 genes: 1 (or more) concerned with replication of the nucleic acid, 1 (or more) concerned with cell-to-cell movement of the virus and 1 (or more) encoding a structural protein that is assembled into the virus particle (usually called the "coat" or "capsid" protein). There may also be additional genes that have a regulatory function or which are required for transmission between plants (association with a vector).

(d) Translation Strategy

A variety of strategies are employed to translate the genes from the genome components either directly or via mRNA intermediates and (in some cases) to permit different amounts of protein to be produced from the different genes. These are summarised for each genus in the genus description pages but 3 examples here serve to illustrate some of the variety:

Genus *Potyvirus*

In this very large genus, there is one ssRNA component that encodes one large (*c.* 350 kDa) polyprotein. This is cleaved by 3 different proteases (all encoded by the virus itself) into 10 different mature proteins. The two proteins at the C-terminus of the polyprotein are respectively an RNA-dependent RNA polymerase (NIb, involved in replication of the virus) and the (single) coat protein (CP). Many of the proteins have multiple functions. The genome organisation of a typical member is shown here, indicating the 10 mature proteins and the nine cleavage sites (arrowed) (Figure 21).

Genus *Furovirus*: in this genus there are two ssRNA components. The 5'-proximal gene on each RNA is translated directly from the genomic RNA: on RNA1 (the larger RNA component) this gene encodes a replication protein and on RNA2 it is the coat protein. The stop codons of both of these genes are "leaky" and in a small percentage of cases, translation continues to produce a larger ("readthrough") protein. On RNA1, the replication protein is extended to include an RNA-dependent RNA polymerase (RdRp) while the readthrough region of the coat protein is probably required for particle assembly and for transmission by the plasmodiophorid vector. There is a further (3'-proximal) gene on each of the RNAs and these are translated from shorter RNA molecules transcribed from the 3'-end of the genomic RNA ("subgenomic" mRNAs). That from RNA1 is a cell-to-cell movement protein (MP) that enables the virus to move between adjacent plant cells via the plasmodesmata while the function of the product from RNA2 is uncertain but may involve supression of the host plant defence reaction. The genome organisation of a typical member is shown here (Figure 22).

Genus *Fijivirus*

In this genus there are 10 components of dsRNA. Most of the components encode a single protein and at least 3 of these are structural proteins assembled into the complex virion.

Genome Relatedness

The degree of nucleotide identity (or amino acid identity in the protein sequence) between sequences is often used to examine the relationship between different viruses

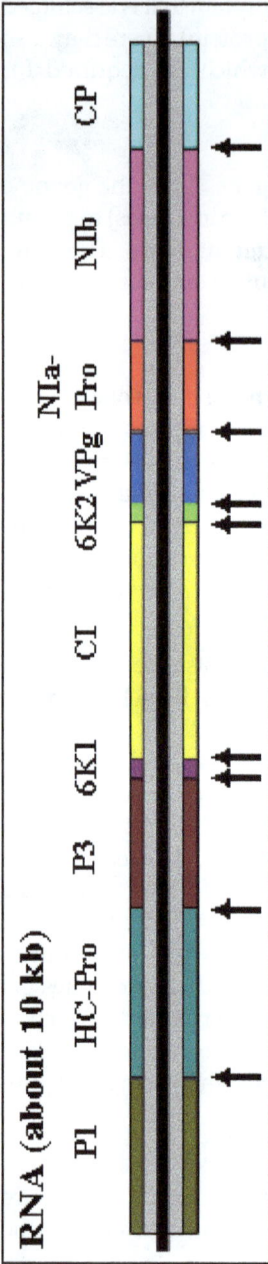

RNA (about 10 kb)

P1   HC-Pro   P3   6K1   CI   6K2 VPg   NIa-Pro   NIb   CP

Figure 21. Genus Potyvirus

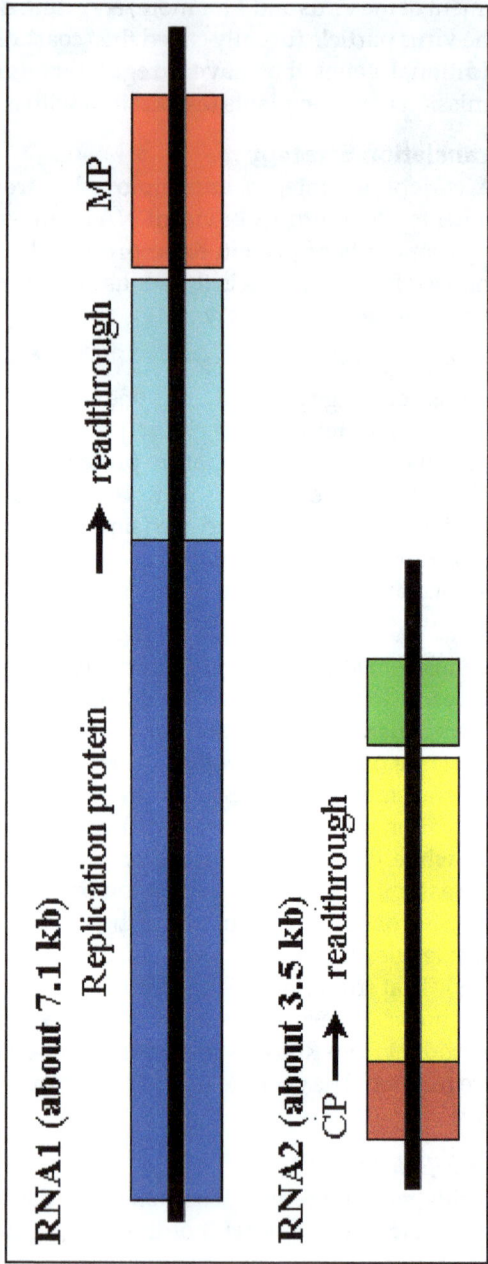

RNA1 (about 7.1 kb)
Replication protein → readthrough   MP

RNA2 (about 3.5 kb)
CP → readthrough

Figure 22. Genus *Furovirus*

or isolates. For example, recent studies in the genus *Carlavirus* show that when different species are compared, they have less than 73 per cent nucleotide identity (or 80 per cent amino acid identity) in their coat proteins.

Biological Properties

☆ In some families, the type of host is a useful feature for classification. For example, in the family *Reoviridae*, there are currently 3 genera with plant-infecting members (*Fijivirus, Oryzavirus, Phytoreovirus*), 1 genus of mycoviruses (*Mycoreovirus*), 1 genus containing viruses of fish and cephalopods (*Aquareovirus*), two genera that are restricted to insects (*Cypovirus* and *Entomoreovirus*) and 5 genera of vertebrate viruses that sometimes also infect insects.

☆ The mode of transmission is also a useful characteristic of some groups of plant viruses. For example in the family *Potyviridae*, members of the largest genus (*Potyvirus*) are transmitted by aphids, while viruses in the genera *Rymovirus* and *Tritimovirus* are transmitted by mites of the genus *Abacarus* or *Aceria* respectively, those in the genus *Ipomovirus* are transmitted by whiteflies and those in the genus *Bymovirus* by plasmodiphorids (root-infecting parasites once considered to be fungi but probably more closely related to protists).

Serological Properties

Many viruses are good antigens (elicit strong antibody production when purified preparations are injected into a mammal) and this property has been widely exploited to produce specific antibodies that can be used for virus detection and for examining relationships between viruses. Earlier studies used agar diffusion plates but in the last 20 years these have been largely superseded by ELISA (enzyme-linked immunosorbent assay) procedures. Although serological properties are still important, their significance in taxonomy has declined to some extent now that nucleotide sequence data are available.

Symptoms of Viral Diseases in Plants

How a plant looks after infection with a virulent virus is called *disease symptom*. Virus may affect the metabolism and growth of a susceptible plant and disturb the normal function of the infected hosts. Host-virus interaction is mainly detected in the form of symptoms. Any modification in natural structure or function of plants is called symptom. Infection may result in the formation of a group of symptoms called *syndrome*.

Viruses cause several kinds of disease symptoms in plants. They may cause systemic infection, associated with each stage of life cycle of plant or localized infection, being restricted to site of infection only. In systemic infection, the virus infects the plant at its early stages of development and persists throughout the plant. Field crops show such systemic symptoms. If the viruses are artificially inoculated, plants show lesions at the site of infection alone and the symptoms are called *localized symptoms*.

In the production point of view, viral infection reduces growth, leading to dwarfing or stunting of the entire plant and reduction in the total yield. Many viruses may infect some hosts without causing visible symptoms. Such viruses are called latent viruses and the hosts are called *symptomless carriers*. Plants may show acute or severe symptoms soon after infection.

The most common type of symptoms in plants due to virus infections is mosaic pattern in leaves. The other common symptoms are stunt, dwarf, leaf roll, chlorosis, necrosis, yellows, streak, enation, tumours, pitting of sterm, pitting of fruit, flattening, distortion of stem, etc.

Virus infections produce *morphological changes (external symptoms), histological changes (internal symptoms) and physiological changes* in plants. These three types of Symptoms are discussed hereunder:

I. Morphological Changes or External Symptoms

Disease symptoms that can be seen on the plant surface with naked eye are called external symptoms or morphoIohical changes.

The following are important morphological changes which are expressed in the form of external symptoms in diseased plants:

1. Stunting

Reduction in the growth of different parts of a plant is called stunting. Viral infection reduces the growth rate. It affects all parts of plant such as leaves, flowers, fruits, petioles and internodes. Stunting results in reduction of yield some parts of the plant show much more pronounced effect of stunting than others. For example, in little cherry disease, fruit remains small due to reduced cell division in spite of good plant growth.

2. Mosaic

A pattern of alternate light green patches with dark green areas on leaves of plants is known as a mosaic. Such symtoms are seen on leaf blade, leaf sheath, floral parts and other green parts of plants. The mosaic pattern varies with the virus host interaction. In most of the dicotyledons, the mosaic is composed of irregular outlines of dark green and pale or yellow green patches. But in monocots, it appears to be streaks.

3. Vein Clearing and Vein Banding

Chlorosis of tissues adjacent to veins is known as clearing. A decolourized line is therefore seen on either side of the veins of leaves. Vein clearing occurs in before the mosaic or mottle. Formation of dark-green line on either side of vein of leaves is called vein banding. This occurs due to accumulation of chlorophylls in parenchyma cells adjacent to veins. It may be a transient stage of viral infection or formed fore mosaic formation.

4. Stripes and Streaks

Long narrow bands, which are parallel to each other, found on the surface of leaves are called stripes. They may be due to chlorosis of infected leaves. Stripes may be yellow or brown or dark in colour in mature leaves, but they become necrotic at older stage. Broken stripes are called streaks. In general, stripes are confined to monocot plants, whereas, streaks are common in dicot plants.

5. Variegation

A pattern of white patches in leaves or flowers or other green parts of plants is called variegation. It occurs due to failure of infected tissue to produce chlorophylls. The white patches seem to be flecks or streaks. In flowers, white patches are developed due to halting of anthocyanine production.

6. Fruit Abnormalities

A variety of symptoms may develop on the fruit in certain cases. Fruit of cucumber infected with (cucumber Mosaic Virus (CMV) are small and deformed. Tomato fruits show mottling when infected with tomato mosaic virus and concentric rings when infected with tomato spotted wilt virus (or bushy stunt virus. Papaya and apple fruits show mosaic or ring spot when plants are infected with papaya ring spot and apple ring spot viruses respectively.

7. Yellows

The initial symptom usually consists of clearing and yellowing of the veins in the younger leaves. The yellowing may be complete or partial or limited to chiorotic spots without mosaic pattern. The yellowing may be confirmed to leaf margin.

8. Necrosis

Rapid death of cells or tissue at the site of infection is called necrosis. It is usually accompanied by browning or blackening of the infected area. It may be in the form of superficial lesion or of extensive type. Some viruses cause vein necrosis or necrotic streaking of petioles and stem as in potato infected with Potato Virus Y (PVY).

Necrosis may affect stem tips or buds (tip necrosis/bud necrosis) and may result in degeneration of phloem in vascular bundle (phloem necrosis).

In some potato cultivars, the streaking spreads to the growing point. This results in death of the plant, when infected with Potato Virus X (PVX) and Potato Virus Y (PVY). Some potato virus infection may lead to tuber necrosis.

9. Leaf Rolling

Margin of infected leaves folds upward or rarely down.wards and turns on its axis to form a leaf role. A leaf has to such rolls on either side of the midrib.

10. Curling

Abnormal bending of shoots or leaves due to virus infection is called curling. This is mainly due to localized over growth of cells or tissues at the site of infection.

11. Ring Spots

These are circular areas in the green parts of the plant in some cases, the ring spot may be irregular in outline, but in some others it may have more than one concentric ring. If the ring spot appears to be a chlorotic area, then it is known as chlorotic ring spot and if it is formed of necrotic tissue, it is called necrotic ring spot. Infections with ring spot viruses usually lead to ring spot symptoms.

12. Growth Abnormalities

The growth abnormalities lead to distortions and malformation of leaves, stems, roots and shortemng of internodes.The common growth abnormalities are as follows

(i) Witche's Broom

The leaves become much reduced and inter-nodes become shortened. This is an abnomal growth of leaves, turning to a densely packed broom like structure

(i) Little Leaf

Here the leaves are reduced in size. In fern-leaf, there is much suppression of the lamina.

(iii) Blistering

The uneven growth of leaf larnina with dark green raised area is referred as blistering.

(iv) Fern Leaf Effect I Shoe String/Fill Form

Twisting, elongation and turning of leaf to needle shape is known as fern leaf effect/shoe string /fili form.

(v) Enation

It is the out growth from the lower surface of the leaf veins.

(vi) Stem Abnormalities

Some viruses cause abnormalities such as cocoa swollen shoot, apple stem pitting and apple flat limb.

(vii) Cracking and Scaling

Cracking and scaling of stem bark are the characteristics symptoms of citrus, which is infected with a viroid.

(viii) Roots

Tumour as in sweet clover plant infected by wound tumour virus and death of roots in citrus due to *tristeza virus* are abnormalities due to virus infections.

II. Histological Changes or Internal Symptoms

Changes in internal structures of plants due to virus infections are called histological changes or histopathobogical changes. Various histological changes are observed in leaf, petiole and stern. They are:

1. Necrosis

This histopatholOgical change is primarily observed in phloern cells, it resulting in necrosis of phloern vessels. The top necrosis (or streak in potato is caused by PVY involves the death of cholenchyma and tissue spreading along the vein without affecting vascular elements.

2. Hyperplasia

Excessive growth of virus infected tissue in the form of tumour or enation is known as *hyperplasia*. Sugar beet leaf infected with curly virus develops a large number of sieve elements and forms hyperplasias in veins. In swollen shoot of cocoa due to wound tumour virus, the cells are generally normal but swelling is due to excessive production of xylem cells.

3. Hypoplasia
Weakening of infected tissue is called to be a *hypoplasia*. It is more evident in yellow sector of mosaic patterns.

4. Lignified Strands
In grape infected with *Grape vine fan leaf virus*, xylem elements are lignified at their inner surface; these thickened walls are called endocellular cordons.

5. Tyloses
In barley infected with *Barley yellow dwarf virus*, loses are formed in the xylem tubes. Many cytopathological effects of viral infection have also been observed in infected plants. They are:

☆ In mosaic affected leaf, the size and number of starch grains are reduced.

☆ Infections with Tymovirus form marginal vesicles chloroplasts.

☆ In infections with Tobacco rattle virus, mitochondria are aggregated into inclusion bodies.

☆ Necrosis or streak symptoms cause drastic cytological changes in cells as they approach death.

☆ The development of inclusion bodies in the virus fected cells is the major cytopathic effect found either in to cytoplasm or in the nucleus. They are aggregates of virus particles in more or less crystalline form. Virus induces proteins or other materials to form inclusion bodies, which may be *amorphous granule* or *crystalline*. Examples:

Angled layer aggregates, Complex inclusions, Crystalloid inclusions, Cylindrical inclusions, Hexagonal crystals, Laminated aggregates, Laminated inclusion components, Paracrystals, Pinwheel, Rounded plates, Scroll bodies.

III. Physiological Changes
There are several physiological changes in virus infected plants, that have not been noticed in healthy plants. Diener (1963) has pointed out the following physiological changes in plants infected with viruses:

1. Decreased photosynthetic activity
2. Increased rate of respiration.
3. Accumulation of soluble nitrogen compounds
4. Increased phenol oxidase activity
5. Decreased activity of growth hormones
6. Translocation of viruses

1. Decreased Photosynthetic Activity
In several cases, virus infection decreases the phyll synthesis. Besides this, even

in the existing chlorolasts photosynthetic phosphorylation activity and hill reaction are very slow.

2. Increased Rate of Respiration

In many plants, the rate of respiration increases immediately after virus infection. The rate of respiration however depends on *duration of incubation, physiological state of the plant, age of plant, environmental conditions during growth and nature of leaves.*

3. Accumulation of Soluble Nitrogen Compounds

The amount of soluble nitrogen, especially amides, is higher in virus infected plants than in healthy plants. On contrary to it, total nitrogen and protein contents are lower than those in healthy plants.

4. Increased Phenol Oxidase Activity

Oxidation of polyphenols by phenol oxidase is one of the self-defense mechanism in virus infected plants. This phenol oxidase activity increases immediately after the virus infection.

5. Decreased Activity of Growth Hormones

In virus infected plants, the concentration of growth hormones such as IAA is very low but that of growth inhibiting substances such as antiauxin and oxidase enzymes is high.

☆ Tomato leaf infected with *Tomato spotted wilt virus* has low amount of auxin (IAA).

☆ Scopoletin that inhibits IAA synthesis is high in some virus infected plants.

☆ Certain viral infections reduce the gibberellic production in plants.

6. Translocation of Viruses in Plants

Movement of virus particles from the site of infection to other parts of the plant is a prominent feature in infected plants. There are two types of movements:

☆ Local spreading of viruses takes place by cell to cell movement of viruses through plasmodesmata.

☆ Movement of viruses through phloem occur for spreading to tissue systems or organs.

Importance of Virus

Viruses also cause many important plant diseases and are responsible for huge losses in crop production and quality in all parts of the world. Infected plants may show a range of symptoms depending on the disease but often there is leaf yellowing (either of the whole leaf or in a pattern of stripes or blotches), leaf distortion (*e.g.* curling) and/or other growth distortions (*e.g.* stunting of the whole plant, abnormalities in flower or fruit formation).

Virus Transmission

Viral diseases are always contagious. They multiply in host cells and spread from cell to cell and from plant to plant. Epidemics lead to wide spread damage to the crop and economic loss. Plant viruses rarely come out of the plant spontaneously and transmitted through air. They are carried in debris or plant sap. They would cause infections only when come in contact with the contents of wounded living cell. Viruses are transmitted from plant to plant in many ways:

1. Experimental transmission
2. Transmission by vegetative propagation
3. Transmission by nematodes and soil fungi
4. Transmission by vectors
5. Transmission through seed and pollen
6. Mechanical transmission
7. Dodder transmission

1. Experimental Transmission

Experimental transmission is performed for virological studies in laboratory. It may be done by grafting the diseased scions on healthy stock. Cleft, wedge, bud and approach grafting is useful for virus transmission from diseased plants to healthy plants.Tobacco mosaic virus is injected into healthy plants by pricking with a capillary glass needle. Virus transmission is enhanced by using abrasives such as carborundum powder on the surface of plant.

2. Transmission by Vegetative Propagation

In vegetatively reproducing plants, virus transmission takes place through infected propagules. When such diseased Propagules are used as planting materials, the disease first appears in plants raised from them and then it spreads to other plants. Thus viruses are transmitted by *buds, grafts, cuttings, tubers, corns, bulbs, rhizomes,* etc. This mode of transmission is most important for ornamental trees, shrubs and tile field crops. For example, bunchy top of banana is transmitted through infected suckers.

3. Transmission by Nematodes and Soil Fungi

Free-living ectoparasitic nematodes play an important ole in virus transmission. *Xiphineria sp.* transmits polyhedral nepoviruses. *e.g. Grapevine fan leaf virus.* Nematode feed on roots of infected plant and then move to roots of healthy plants for feeding. This larvae as well as adults can acquire the viruses from infected plant and deliver the virus in healthy plants (Figure 23)

Several soil fungi also transmit viruses from diseased plants to healthy plants because of growth process. The fungal hyphae grow to occupy large areas and carry viruses in healthy plants.

☆ *Trichoderma viridae* transmits tubular tobra virus of pea

☆ Chytrids transmit *Tobacco necrosis virus, cucumber necrosis virus and Red clover necrotic mosaic virus*

☆ Phycomycetes transmit *Beet necrotic yellow vein virus* and *Potato top virus.*

4. Transmission by Vectors

Organisms that act as carrier of pathogen and are involved in the spreading of disease from one plant to another are called vectors. The most common method of viral transmission in the field is by insect vectors. Aphids, leaf hoppers, white flies, the mealy bugs, scale insects, tree hoppers, thrips, beetles, grasshoppers and true bugs are important vectors of plant viruses. On the basis of behaviour of virus within the vector viruses are also categorized into three types:

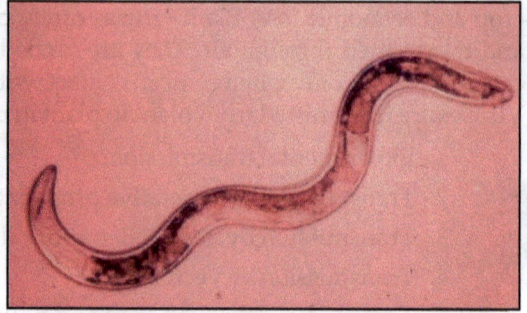

Figure 23. Nematodes: These are Root-Feeding Parasites, some of which Transmit Viruses in the Genera *Nepovirus* and *Tobravirus*. The picture shows an adult female of *Paratrichodorus pachydermus*, the vector of *Tobacco necrosis virus*.

(a) Non-persistent or Stylet-borne Viruses

These viruses carried by insects with their sucking mouthparts on the stylets. *Alfamovirus, Caulimovirus, Cucumovirus, Polyvirus*, etc. are examples of this type. Acquisition very short and fast, but rate of transmission is affected by environmental factors.

(b) Persistent or Circulative Viruses

There are the viruses accumulated in the insect's mouthparts and the internal tissues. They are circulated throughout the insect body and introduced again into plants through the mouth parts. Transmission of RTSV by the aphid *Nephotettix virescens* is example for this type.

(c) Propagative Viruses

There are some of the circulative viruses that multiply within the vectors. They are called propagative viruses. Viruses transmitted by insects with chewing mouthparts are of propagative type may be carried on the mouthparts (stylet-borne) Aphids are the most important vectors for propagative plant viruses. BVDV is transmitted in this way by aphid *Schizaphis graminum*.

Aphids

Aphids are members of the family *Aphidaceae* of the class *Homoptera*. Several aphids are known to transmit viruses from diseased plant to healthy plants (Figure 24). Of these, *Myzus persicae* is very important one that can transmit about 60 viruses in different plants. Egs. *Tobacco ring spot virus, Rosette disease virus, Tobacco yellow mosaic virus, Potato roll virus*, etc.

Aphids insert their stylct into the plant to absorb phloem sap from the plant. At that time, some cells in the path of stylet are injured and the virus in the stylet is left in the Injured cells. As a result, the plant gets infected with the virus.

Leaf Hoppers

Active jumping plant bugs belonging to the families and *Fulgoridae* and *Jassidae* are known as leaf hoppers. Nymphs and adults are found in large swarms on tender foliages and inflorescences and sucking the plant sap. While feeding on diseased as well as healthy plants, virus contained in the sap is transmitted from plant to plant.

Figure 24. Aphids: Transmit Viruses from Many Different Genera. The picture shows the green peach aphid *Myzus persicae*, the vector of many plant viruses, including *Potato virus Y*.

☆ *Nephotettix implicticeps* transmits *Rice tungro virus*

☆ *Nephotettix nigripictus* takes part in the transmission of rice dwarf disease.

☆ Beet curly top is transmitted by *Circulifer tenellus*.

Flies

White flies are members of the family *Aleurodidae* of the class *Homoptera*. They feed by sucking the plant sap from green tender parts of the plant. When they feed on healthy and diseased plants repeatedly, virus contained sap is exchanged between the plants. As a consequence, th virus is transmitted from diseased plant to healthy plant (Figure 25).

☆ *Bemesia tabaci* transmits leaf curl of tobacco and yellow mosaic of *Acalypha indica* and *Phaseolus aureus*.

Thrips

Thnps are tiny insects of the family *Thripidae* of class *Homoptera*. They feed on plant saps by their mouth parts adopted for sucking. While feeding the saps of healthy plants and diseased plants, a fraction of sap infected with a exchanged between the plants. Thus the virus is transmitted from diseased plant to healthy plants (Figure 26).

Mealy Bugs

These are small bugs included in the family *Coccidae* of the class *Homoptera*. They cluster in large numbers on the under surface of leaves, sucking sap. While feeding on sap of diseased and healthy plants, virus is transmitted through exchange of virus contained sap. Nymphs and bugs transmit *Badna virus* and *Terovirus*.

Mites

Some eriophyid mites transmit certain viruses from diseased plant to healthy plants while moving from plant to plant. Viruses causing *Fig Mosaic disease* and *pigeon pea sterility* are transmitted by mites, *Aceria tosichella* (Figure 27).

Figure 25. Whiteflies: Transmit Viruses from Several Genera but Particularly those in the Genus *Begomovirus*. The picture shows *Bemisia tabaci*, the vector of many viruses including *Tomato yellow leaf curl virus* and *Lettuce infectious yellows virus*.

Figure 26. Thrips: Transmit Viruses in the Genus *Tospovirus*. The picture shows *Frankinellaoccidentalis*, the western flower thrips that is a major vector of *Tomato spotted wilt virus*.

Grass Hoppers

The grass hopper *Melanoplus differentialis* is known to transmit *Turnip yellow mosaic virus* (TYMV) and *Turnip crinkle virus* from plant to plant

5. Transmission through Seed and Pollen

Seed Transmission

Although seed transmission is not a common method of viral transmission more than 100 viruses are reported to be transmitted by seeds. Only a small portion (1-30 per cent) of the seeds derived from virus-infected plants transmit viruses. About 100 per cent of the tobacco ring spot virus in Soyabean is transmitted through seeds. Barley stripe mosaic virus also shows high transmission ability through seeds (50-100 per cent).

Figure 27. Mites: These Transmit Viruses in the Genera *Rymovirus* and *Tritimovirus*. The picture shows *Aceria tosichella*, the vector of *Wheat streak mosaic virus*.

Pollen Transmission

Some viruses are transmitted through infected pollen grains. When the infected pollen fall on pistil of flower of healthy plant, the flower contracts the virus. Pollen transmission may result in the production of virus infected seed. *e.g.* Transmission of *Alfalfa mosaic virus* and *Bean common mosaic virus*.

6. Mechanical Transmission

The mechanical transmission involves the transfer of plant sap from diseased plant to healthy plant by natural or artificial means Mechanical transmission may take place between closely planted plants by direct contact. Strong wind may cause the leaves of adjacent plants to rub together, leading to wound formation. Sap may be exchanged between the plants leading to the virus transmission. *Potato Virus X* (PVX) is most usually transmitted in this way.

7. Dodder Transmission

Dodder (*Cuscuta* sp.) is a parasitic plant that obtains nutrition from its host by root like *haustoria*. Meantime, it transmits the virus to host plants. *e.g.*: *Sugarbeet curly top virus* is transmitted by *Cuscuta subinclusa*.

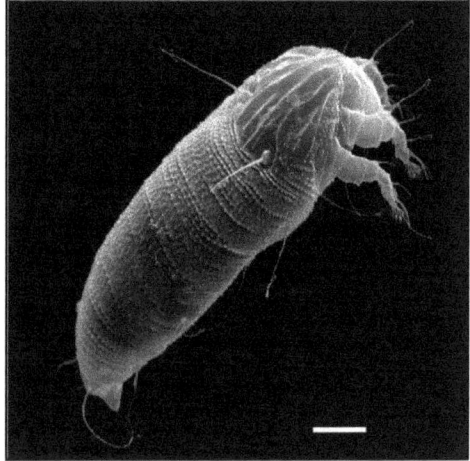

Control of Plant Viral Diseases

Plant viruses cause severe damage to agricultural crops such as cereals, pulses, vegetables, etc. which reduces the crop growth and net yield returns. In order to avoid the crop loss due to virus attack, it is necessary to control viral diseases in the field.

Several methods have been employed to prevent the viral attack as well as to cure viral infections in the crop field. As in human health, in agriculture also, prevention of a disease is better than its cure. The following methods are used in the control of virus disease in crops.

1. Removal of sources of infection
2. Hygiene
3. Virus free seed
4. Virus free vegetative stocks
5. Modified planting and harvesting procedures
6. Change of planting Date
7. Prevention of long distance spread
8. Growing resistance crop varieties
9. Control of vectors
10. Antiviral chemicals

1. Removal of Sources of Infection

Virus infected plants should be uprooted and removed from the crop field before cultivation of new crop to avoid virus transmission. This would help to reduce the source of infection in crop fields. Rouging and eradication procedure is adopted to remove source of infection. Burning of crop residues after harvest is yet another method to avoid the source of infection.

2. Hygiene

Workers should maintain hygienic conditions to avoid virus transmission during cultural operations. Some of the viruses are easily spreaded through worker's clothing and body. For example, in dark, TMV persisted for over 3 years on clothing. To reduce this type of transmission workers should wash their hand with a soap after works.

3. Virus Free Seed

Many viruses are spreacled in the crop field through infected seeds. If the seed is infected with a virus, it infects the crop at a very early stage, so the disease spreads to other plants too. Thus seed transmission introduces the infection throughout the crop field. Virus free seeds may provide a very effective control for such diseases. Seed certification scheme may be necessary for controlling virus diseases.

4. Virus Free Vegetative Stocks

The planting materials such as seeds, suckers, cuttings, grafts, etc. should be taken from disease free mother plants. It would reduce the risk of disease development through infected planting materials.

5. Modified Planting and Harvesting Procedures

Application of antiviral substances and manures to crop field before planting keep some virus infection in control. Therefore, such farming practices should be encouraged.

6. Change of Planting Date

Date and stage of planting a crop may influence the time and amount of infectiori. The best time to sow depends on the time of migration vector. If it migrates early, late sowing may be advisable. Time of planting of particular crop should be changed depends upon the season of the insect vector availability. This will reduce the spreading of the virus in the field.

7. Prevention of Long Distance Spread

Most developed countries have regulations to control the entry of infected plant materials to other areas to prevent the entry of diseases and pests. Prevention of free movement of plant materials to far away areas reduces the risk of disease spreading. The setting up of quarantine measure is important to avoid certain viruses.

8. Growing Resistance Crop Varieties

Existing crop varieties should be screened for virus resistance under laboratory conditions and the virus resistant varieties should be grown in areas where incidence of the disease is high. Now-a-days virus resistant plants are developed by transgenic process, which has shown a considerable promise for virus resistance in crops. By adopting transgenic methods, papaya varieties resistant to papaya ring spot virus has been developed and grown in the fields. This has been done for several other crops also.

9. Control of Vectors

Control or avoidance of invertebrate vectors is of prime importance for limitation of crop damage by viruses. Spraying photo-stable synthetic pyrithroid reduces the burden of air borne vectors. Chemical pesticides are also found to be useful to control vectors.

Oil sprays have given useful results of field trail against a range of vectors of non persistent viruses. *e.g.* Mineral oil.

The oil coated on the leaf surface interferes with feeding behaviour of insects and hence they die off because of starvation.

10. Antiviral Chemicals

Synthetic analogs of the purine and pyrimidine are useful to reduce the burden of viral infection. *e.g.* Gentamycin, Actinomycin-D, phenolic compounds, etc.

Methods for Obtaining Virus Free Plants

The following methods are adopted to get virus free planting materials. They are

1. Heat therapy
2. Meristem tip culture
3. Tissue culture
4. Transgenic technique

1. Heat Therapy

Keeping the infected plants at 30-45°C jbr 3 hrs to 4 weeks in glass house is called heat therapy. It is the most useful method for obtaining virus free planting material from infected plants. Hot air is used to treat the plants. High ternperature stops the virus growth so that the virus does not move to daughter cells at the meristem tip. Thus the meristern becomes free from viruses.

Heat therapy is of much use for getting virus free meristern for in vitro culture of plantlets. It eliminates viruses causing peach yellow, little peach, ring patterns in chrysanthemum, bayberry yellows, etc.

2. Meristem Tip Culture

Apical meristem obtained from infected plant subjected to heat therapy is free from viruses. Therefore, virus free plantlets are cultured from pieces of such meristerns. The meristem excised from the plants is sterilized with mercuric, chloride and then cut into small pieces called explants. These explants are inoculated into *Murashige and Skoog medium,* in sterile flasks and the flasks are incubated until calli are formed. The calli are then transferred to another medium for organogenesis (root and shoot development). Finally, these calli are grown into plantlets free from viruses. Culture of meristem tips has proved to be an effective way of getting virus free plants for cultivation from diseased plants.

Scheme for virus free plant production by meristem tip culture is given in the following flow chart (Figure 28).

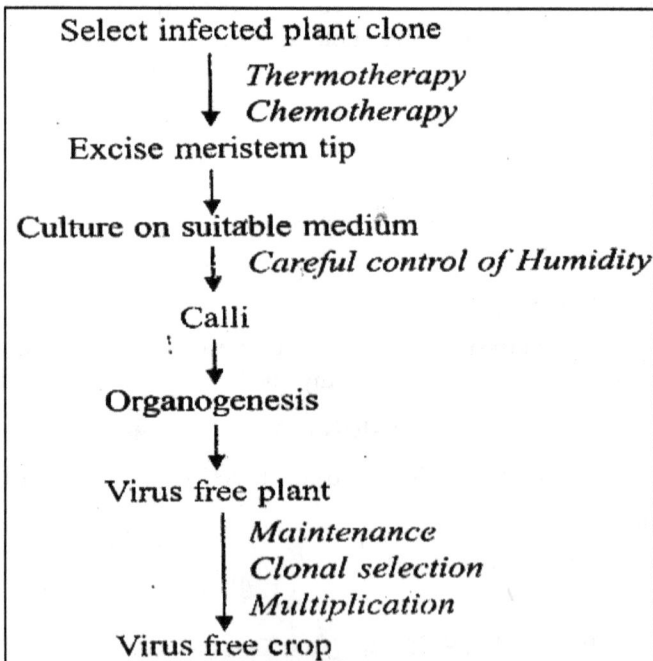

Figure 28. Flow Chart Illustrating Different Stages of Meristem Culture

3. Tissue Culture

Nucellus present in the immature seeds is free from viruses. The nucellus of seeds froni virus infected seeds is taken and grown into virus free plantlets. Virus free Citrus plantlets are thus obtained for cultivation.

4. Transgenic Techniques

Virus resistant plants are developed by transgenic techniques. Genes coding for viral coat protein are cloned in virus susceptible plants using suitable vectors. The transgenic plant thus obtained is resistant to the virus. Transgenic plants have shown considerable promise in disease resistance. Using transgenic technique, papaya has developed for resistance against viral attack.

Diagnosis of Plant Viruses in Seedstocks and Diseased Plants

Seeds and seedlings are frequently found infected with different viruses. If such infected planting materials are used to raise a crop, the disease would spread rapidly throughout the field and cause a heavy loss of yield. Keeping this in mind, a seed certification program has been performed for preserving seeds in granary and releasing them for sowing. Similarly, seedlings in nurseries are released after, testing for plant pathogens. Diagnosis of plant virus infections in planting materials is one of the most important step before releasing seed stocks for sowing and seedlings for planting.

The following methods are used in the diagnosis of plant viruses in seed stocks and diseased plants:

1. Seed morphology
2. Seedling symptornatology
3. Indicator plant test
4. Serological tests
5. Histochemical test
6. Fluorescent microscopy

1. Seed Morphology

In genera!, viruses are not transmitted through seeds, but about 30 plant viruses are known to be transmitted through seeds. *Lettuce mosaic virus, Tobacco ring spot virus in soybean, Cucumber mosaic virus, Barley stipe mosaic virus*, etc. are examples for seed transmission. Unlike the fungal pathogen that produces changes in seed morphology, viruses do not change the seed morphology. So the infected seeds and normal healthy seeds look alike, but infected seeds are somewhat smaller than healthy seeds. In severely infected seeds, whitish patches or discolouration occur in the seed coat. Seeds with such symptoms are not preserved in granary.

2. Seedling Symptomatology

Presence of a virus in plants can be determined by disease symptom produced in those plants. The nature of disease symptom is characteristic of virus infected the host plant and severity of the infection. Identification of diseased plants by looking at seedlings for disease symptoms is known as seedling symptomatology. The following are symptoms of Virus infections in seedlings:

- ☆ Stunted growth of seedlings
- ☆ Yellowing of leaves
- ☆ Chlorosis in leaves
- ☆ Mottling of leaves
- ☆ Spotted wilt at the site of infection
- ☆ Vein banding and vein clearing
- ☆ Little leaves
- ☆ Leaf rolling

Not all these symptoms occurs in a seedling infected with a particular virus. It is essentially depending on the virus and its target host.

3. Indicator Plant Test

Plant viruses found in infected specimens such as seeds and seedlings can be detected with indicator plant test. The indicator plant is nothing but the assay host in which the virus, on inoculation, produces its characteristic, symptom and thereby indicates the presence of the particular virus. Any change in the morphology and physiology of the plant is indicated by the plant. So it is also known as *index plant*. Extract of suspected seeds or plants is artificially inoculated into or on the indicator plant and the plant is covered with a mesh to prevent contaminations. After keeping this plant for a considerable period of time, the plant is visualized for the presence of characteristic disease symptom of the virus. If the plant shows the suspected disease symptom, then it is considered that the seed or plant is infected one. Such infected seeds are usually destroyed by incineration. To test the corn seed suspected to be infected with maize mosaic virus, corn plant is used as the indicator plant (Figure 29).

4. Serological Tests

Proteins in plant viruses are good immunogens (antigens) which stimulate the production of antibodies when injected into a mammal (*e.g.* Mouse or rabbit). These antibodies in the blood serum react with the antigens and precipitate them. This is the basis of serological tests. The serological tests involve the following steps:

- ☆ Plant virus is isolated from the infected seeds or diseased plant.
- ☆ It is injected into a mouse or rabbit.
- ☆ After a suitable duration of incubation, known volume of blood is taken from the animal using a syringe.
- ☆ The blood containing antibodies is called antiserum.
- ☆ The antiserum is diluted to a suitable concentration by adding a buffer solution.

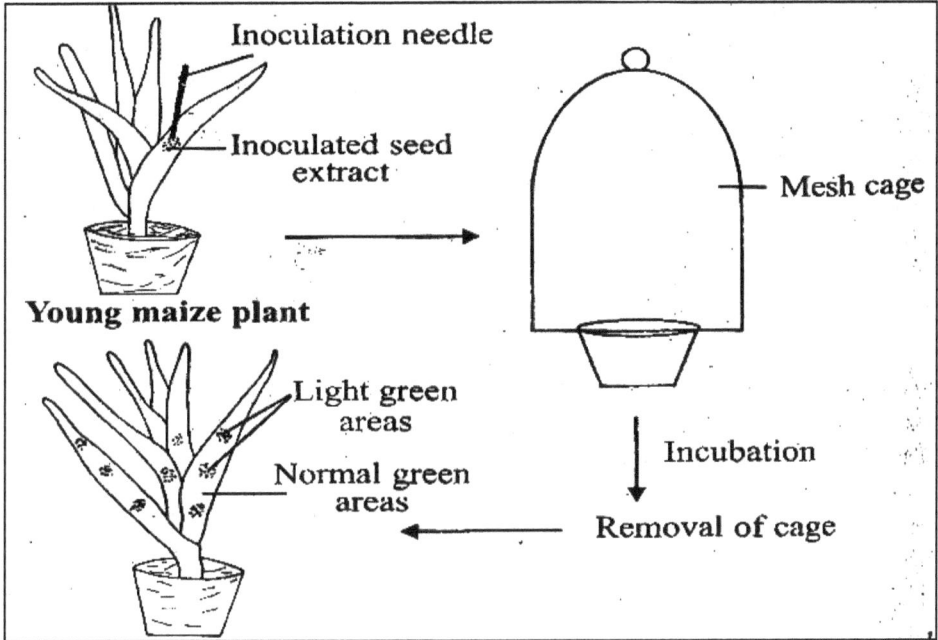

Figure 29. Indicator Plant Test with Corn Seed Suspected to be Infected with *Maize Mosaic Virus*

☆ The viruses to be tested serologically are taken in separate tubes and each tube is treated with a known volume of the antiserum.

☆ These tubes are incubated for 10 minutes.

☆ After incubation, these tubes are viewed for presence of precipitates.

Presence of precipitate indicates that the virus is related to the virus injected into the animal. As antibodies are specific towards their antigens, this test would identify the strains of plant viruses. Further, sensitivity of antigen-antibody reaction is used to estimate the concentration of virus in the extracts.

☆ Monoclonal antibodies may be used in this test to identify antigens in plant viruses.

☆ ELISA technique is also useful for identification and confirmation of plant viruses in extracts.

5. Histochemical Tests

In infected seeds and seedlings, the virus may occur in almost all regions. Hence, presence of virus in infected materials can be demonstrated in cells of the infected region by histochemical method. In this method, thin sections are cut from the infected or suspected specimen and stained with a suitable stain to visualize the virus in the cells. The dye *Calcomine orange-Luxol brilliant green* stains the virus inclusions green, leaving cell proteins as such. *Azure-A* gives magenta colour to viruses and viral inclusions. After staining the fine sections with anyone of these stains, they are viewed under a microscope. Presence of stained bodies indicates the exact location of virus particles in cells of the tissue sections. If there is no such bodies, it is confirmed that the seed or plant is free from virus infection.

6. Fluorescent Microscopy

☆ It is a modification of plaque assay

☆ It is used to assay viruses that do not kill cells.

☆ After adsorption and propagation, cells are treated with methanol or acetone and incubated with an antibody raised against the virus.

☆ A second antibody coupled with an indicator such fluorescein is added to it. This recognizes the first antibody

☆ The cells are then examined under UV microscope.

☆ Infected cells fluoresces against dark background.

☆ Viruses can be expressed as fluorescent–focus-forming units/millimetre.

Cauliflower Mosaic Virus (CaMV)

Group	–	*Group VII (dsDNA-RT)*
Family	–	*Caulimoviridae*
Genus	–	*Caulimovirus*
Species	–	*Cauliflower mosaic virus*

CaMV is included under the genus- caulimovirus. these are widespread and they are found where their host plants are grown, most of them have a limited host range.

The CaMV are the only plant viruses known to have dsDNA genome. Due to the presence of dsDNA, they serve as vectors in plant genetic engineering.

Structure

☆ CaMV is an isometric particle about 50 nm in diameter with a holo cetre of about 15 to 20 nm.

☆ Holo centre contains dsDNA which is encapsulated in subunits of 58 KD protein precursors

☆ The 2 proteins comprises more than 84 per cent of the virus associated protein.

☆ Its molecular weight is about 22.8×10^6

☆ Each viral particle is composed of 420 molecules of the smaller protein and 50 to 60 molecules of larger protein.

☆ The protein has icosahedral structure.

Genome Structure

☆ The genome size is 80 to 4 base pair.

☆ The genome is a circular dsDNA with one discontinuity in one strand and one or more discontinuity in the other strand.

☆ Gaps are produced because of strands displacement.

☆ These gaps are associated with the replication of the virus.

☆ The genome contains six major coding regions (I, III, IV, V,VI,VIII) and 2 minor coding regions. (II, VII).

☆ The minor coding regions are the store house of non essential genes (Figure 30).

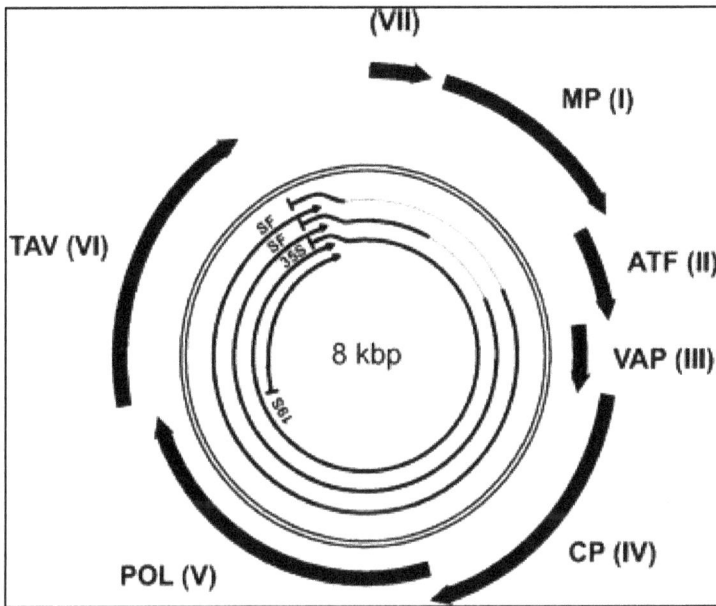

Figure 30. Geneme Structure of CaMV

☆ *Coding region I*–codes for 38 KD protein which helps in transfer of virus from one cell to another.

☆ *Coding region II*–codes for 19 KD protein which attracts aphids.

☆ *Coding region III*–codes for viral 15 KD viral capsomeres.

☆ *Coding region IV*–codes for 57 KD coat protein.

☆ *Coding region V*- codes for 79 KD reverse transcriptase

☆ *Coding region VI*–codes for 58 KD proteins which is used in building inclusion bodies.

☆ *Coding region VII*–functions not known.

☆ *Coding region VII*–codes for protein which targets newly formed virus to inclusion body.

Transmission

CaMV is transmitted by mechanical methods, insect vectors and through vegetative propagules. The cultural operations being done for diseased plants along with healthy plants bring out the transfer of plant sap from plant to plant. Virus in the sap is thus transmitted from diseased plant to healthy plants.

CaMV is also transmitted by aphids. The important aphids that transmit CaMV are *Brevicoryne brassicae, Rhopalosiphan pseudobrassicae* and *Myzus persicae*. CaMV does not multiply in the vector, so the progenies of these vectors do not transmit the virus unless they feed on infected plant sap. Aphids retain the virus for 3-20 hrs. Planting of infected materials is yet other mode of transmission of CaMV to plants.

Replication

CaMV is transmitted to plants mainly by Aphids. After entry into the host cell, uncoating and release of CaMV DNA occur within the cytoplasm (Figure 31).

- ☆ Uncoated viral DNA is transported to the cell nucleus. There occurs removal of overhangs by host encoded DNAse and closing of gaps by host encoded DNA ligase.
- ☆ Viral DNA becomes associated with the histone proteins of the host cell to form mini chromosome configuration.
- ☆ Two RNA transcripts are generated as a result of transcription and transported to the cytoplasm of the host cell.
- ☆ 35S RNA codes for structural proteins.
- ☆ 19S RNA is involved in replication of genome.
- ☆ ds DNA virus is synthesized from RNAs by the enzyme reverse transcriptase.
- ☆ Host RNA acts as a primer for the synthesis of viral DNA,
- ☆ Encapsidation occurs in the cytoplasm to form progeny virions.
- ☆ Newly formed virus particles aggregate together form inclusion bodies in the cytoplasm. Inclusion body formation may be regulated by gene II and gene VI.
- ☆ Mature virions escape from the infected cell and infect new cells.
- ☆ As a result of replication, disease symptom appears in the host.

Symptoms

Mosaic disease of Cauliflower is characterized by the following symptoms:

- ☆ Vein clearing or chlorotic vein banding occurs in *Arabidopsis thaliana, Brassica* sp. etc.
- ☆ Mosaic symptoms occur in *B. campestris*.
- ☆ Stunting of the entire plant.
- ☆ Electron microscopic studies have revealed that CaMV particles are found in cytoplasm and plasmodesmata.

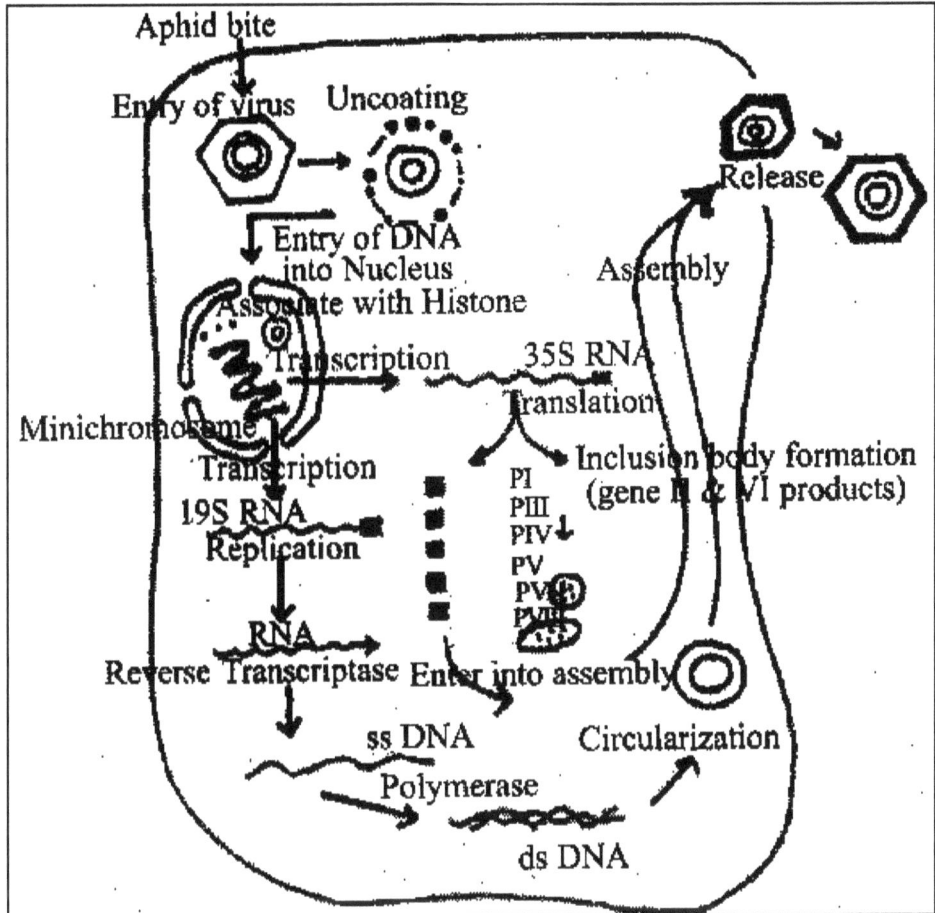

Figure 31. Replication of CaMV

☆ Inclusion bodies containing virus particles, viral proteins and host proteins are seen in the cytoplasm.

☆ Nucleus of host cell may become lobed one.

☆ Certain vesicles protrude out from the cell wall.

Control

Control of insect vectors using insecticides, Elimination of diseased plants, Cultivation of virus-resistant varieties, Selection of disease free planting materials., Selection of disease free soil for cultivation.

Tobacco Mosaic Virus (TMV)

Group   –   Group IV ((+)ssRNA)

Genus   –   *Tobamovirus*

Species   –   *Tobacco mosaic virus*

Tobacco mosaic virus (TMV) is an RNA virus that infects plants, especially tobacco and other members of the family Solanaceae. The infection causes characteristic patterns (mottling and discoloration) on the leaves (thence the name). TMV was the first virus to be discovered. Although it was known from the late 19th century that an infectious disease was damaging tobacco crops, it was not until 1930 that the infectious agent was determined to be a virus.

History

In 1883, Adolf Mayer first described the disease that could be transferred between plants, similar to bacterial infections.. Dimitri Ivanovski gave the first concrete evidence for the existence of a non-bacterial infectious agent, showing that infected sap remained infectious even after filtering through Chamberland filter candles, in 1892. However, he remained convinced despite repeated failures to produce evidence, that bacteria were the infectious agents. In 1898, Martinus Beijerinck showed that a filtered, bacteria-free culture medium still contained the infectious agent. Wendell Meredith Stanley crystallized the virus in 1935 and showed that it remains active even after crystallization. For his work, he was awarded 1/4 of the Nobel Prize in Chemistry in 1946, even though it was later shown some of his conclusions (in particular, that the crystals were pure protein, and assembled by autocatalysis) were incorrect. The first electron microscopical images of TMV were made in 1939 by Gustav Kausche, Edgar Pfankuch and Helmut Ruska–the brother of Nobel Prize winner Ernst Ruska. In 1955, Heinz Fraenkel-Conrat and Robley Williams showed that purified TMV RNA and its capsid (coat) protein assemble by themselves to functional viruses, indicating that this is the most stable structure (the one with the lowest free energy), and likely the natural assembly mechanism within the host cell.

Structure

Schematic model of TMV: 1. nucleic acid (RNA), 2. capsomer (protomer), 3. capsid (Figure 32).

Tobacco mosaic virus has a rod-like appearance. Its capsid is made from 2130 molecules of coat protein (see image above) and one molecule of genomic RNA 6390 bases long. The coat protein self-assembles into the rod like helical structure (16.3 proteins per helix turn) around the RNA which forms a hairpin loop structure (Figure 33). The protein monomer consists of 158 amino acids which are assembled into four main alpha-helices, which are joined by a prominent loop proximal to the axis of the virion. Virions are ~300 nm in length and ~18 nm in diameter. Negatively stained electron microphotographs show a distinct inner channel of ~4 nm. The RNA is located at a radius of ~6 nm and is protected from the action of cellular enzymes by the coat protein. There are three RNA nucleotides per protein monomer. TMV is a thermostable virus. On a dried leaf, it can withstand up to 120 degrees Fahrenheit (50 °C) for 30 minutes.

Transmission

☆ TMV is sap transmissible and enters the host through wounds. No vector is known to transmit this virus from plant to plant.

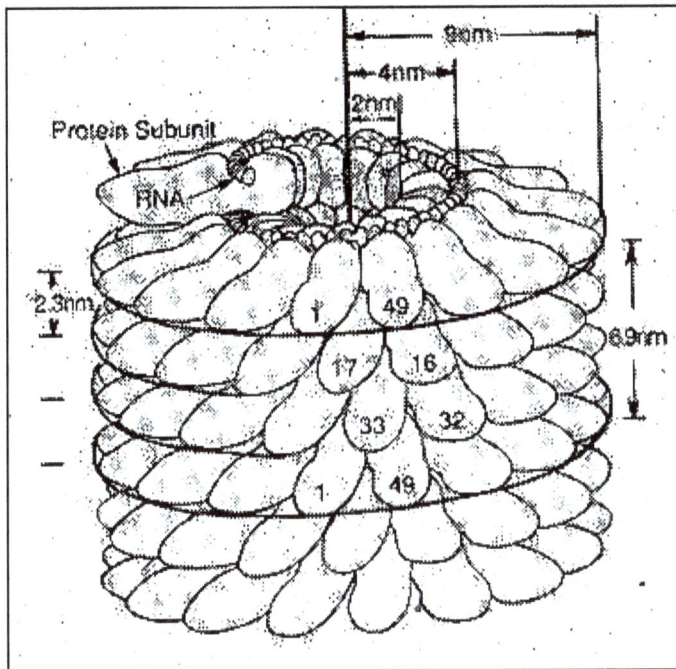

Figure 32. Schematic Model of TMV

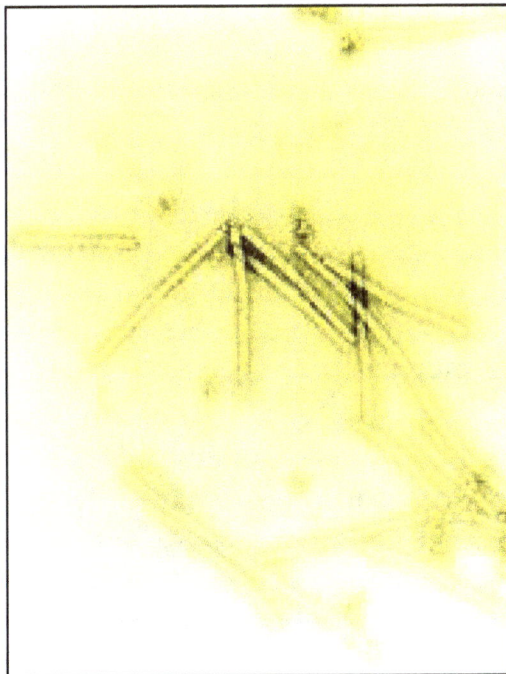

Figure 33. Electron Micrograph of TMV Particles Stained to Enhance Visibility at 160,000x Magnification

☆ TMV can also be transmitted by mechanical means, wind and water.

☆ The cultural operations such as topping and clipping of shoots make wounds on the plants, which become portals of entry of this virus. When these operations are done for infected plants and then for healthy plants, the virus spreads to the healthy plants.

☆ Virus is resistant to adverse environmental and climatic conditions and is infective at any period. Seed transmission of TMV has been reported in tomato.

Replication

Plant viruses like TMV penetrate and enter the host cells in to and their replication completes within such infected host cells. Inside the host cell, the protein coat dissociates and viral nucleic acid becomes free in the cell cytoplasm. Although the

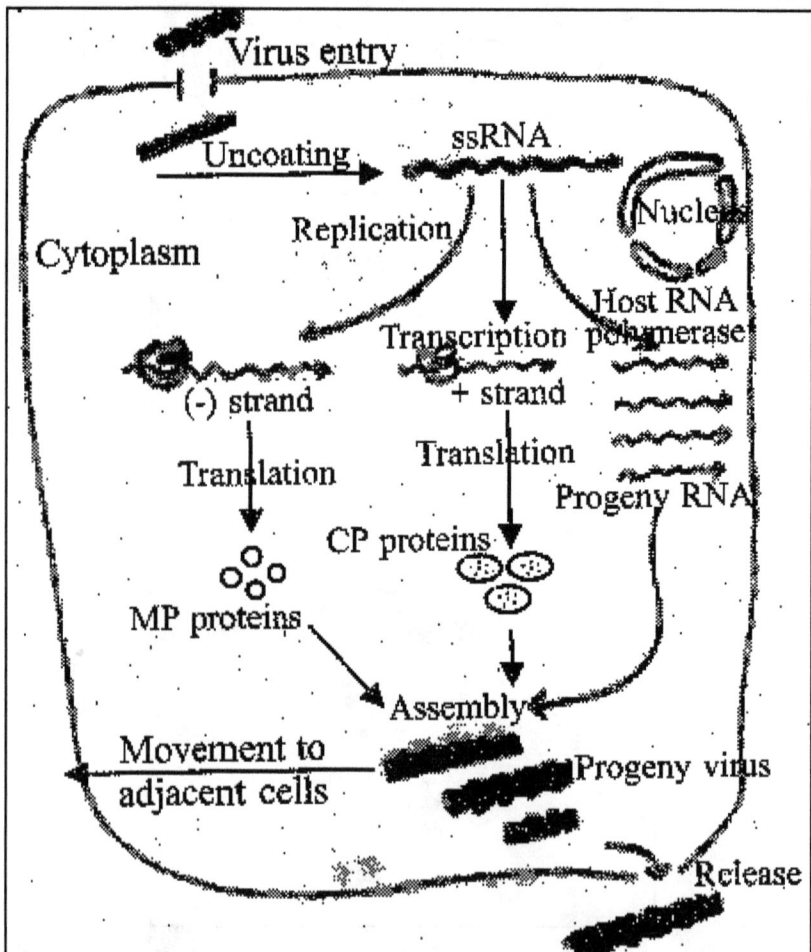

Figure 34. Replication of TMV

sites for different steps of the viral multiplication and formation of new viruses have not yet been determined with absolute certainty, the studies suggest that after becoming free in the cell cytoplasm the viral-RNA moves into the nucleus (possibly into the nucleolus).

The viral-RNA first induces the formation of specific enzymes called 'RNA polymerases', in the presence of which the single-stranded viral-RNA synthesizes an additional RNA strand called 'replicative RNA'. This RNA strand is complementary to the viral genome and serves as 'template' for producing new RNA single strands which are the copies of the parental viral-RNA. The new viral-RNAs are released from the nucleus into the cytoplasm and serve as messenger-RNAs (m-RNAs). Each m-RNA, in cooperation with ribosomes and t-RNA of the host cell directs the synthesis of protein sub-units (Figure 34)

Infection

When TMV infects a tobacco plant, the virus enters mechanically (For example through a ruptured plant cell wall) and replicates. After its multiplication, it enters the neighboring cells through plasmodesmata. For its smooth entry, TMV produces a 30,000 dalton protein called P30 which tends to enlarge the plasmodesmata. TMV most likely moves from cell-to-cell as a complex of the RNA, P30, and replicase proteins. The first symptom of this virus disease is a light green coloration between the veins of young leaves. This is followed quickly by the development of a "mosaic" or mottled pattern of light and dark green areas in the leaves. These symptoms develop quickly and are more pronounced on younger leaves. Mosaic does not result in plant death, but if infection occurs early in the season, plants are stunted. Lower leaves are subjected to "mosaic burn" especially during periods of hot and dry weather. In these cases, large dead areas develop in the leaves. This constitutes one of the most destructive phases of tobacco mosaic virus infection. Infected leaves may be crinkled, puckered, or enlongated.

Scientific and Environmental Impact

In plants, tobacco mosaic virus leads to severe crop losses. It is known to infect members of nine plant families, and at least 125 individual species, including tobacco, tomato, pepper, cucumbers, and a number of ornamental flowers. There are many different strains.

The large amount of literature about TMV and its choice for many pioneering investigations in structural molecular biology, X-ray diffraction, virus assembly and disassembly, and so on, are fundamentally due to the large quantities that can be obtained, plus the fact that it does not infect animals. After growing a few infected tobacco plants in a greenhouse and a few simple laboratory procedures, a scientist can easily produce several grams of virus. As a result of this, TMV can be treated almost as an organic chemical, rather than an infective agent.

Potato Virus

Potatoes can be infected by many different viruses that can reduce yield and tuber quality. Virus diseases can often be diagnosed by mosaic patterns on leaves, stunting of the plant, leaf malformations, and tuber malformations. Symptoms are not always expressed due to interactions between the virus and the potato plant, growing conditions such as fertility and the weather, or the age of the plant when it is infected. Serology and nucleic acid detection techniques are often used to diagnose and characterize suspected virus diseases.

Potato Virus Y

Virus Description and Symptoms

Potato virus Y (PVY) is one of the most prevalent and important viruses in potatoes. Recently, strains of PVY which can cause necrosis (dead spots on leaves and in tubers) have been discovered, creating more concern about this widespread virus. PVY is a *Potyvirus*, the type member of the largest group of plant viruses. It is transmitted by aphids in a nonpersistent manner, by sticking to aphid mouthparts (stylet). The virus can be acquired from the infected plant within seconds, and transmitted to a healthy plant just as fast. PVY can also be transmitted mechanically by machinery, tools, and damaging plants while walking through the field. Aphids are by far the most efficient means of transmission.

Several strains of PVY have been identified that differ by the symptoms they cause in potatoes and tobacco. PVY^O is the common strain, and causes mosaic symptoms. PVY^C causes stipple streak. PVY^N, the necrotic strain, generally causes mild foliage symptoms, but necrosis in the leaves of susceptible potato varieties. Mixed infections of common strains and the necrotic strain are common, and the genomes (genetic material) can mix, producing hybrid strains (*i.e.* $PVY^{N:O}$ and PVY^{NTN}). PVY^{NTN} strains can cause tuber necrosis. Diagnosis can be difficult, because there are antibodies to PVY^O and PVY^N, but immunological methods (ELISA, Enzyme Linked Immunosorbent Assay) cannot distinguish PVY^{NTN} from these two virus strains. Additionally, not all PVY^N isolates will react with PVY^N–specific' antibodies, and some PVY^O isolates will. Symptoms alone cannot distinguish these virus strains, as symptoms vary with age, time of infection, temperature, and the genetics of both the virus and the plant host. PVY strains can interact with other potato viruses such as *Potato virus X* (PVX) and *Potato virus A* (PVA) to result in heavier losses. Necrotic symptoms in tubers often increase after storage. Some varieties such as Russet Norkota and Shepody rarely show symptoms, but can carry the virus and serve as reservoirs for aphid transmission. Yukon Gold is particularly susceptible to tuber necrosis (Figure 35 and 36).

Potato Leafroll Virus

Virus Description and Symptoms

Potato leafroll virus (PLRV) is a phloem-limited *Luteovirus* which is transmitted by aphids in a persistent manner. In contrast to PVY and AMV, PLRV takes longer to be acquired (10-30 minutes) and transmitted (24 to 48 hours) by aphids, since the virus needs to move into the gut, through the body and back out through the salivary system of the aphid. Symptoms of PLRV include a characteristic upright character and rolling of the leaves, chlorosis (yellowing) or reddening, leaves with a leathery feel, phloem necrosis (dead spots along the leaf veins), stunting (reduced height) of

Figure 35. Necrotic Symptoms on Leaves of Potato and
Tobacco (Lower R) Caused by PVY^{N}

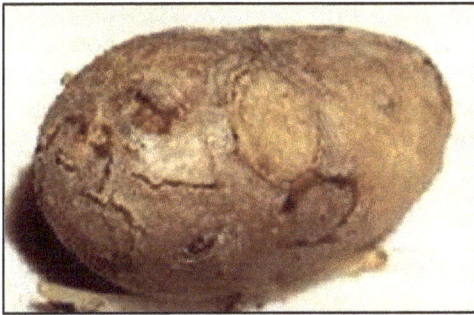

Figure 36. Necrotic Symptoms on Potato Tubers of the
Variety Nicola Caused by PVY^{NTN}

the plant, and net necrosis in tubers. The severity of net necrosis will vary depending on when the plant was infected, and may increase during storage. Some varieties are more susceptible than others, including Russet Burbank, one of the most commonly grown commercial potato varieties in the western US (Figure 37).

Management

Since the spread of PLRV takes more time than PVY, insecticide application can be effective if the aphid populations are closely monitored. Colonizing aphids are the most important vectors for this virus because transmission requires an extended

Figure 37. Symptoms of PLRV on Foliage (Rolling of leaves, Stunting of plants) and Tubers (Net necrosis)

feeding period. The green peach aphid, *Myzus persicae*, is one of the most important vectors. The use of clean seed is critically important. Roguing of infected plants helps prevent the spread of PLRV and early harvest can help prevent late-season infection. Handling plants will not spread the virus, since PLRV is not mechanically transmissible.

Potato Virus S

Virus Description and Symptoms

Potato virus S (PVS) is of increasing importance in potato. It remained unknown until the 1950's because its symptoms are very inconspicuous. PVS can cause yield loss up to 20 per cent. Seed potatoes are not yet certified for PVS, which contributes to its widespread distribution.

Most potato cultivars are symptomless. On some cultivars, if infected early in the season, will show a slight deepening of the veins, rough leaves, more open growth, mild mottling, bronzing, or tiny necrotic spots on the leaves. PVS is a *Carlavirus*, and is nonpersistently transmitted by aphids, including *Myzus persicae*, the green peach aphid. It is also mechanically transmissible, and transmissible through tubers.

Management

PVS is very difficult to detect using visual cues. Insecticides are ineffective in controlling nonpersistently transmitted viruses. Crop oils may be used early in the season. Plants tend to be resistant to infection by PVS later in the season. Prevent mechanical spread within the field by sanitizing tools and minimizing movement through the field. Rogue (remove) any symptomatic plants.

Potato Virus X

Virus Description and Symptoms

Potato virus X is the type member of the *Potexvirus* family of plant viruses. Plants often do not exhibit symptoms, but the virus can cause symptoms of chlorosis, mosaic, decreased leaf size, and necrotic lesions in tubers. PVX can interact with PVY and

PVA to cause more severe symptoms and yield loss than either virus alone. The source of this virus is infected tubers. It is transmitted mechanically, not by an insect vector. Tobacco, pepper, and tomato can also serve as hosts of PVX.

Host Range

Potato virus-X mainly infects potato and other Solanaceous plants. It causes mild mosaic disease in these plants. This virus also infects *Allium cepa, Antirrhinum, Amaranthus, Anthriscus* and *Apium.* Members of alliaceae, Amaranthaceae, Amaryllidaceae, and Asparagaceae are frequently infected with PVX. This virus infects some fungi aIso.

Morphology and Structure

Potato virus-X is a filamentous virus belonging to genus *Potexvirus* of the family *Closteroviridae.* The following are distinctive features of PVX:

PVX is a *flexuous rod-shaped virus* measuring 445–775 nm length and 11–15 nm diameter. The viral capsid shows helical symmetry. There is a central hollow space called axial canal, whose size is about 3.4-6.3 nm diameter. Virus capsid occupies 92–94°/s of total weight of the virion. The capsid is composed of two types of proteins. Of which one protein is a 30,000 Da protein and the other is a 35,000 Da protein. This virion also contains five non-structural proteins coded by the viral genome. The virion contains 5-8 per cent of nucleic acid in its total weight. The nucleic acid is linear and single stranded RNA. It is of minus sense strand type. 5' end of the ssRNA is caped with a methionine and the 3' end has a poly-A sequence. Sub-genomic RNA is often seen in infected cells. Lipid is absent. Genornic nucleic acid is infective type.

Transmission

This virus is usually transmitted by sap inoculation. The mode of sap inoculation may be any one of the following:

☆ It spreads in the field by direct contact between disand healthy plants.

☆ Mechanical inoculation occurs during various cultural operations in the field.

☆ Grafting of infected specimens is the more direct way of sap inoculation.

☆ Natural source of perennation of this virus is infected seed stocks.

☆ Vector mediated transmission is seen in some plants in the field, but not at all reported in potato.

☆ Helper virus is necessary for infection of PVX.

Replication

The replication of PVX is very similar to that of *TMV*

Symptoms

Plants infected with PVX show the following disease symptoms:

Interveinal mottling of young leaves, Chlorotic patches in the leaves, Slight dwarfing of infected plants, Necrosis in old leaves and tubers, Deformation of foliages, Asymptomatic at high temperatures above 21°C 1 the field, X-bodies are found in cells of mottled region of leaves, Virus is seen in cytoplasm and plasmodesmata, Paracrystal inclusions occur in cytoplasm, vacuole and even inside the nucleus of infected cells.

Control

PVX is resistant to ageing at room temperature. It remains infective even after one year. However, it is inactivated when it is exposed to 74°C for 10 minutes. The following methods are adopted to control PVX in plants.

 ☆ Disease-free seed tubers should be used for planting.

 ☆ infested plants must be extinguished in fire.

 ☆ Movement in infested fields should be avoided.

 ☆ Passage of machinery among the young plants should be kept in limit if necessary.

 ☆ Favourable environment should be maintained.

 ☆ Rouging of virus-infected plants.

Potato Mop Top Virus

Virus Description and Symptoms

Potato mop top virus (PMTV) is a *Pomovirus*, and is transmitted by the pathogen (a protozoan with zoospores for infection) which causes powdery scab disease, *Spongospora subterranea*. PMTV is one of the causes of spraing disease, the other being *Tobacco rattle virus* (see below). "Spraing" means the tuber has a rust-brown necrosis which occurs in rings or flecks (Figure 38). PMTV occurs more often in heavy, wet soils. Cool temperatures and wet conditions which are favorable for the germination of *S. subterranean* spores favor spread of the disease. The virus can survive with its vector in the soil for decades.

Management

Conditions that favor the spread of powdery scab, cool and wet conditions, also favor the spread of this virus. Viral infection may be reduced in some cases by improving drainage, reducing irrigation, or delaying planting until soils are warmer and drier. Spores infected with virus can survive in the soil for decades. Longer rotations may help reduce the incidence of powdery scab. In the absence of its vector, the virus is rapidly diluted.

Tobacco Rattle Virus

Virus Description and Symptoms

Tobacco rattle virus (TRV) is a *Tobravirus* which causes corky ringspot or spraing disease. It occurs more often in coarse sandy soils. Symptoms often do not appear on the foliage, but the tuber contains corky layers of tissue interspersed with rings of healthy tissue and brown flecks distributed throughout the tuber (Figure 39). This virus is transmitted by 'stubby-root' nematodes in the *Paratrichodorus* or *Trichodorus* genus, and can also be transmitted mechanically.

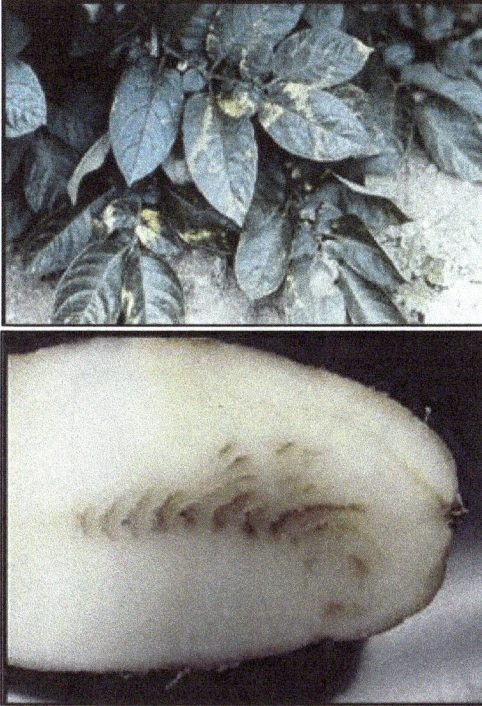

Figure 38. Symptoms of PMTV in Foliage and Necrosis of Tubers

frequently. Plant varieties with partial resistance.

Potato Spindle Tuber Viroid

Virus Description and Symptoms

Potato spindle tuber viroid (PSTV) is not a virus. It is a viroid, which is essentially a self-replicating RNA, without a protein coat. PSTV is an important disease in breeder stock, where it is often transmitted mechanically, as well as through pollen and true seed. It causes mild foliar symptoms including smaller leaves that curl downward, giving the plant a more upright growth habit. Plants can also be stunted, and leaves can be grey and distorted. The stems are often more branched, with the branches having sharp angles on the

Management

The most important control measure is to avoid introducing inoculum into your field by planting certified seed potatoes. The soil should be sampled for the identification of nematodes which can transmit TRV. If the nematode vectors are not present, the virus will not spread. Nematicides are available for use to destroy the vector in the soil. TRV can survive in dormant nematodes for 2 to 4 years.

Nematode populations can increase on cereal crops, so rotations should not include them. Several weeds such as shepherd's purse and chickweed are reservoirs, so good weed control can limit viral increase. Limit handling of plant material and sanitize tools and equipment

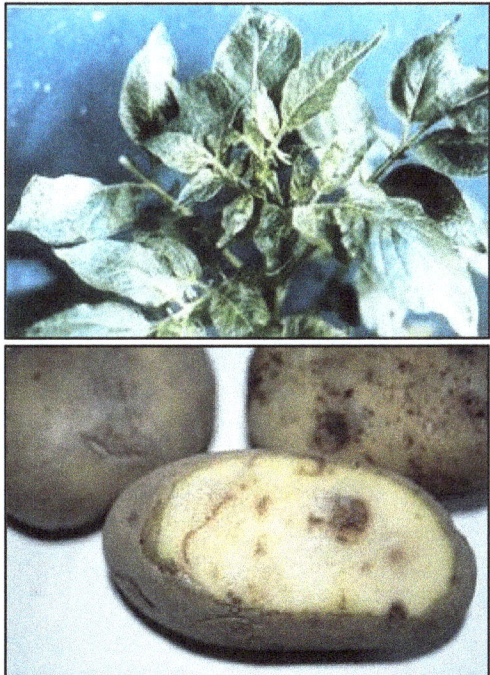

Figure 39. Foliar Symptoms of TRV Infection and Tuber Symptoms of Corky Ringspot Disease

stem. Tubers are narrow and spindle or oblong in shape, or more rounded than expected for a particular variety, and have prominent eyebrows (Figures 40 and 41). Tubers can also become cracked or develop knobs and swellings. PSTV can also infect tomato and nightshade.

Figure 40. Foliar Symptoms of PSTV.
Note the small leaves and upright growth habit.

Figure 41. PSTV Causes Elongation of Tubers (Right Side).
H. David Thurston, Cornell University.

Management

The most important control measure for PSTV is to plant certified seed. Various insects have been reported to transmit PSTV, but the most important disseminator is human movement through the field. Sanitize tools and machinery. Rogue infected plants before they can serve as a source of inoculum.

Alfalfa Mosaic Virus

Virus Description and Symptoms

Alfalfa mosaic virus (AMV) is a *Potyvirus*, like PVY. It is nonpersistently transmitted by aphids. AMV causes a distinct calico pattern (yellow blotching) on leaves. Some strains of the virus can cause severe stunting and stem and tuber necrosis. It is not considered an economically important virus. Aphids carry the virus in to potato crops from nearby alfalfa or clover fields.

Management

Control of AMV is similar to PVY. Insecticides are not effective. Avoid planting potatoes near alfalfa or clover.

General Virus Management Guidelines

☆ Buy seed potatoes that have been certified virus-free. Saving potatoes from a field which was infected by virus will increase the number of plants serving as sources of virus in the following season.

☆ Remove "volunteer" potatoes in the spring (potato plants coming up from tubers left over from the previous season), as these may be virus reservoirs.

☆ Rogue (remove) symptomatic plants — these serve as excellent sources for virus spread within the field. Do not leave rogued plants in the field — remove and trash or burn them. Proper composting may be effective to remove potatoes with *Potyviruses*, *Carlaviruses*, or *Potexviruses*, but not those with protozoan or nematode vectors.

☆ Rogue weeds which may serve as reservoirs of viruses.

☆ Plant early to avoid aphid-transmitted viruses, scout for aphids regularly and apply insecticide (not effective for PVY) or crop oils as appropriate.

☆ Avoid planting potatoes next to alfalfa or red clover crops to reduce the risk of AMV.

☆ Symptoms vary by time of infection, temperature, variety, and virus strain– It is not possible to tell what virus is present in potato plants by symptoms alone. In order to choose the best management technique, get symptomatic plants tested:

☆ Plant varieties which have reported tolerance or resistance to viruses.

☆ Sanitize all tools, planters, and cultivators frequently, especially when moving equipment into a new area. Avoid the spread of soil which could harbor the vectors of some potato viruses between fields.

Chapter 12
Animal Viruses

General Features of Viruses

Viruses consist of nucleic acid (DNA or RNA) surrounded by a protein coat called a *capsid*. The capsid is made up of individual structural subunits called *capsomeres*. The combination of the nucleic acid genome enclosed in the capsid is called the *nucleocapsid*. In addition, many animal viruses have an *envelope*, which is a membranous lipid structure that surrounds the nucleocapsid.

The structural components of a Herpes virus are illustrated in Figure 42.

Viruses are quite different from cells. They contain only one type of nucleic acid, DNA or RNA, never both. They lack membranes and a cytoplasm, as well as ribosomes and any means to produce energy. Although viruses can replicate, mutate and maintain genetic continuity, which are features of all cells, they depend entirely upon a host cell to supply a habitat, energy and raw materials (precursors) for viral replication. Thus, viruses must exist as *obligate intracellular parasites* of cells.

Viruses are very small in size. Some are not as large as a cell ribosome. Their size is so small that individual virus particles cannot be visualized with the light microscope. The range of particle size is from about 20 nanometers for a small virus (*e.g.* poliovirus) to about 0.3 micrometers for a very large virus [*e.g.* smallpox (variola) virus] (Figure 43).

Animal viruses have many shapes ranging from cubical, bullet-shaped, polygonal, spherical, filamentous or helical, to a complex layered morphology. One of the most common morphologies of the viral capsid is the *icosahedron*, which consists of 20 triangular faces (capsomeres) that coalesce to form a roughly spherical structure enclosing the viral nucleic acid. The herpes virus illustrated above has the icosahedral shape.

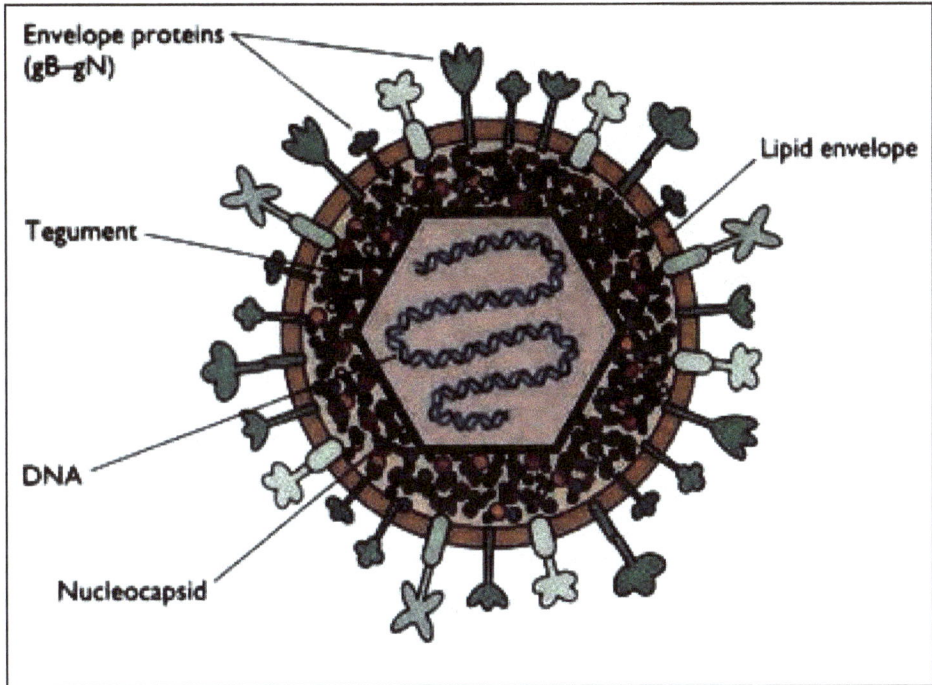

Figure 42. Herpes Simplex Virus 1, Illustrating the Basic Structural Features of a Virus. HSV1 is an enveloped, icosahedral DNA virus. The region between the outer lipid envelope and the nucleocapsid is called the tegument. The DNA of the virus resides in the core. The envelope proteins ("Glycoprotein Spikes") are unique viral proteins, but the envelope itself is derived from the virus host cell.

Classification of Viruses

The primary criteria for taxonomic classification of animal viruses are based on morphology (size, shape, etc.), type of nucleic acid (DNA, RNA, single-stranded, double-stranded, linear, circular, segmented, etc.), and occurrence of envelopes. ssRNA viruses possess either (+)RNA (if it serves as messenger RNA) or (-)RNA (if it serves as a template for messenger RNA). Host range is not a particularly reliable criterion for classification. Although some animal viruses exhibit a very narrow or specific host range, such as HIV in humans or canine distemper virus (CDV) in dogs. But for classification purposes, host range cannot be a criterion because each animal species is subject to infection by a wide variety of viral agents, and numerous viruses infect several different animal species. For example, West Nile virus has a primary host of birds, but it infects and causes disease in horses and humans. Some viruses, such as the influenza virus, are able to change their structure in such a way that they can shift from one primary host to another, for example birds to humans.

Morphologic similarity among animal viruses correlates closely with similarity of viral components, particularly with the type and size of the viral nucleic acid (genome). For example, all viruses with the morphology of adenoviruses contain

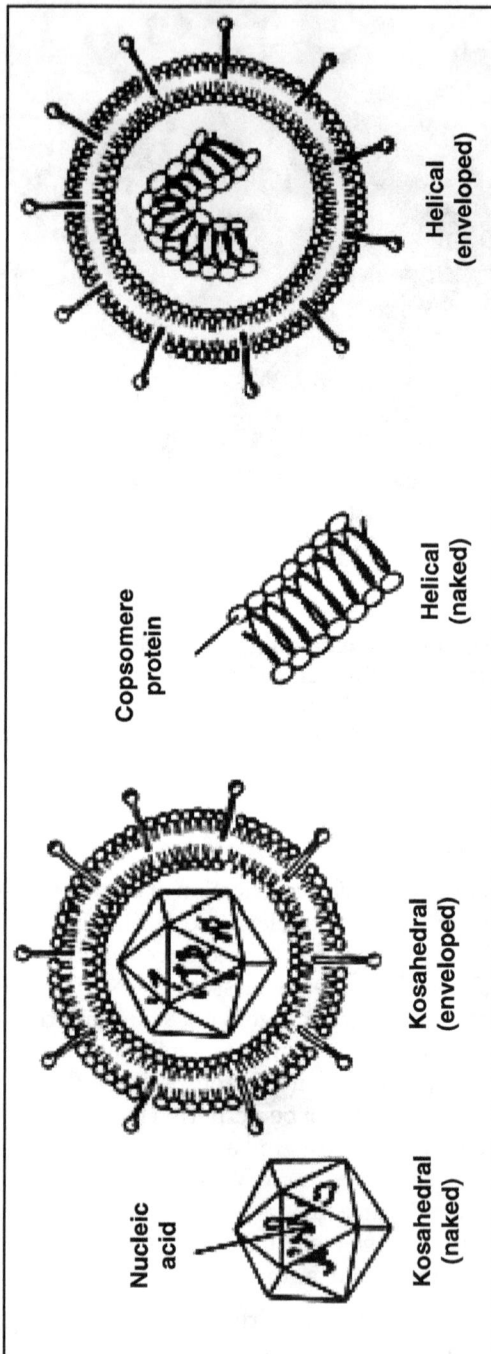

Figure 43. Common Morphologies Seen in Animal Viruses.
Left to Right. A naked icosahedral virus (*e.g.* poliovirus), an enveloped icosahedral virus (*e.g.* herpes virus), a naked helical virus, and an enveloped helical virus (*e.g.* influenza virus). Individual capsomeres are arranged to form a capsid which encloses the nucleic acid (DNA or RNA) of the virus.

dsDNA genomes with a molecular weight of about 23 million daltons; all reoviruses contain segmented dsRNA genomes. In fact, a system of virus classification based on structure and size of viral genomes yields that same grouping as one based on morphology. This information is organized in two ways.

According to the *Baltimore method of classification*, animal viruses are be separated into several classes, grouped by type of nucleic acid. Class I. dsDNA viruses; Class II. ssDNA viruses; Class III. dsRNA viruses; Class IV. (+)RNA viruses; Class V. (-)RNA viruses: Class VI. RNA reverse transcribing viruses; Class VII. DNA reverse transcribing viruses. The Baltimore method of classification is illustrated the Table 6.

Table 6: Classes of Animal Viruses, Grouped by Type of Nucleic Acid

Sl.No.	Class*	Examples/Diseases
I.	dsDNA**	
	Papovavirus	Papilloma (human warts, cervical cancer); polyoma (tumors in certain animals)
	Adenovirus	Respiratory diseases; some cause tumors in certain animals
	Herpesvirus	Herpes simplex I (cold sores), herpes simplex II (genital sores); vericella zoster (chicken pox, shingles); Epstein-Barr virus (mononucleosis, Burkitt's lymphoma)
	Poxvirus	Smallpox, vaccinia, cowpox
II.	ssDNA	
	Parvovirus	Roseola; most parvoviruses depend on co-infection with adenoviruses for growth
III.	dsRNA	
	Reovirus	Diarrhoea; mild respiratory diseases
IV.	ssRNA that can cerve as mRNA	
	Picornavirus	Poliovirus, rhinovirus (common cold); enteric (intestinal) viruses
	Togavirus	Rubella virus; yellow fever virus; encephalitis viruses
V.	ssRNA that is a template for mRNA	
	Rhabdovirus	Rabies
	Paramyxovirus	Measles; mumps
	Orthomyxovirus	Influenza viruses
VI.	ssRNA that is a template for DNA synthesis	
	Retrovirus	RNA tumor viruses (*e.g.* leukemia viruses); HIV (AIDS virus)
VII.	dsDNA with an RNA intermediate that is a template for DNA synthesis	
	Hepadnavirus	Hepatitis B virus

* The subclass within each class differ mainly in capsid structure and in the presence or absence of a membraneous envelope.

** ds: Double-stranded; ss: Single-stranded.

Table 6. Baltimore Method of classification of animal viruses, grouped by genome structure. This method classifies viruses with regard to the various mechanisms of viral genome replication. The central theme is that all viruses must generate positive strand mRNAs [(+) RNA] from their genomes, in order to produce proteins and replicate themselves. The precise mechanisms whereby this is achieved differ for each virus family. These various types of virus genomes can be broken down into seven strategies for their replication. For a more complete listing of family groups of viruses classified by the Baltimore method.

On the basis of morphology alone, animal viruses are organized into a hierarchical scheme consisting of virus families and constitutive genera based on size, shape, type of nucleic acid and the presence or absence of an envelope. Some families of viruses generated in this scheme are described and illustrated below (Table 7).

Replication of Animal Viruses

Outside its host cell a virus is an inert particle. However, when it encounters a host cell it becomes a highly efficient replicating machine. After attachment and gaining entry into its host cell, the virus subverts the biosynthetic and protein synthesizing abilities of the cell in order to replicate the viral nucleic acid, make viral proteins and arrange its escape from the cell. The process occurs in several stages and differs in its details among DNA-containing and RNA-containing viruses.

The Stages of Replication

1. The first stage in viral replication is called the *attachment (adsorption) stage*. Like bacteriophages, animal viruses attach to host cells by means of a complementary association between attachment sites on the surface of the virus and receptor sites on the host cell surface. This accounts for specificity of viruses for their host cells. Attachment sites on the viruses (usually called *virus receptors*) are distributed over the surface of the virus coat (capsid) or envelope, and are usually in the form of glycoproteins or proteins. Receptors on the host cell (called the *host cell receptors*) are generally glycoproteins imbedded into the cell membrane. Cells lacking receptors for a certain virus are resistant to it and cannot be infected. Attachment can be blocked by antibody molecules that bind to viral attachment sites or to host cell receptors. Since antibodies block the initial attachment of viruses to their host cells, the presence of these antibodies in the host organism are the most important basis for immunization against viral infections.

2. The *penetration stage* follows attachment. Penetration of the virus occurs either by engulfment of the whole virus, or by fusion of the viral envelope with the cell membrane allowing only the nucleocapsid of the virus to enter the cell. Animal viruses generally do not "inject" their nucleic acid into host cells as do bacteriophages, although occasionally non enveloped viruses leave their capsid outside the cell while the genome passes into the cell.

3. Once the nucleocapsid gains entry into the host cell cytoplasm, the process of *uncoating* occurs. The viral nucleic acid is released from its coat. Uncoating processes are apparently quite variable and only poorly understood. Most

Table 7. Some Families of Animal Viruses

Family Name	Morphology	Enveloped (E) or Naked (N)	Approximate Size (nm)	Nucleic Acid
Poxviridae (poxviruses)		E	350 × 250	Linear ds DNA
Herpesviridae (herpesviruses)		E	200	Linear ds DNA
Adenoviridae (adenoviruses)		N	75	Linear ds DNA
Parvoviridae (parvoviruses)		N	20	Linear ss DNA
Papovaviridae (papovaviruses)		N	50	Circular ds DNA
Baculoviridae (baculoviruses)		E	300 × 40	Circular ds DNA
Picornaviridae (picornaviruses)		N	27	Plus-strand RNA
Togaviridae (togaviruses)		E	50	Plus-strand RNA
Retroviridae (retroviruses)		E	50	Plus-strand RNA
Orthomyxoviridae (orthomyxoviruses)		E	110	Segmented: 8 minus-strand RNA molecules
Paramyxoviridae (paramyxoviruses)		E	200	Minus-strand RNA
Rhabdoviridae (rhabdoviruses)		E	170 × 70	Minus-strand RNA
Reoviridae (reoviruses)		N	65	Segmented: 10–13 ds RNA molecules

viruses enter the host cell in an engulfment process called receptor mediated endocytosis and actually penetrate the cell contained in a membranous structure called an endosome. Acidification of the endosome is known to cause rearrangements in the virus coat proteins which probably allows extrusion of the viral core into the cytoplasm. Some antiviral drugs such as amantadine exert their antiviral effect may preventing uncoating of the viral nucleic acid.

4. Immediately following uncoating, the *viral synthesis stage* begins. Exactly how these events will unfold depends upon whether the infecting nucleic acid is DNA or RNA.

In DNA viruses, such as Herpes, the viral DNA is released into the nucleus of the host cell where it is transcribed into early mRNA for transport into the cytoplasm where it is translated into *early viral proteins*. The early viral proteins are concerned with replication of the viral DNA, so they are transported back into the nucleus where they become involved in the synthesis of multiple copies of viral DNA. These copies of the viral genome are then templates for transcription into late mRNAs which are also transported back into the cytoplasm for translation into *late viral proteins*. The late proteins are structural proteins (*e.g.* coat, envelope proteins) or core proteins (certain enzymes) which are then transported back into the nucleus for the next stage of the replication cycle.

In the case of some RNA viruses (*e.g.* picornaviruses), the viral genome (RNA) stays in the cytoplasm where it mediates its own replication and translation into viral proteins. In other cases (*e.g.* orthomyxoviruses), the infectious viral RNA enters into the nucleus where it is replicated before transport back to the cytoplasm for translation into viral proteins.

5. Once the synthesis of the various viral components is complete, the *assembly stage* begins. The capsomere proteins enclose the nucleic acid to form the viral nucleocapsid. The process is called *encapsidation*. If the virus contains an envelope it will acquire that envelope and asssociated viral proteins in the next step.

6. The *release stage* is the final event in viral replication, and it results in the exit of the mature virions from their host cell. Virus maturation and release occurs over a considerable period of time. Some viruses are released from the cell without cell death, by *egestion*, whereas others are released when the cell dies and disintegrates. In the case of enveloped viruses, the nucleocapsid acquires its final envelope from the nuclear or cell membrane by a budding off process (*envelopment*) before *egress* (exit) out of the host cell. Whenever a virus acquires a membrane envelope, it always inserts specific viral proteins into the that envelope which become unique viral antigens and which will be used by the virus to gain entry into a new host cell.

Below are illustrated the modes of replication of viruses that conform to this model. Influenza virus is an enveloped, single stranded (-)RNA virus that contains a segmented genome (Figure 44).

How Viruses Cause Disease

Their are several possible consequences to a cell that is infected by a virus, and ultimately this may determine the pathology of a disease caused by the virus.

Lytic infections result in the destruction of the host cell. Lytic infections are caused by virulent viruses, which inherently bring about the death of the cells that they infect (Figure 45).

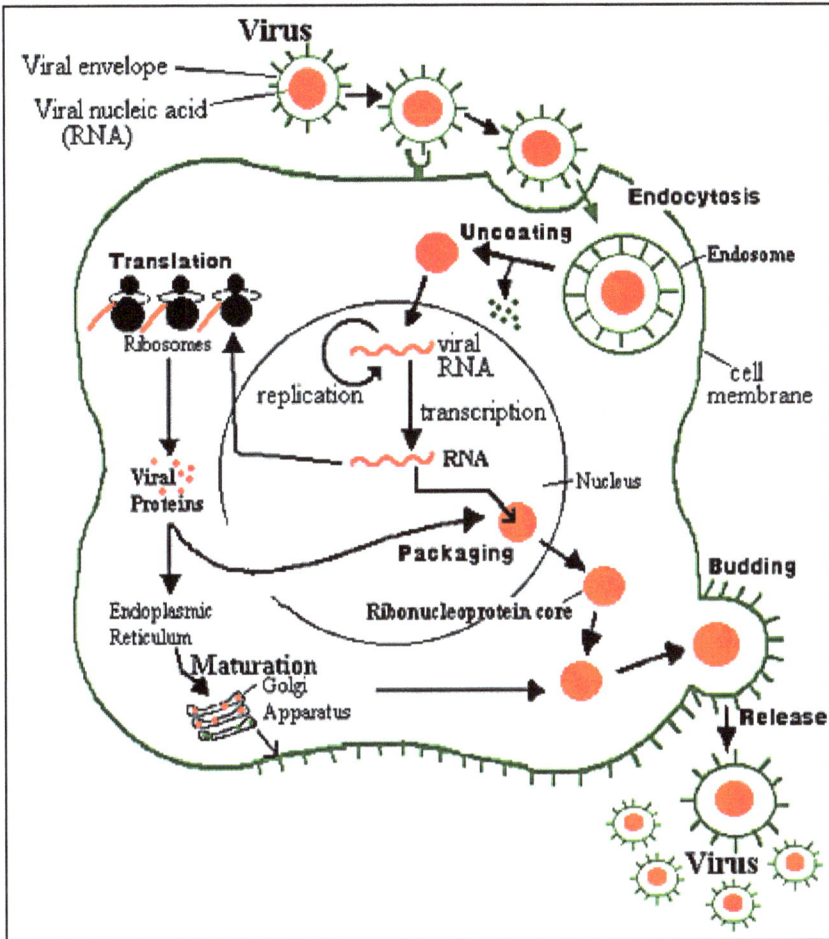

Figure 44.The Replication Cycle of Influenza A Virus

1. The virus adsorbs to the cell surface by means of specific receptors. 2. The virus is taken up in a membrane enclosed endosome by the process of receptor mediated endocytosis. 3. Uncoating takes place in the endosome and the viral RNA (genome) is released into the cytoplasm. 4. The (-) RNA of the viral genome is transported into the nucleus where it is replicated and copied by a viral enzyme into (+)RNA which is both messenger RNA and serves as a template for more (-) RNA. The (+)RNA is transported into the cytoplasm for translation into early and late viral proteins. 5. The viral core proteins are transported back into the nucleus to assemble as the capsid around the viral (-) RNA forming the "ribonucleoprotein core" or the genome-containing nucleocapsid of the virus. The viral envelope proteins assemble themselves in the cell membrane. 6. The nucleocapsid recognizes specific points on cell membrane where viral proteins have become inserted and buds off of the membrane to be released during enclosure in the viral envelope.

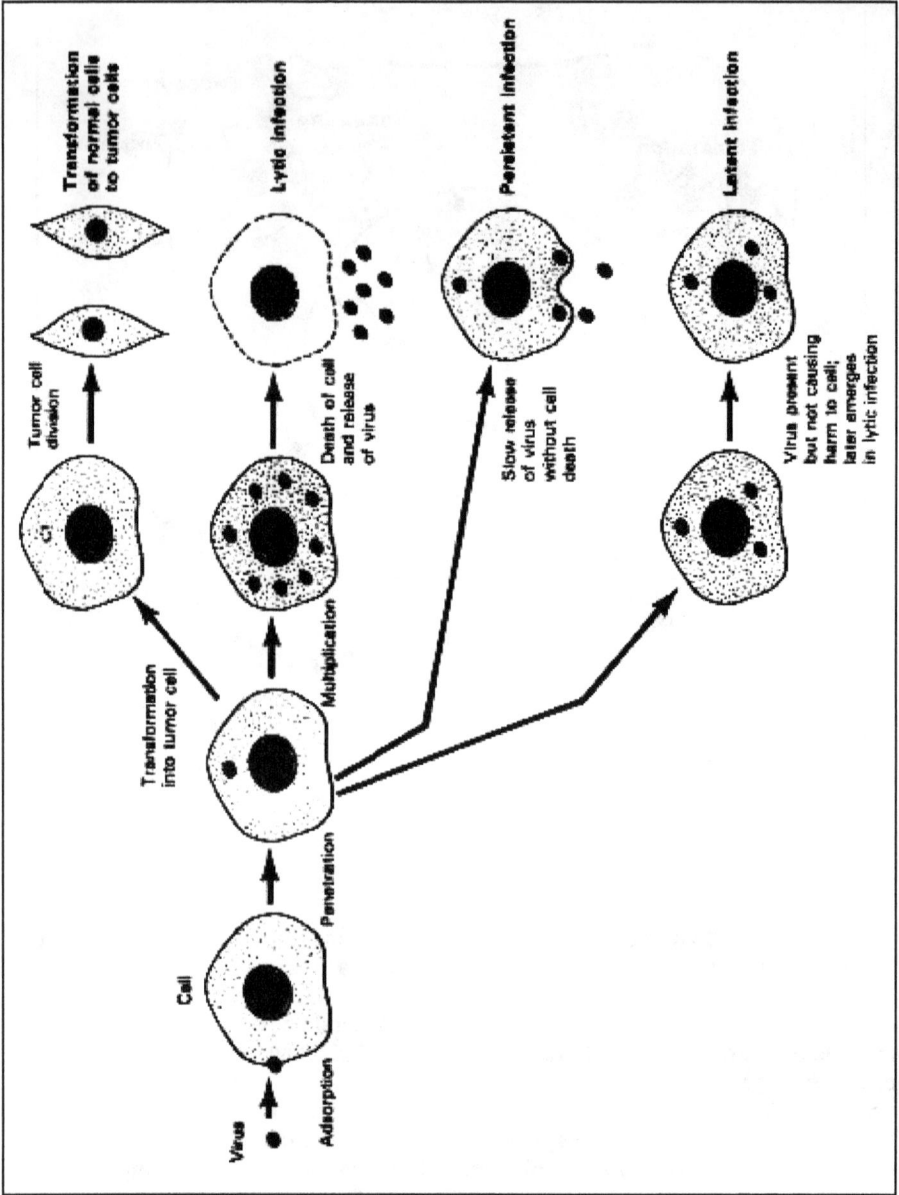

Figure 45. The Possible Effects that Animal Viruses may have on the Cells that they Infect

When enveloped viruses are formed by budding, the release of the viral particles may be slow and the host cell may not be lysed. Such infections may occur over relatively long periods of time and are thus referred to as *persistent infections*. Viruses may also cause *latent infections*. The effect of a latent infection is that there is a delay between the infection by the virus and the appearance of symptoms. Fever blisters (cold sores) caused by herpes simplex type 1 result from a latent infection; they appear sporadically as the virus emerges from latency, usually triggered by some sort of stress in the host.

Some animal viruses have the potential to change a cell from a normal cell into a tumor cell, the hallmark of which is to grow without restraint. This process is called *transformation*. Viruses that are able to transform normal cells into tumor cells are referred to as *oncogenic viruses*.

The vast majority of viral infections in humans are inapparent or asymptomatic. Viral pathogenesis is the abnormal situation and it is of no particular value to the virus, although it typically results in the multiplication of the viruses that can be transmitted to other individuals. For pathogenic viruses, there are a number of critical stages in replication which determine the nature of the disease they produce.

The Stages of Viral Infections

1. Entry into the Host

The first stage in any virus infection, irrespective of whether the virus is pathogenic or not. In the case of pathogenic infections, the site of entry can influence the disease symptoms produced. Infection can occur via several portals of entry.

Skin

Most viruses which infect via the skin require a breach in the physical integrity of this effective barrier, *e.g.* cuts or abrasions. Some viruses employ vectors, *e.g.* ticks, mosquitos, etc. to breach the skin.

Respiratory Tract

The respiratory tract and all other mucosal surfaces possess sophisticated immune defense mechanisms, as well as non-specific inhibitory mechanisms (ciliated epithelium, mucus secretion, lower temperature, etc.) which viruses must overcome. Nonetheless, this is the most common point of entry for most viral pathogens.

Gastrointestinal Tract

A fairly protected mucosal surface, but some viruses (*e.g.* enteroviruses, including polioviruses) enter at this site.

Genitourinary Tract

Less protected than the GI tract, but less frequently exposed to extraneous viruses.

Conjunctiva

An exposed site and relatively unprotected.

2. Primary Replication

Having gained entry to a potential host, the virus must initiate an infection by entering a susceptible cell. Some viruses remain localized after primary infection, but others replicate at a primary site before dissemination and spread to a secondary site. Examples are given in the Table 8.

Table 8: Localized Infections of Primary and Secondary Replication

Virus	Primary Replication	Secondary Replication
Localized Infections		
Rhinoviruses	Upper respiratory tract	
Rotaviruses	Intestinal epithelium	
Papillomaviruses	Epidermis	
Systemic Infections		
Enteroviruses (poliovirus)	Intestinal epithelium	Lymphoid tissues, CNS
Herpesvirus (HSV types 1 and 2)	Oropharynx or urogenital tract	Lymphoid cells, peripheral nervous system, CNS
Rabies virus	Muscle cells and connective tissue	CNS

3. Dissemination Stage

There are two main mechanisms for viral spread throughout the host: via the bloodstream and via the nervous system.

The virus may get into the bloodstream by direct inoculation–*e.g.* arthropod vectors, blood transfusion or I.V. drug abuse. The virus may travel free in the plasma (Togaviruses, Enteroviruses), or in association with red cells (Orbiviruses), platelets (HSV), lymphocytes (EBV, CMV) or monocytes (Lentiviruses). The presence of viruses in the bloodstream is referred to as a *viremia*. *Primary viremia* may be followed by more generalized *secondary viremia* as the virus reaches other target tissues or replicates directly in blood cells.

In some cases, spread to nervous system is preceded by primary viremia, as above. In other cases, spread occurs directly by contact with neurons at the primary site of infection. Once in peripheral nerves, the virus can spread to the CNS by axonal transport along neurons (*e.g.* HSV). Viruses can cross synaptic junctions since these frequently contain virus receptors, allowing the virus to jump from one cell to another.

4. Tissue/Cell tropism

Tropism is the ability of a virus to replicate in particular cells or tissues. It is influenced partly by the route of infection but largely by the interaction of a virus attachment sites (virus receptors) with specific receptors on the surface of a cell. The interaction of the virus receptors with the host cell receptors may have a considerable effect on pathogenesis.

5. Host Immune Responses

There are several ways that the host immune responses may contribute to viral pathology. The mechanisms of cell mediated immunity are designed to kill cells which are infected with viruses. If the mechanisms of antibody mediated immunity result in the production of antibodies that cross-react with tissues, an autoimmune pathology may result.

6. Secondary Replication

This occurs in systemic infections when a virus reaches other tissues in which it is capable of replication. For example, polioviruses initiate infection in the GI where the produce an asymptomatic infection. However, when disseminated to neurons in the brain and spinal cord, where the virus replicates secondarily, the serious paralytic complication of poliomyelitis occurs. If a virus can be prevented from reaching tissues where secondary replication can occur, generally no disease results.

7. Direct Cell and Tissue Damage

Viruses may replicate widely throughout the body without any disease symptoms if they do not cause significant cell damage or death. Although retroviruses (*e.g.* HIV) do not generally cause cell death, being released from the cell by budding rather than by cell lysis, they cause persistent infections and may be passed vertically to offspring if they infect the germ line. Conversely, most other viruses, referred to as *virulent viruses*, ultimately damage or kill their host cell by several mechanisms, including inhibition of synthesis of host cell macromolecules, damage to cell lysosomes, alterations of the cell membrane, development of inclusion bodies, and induction of chromosomal aberrations.

8. Persistence versus Clearance

The eventual outcome of any virus infection depends on a balance between the ability of the virus to persist or remain latent (persistence) and the forces of the host to completely eliminate the virus (clearance).

Long term persistence is the continued survival of a critical number of virus infected cells sufficient to continue the infection without killing the host. It results from two main mechanisms:

(a) Regulation of Lytic Potential

For viruses that do not kill their host cells, this is not usually a problem. But for lytic (virulent) viruses, there may be ways to down regulate their replicative and lytic potential so that they can persist in a state of latency without replication and damage to their host cell. This is the case with herpes viruses.

(b) Evasion of Immune Surveillance

This may be due to several conditions that are properties of the host or the virus. Some viruses, such as influenza, can undergo antigenic shifts or antigenic drift that allows them to bypass a host immune response. Some viruses, *e.g.*, measles, may induce a form of immune tolerance such that the host is unable to undergo an effective immune response to the virus. Other viruses, such as HIV, may set up a direct attack against cells of the immune system such that the immune system is compromised in its ability to attack or eliminate the virus.

Types of Animal Viruses
Papilloma Viruses

Virus

☆ Human Papilloma Virus (HPV)

General Concepts

☆ The Papilloma viruses are small, non-enveloped icosahedral particles containing a circular dsDNA genome.

☆ Papilloma viruses are a member of the Papovavirus family, which is divided into 2 genera:

● *Polyoma virus*, which contains a 5200 base pair DNA genome and has been employed as a good molecular model system. There are 2 polyoma viruses that cause disease in humans; BKV and JCV.

● *Papilloma virus*, which contains an 8000 base pair DNA genome, can induce benign tumors of the head and neck, several varieties of skin warts, and cervical cancers.

☆ Human papilloma viruses are trophic for epithelial cells of the skin and mucus membranes. They appear to replicate in the cell nucleus and have two modes of replication:

● Stable replication in basal cells and

● Vegetative replication in more differentiated cells that generates progeny virions.

☆ Human papillomavirus (HPV) is thought to be the most common sexually transmitted disease in the world.

☆ The CDC estimates that there are approximately 6.2 million new cases of sexually transmitted HPV infections annually and that over 20 million people are already infected.

Distinctive Properties

☆ Human papillomaviruses (HPVs) produce epithelial tumors of the skin and mucous membranes. More than 100 HPV types have been detected.

☆ HPV infections may be latent (asymptomatic), subclinical, or clinical. Most HPV infections are latent; clinical infections are usually apparent as warts.

☆ Of the many types of HPV, types 6 and 11 are generally classified as "low risk" because infection with these types has a low potential for producing cancerous lesions. These two types of HPv are thought to be responsible for 90 per cent of all genital warts cases.

☆ On the other hand, HPV types 16 and 18 are classified as "high-risk" because they are responsible for most of the lesions that may progress to cancers, particularly those in the anogenital and/or mucosal category. These two types are thought to be responsible for 70 per cent of cervical cancer cases.

☆ In "low risk" infections, the HPV genome is thought to exist as a separate circular dsDNA molecule, while in "high risk" malignant infections, the genome is incorporated into the host genome. Some of the viral proteins inactivate host cell tumor suppressor proteins, and this may lead to carcinomas.

Pathogenesis

☆ Clinical HPV infections may be described as:
- Nongenital cutaneous,
- Nongenital mucosal and
- Anogenital.

☆ Nongenital cutaneous diseases include common warts, plantar warts, flat warts and other skin lesions.

☆ Nongenital mucosal diseases include resiratory papillomatosis, laryngeal papillomas, conjunctival papillomas, carcinomas and others.

☆ Anogenital diseases include a variety of warts as well as cancers of the cervix, anus, vagina and penis.

Host Defenses

☆ Host defenses against the papillomaviruses are not entirely understood, but a variety of mechanisms probably contribute.

☆ The efficacy of the new vaccine, however, suggests that humoral responses are protective.

Epidemiology

☆ Papillomaviruses are widespread and warts are common in children and young adults.

☆ Humans are the only host for HPV and infections are generally transmitted by direct contact. However, the virus can survive for extended periods (months) outside the host, and this may provide another means of transmission.

☆ While there is a strong correlation between HPV infection and certain forms of cancer (*e.g.* cervical cancer), infection alone does not result in maligancy; rather, additional factors such as radiation, immunosuppression, or tobacco use are involved.

Diagnosis

Clinical
Warts of the skin, oral cavity and genital area are generally diagnosed by appearance.

Laboratory
Microscopy of wart scrapings shows a characteristic histologic appearance.

Molecular techniques (nucleic acid hybridization) can be used for other HPV infections.

Control

Sanitary
Avoidance of contacts but this is not really practical.

Immunological
In 2006, the first vaccine developed to prevent cervical cancer and genital warts in women due to HPV was approved. This vaccine, Gardasil®, can be administered to females aged 9-26 years of age through a series of three shots over a six-month period. The vaccine is a quadrivalent, recombinant viral protein suspension that protects against infection by types 6, 11, 16 and 18. The vaccine has been shown to be safe and nearly 100 per cent effective.

Chemotherapeutic
Warts can be treated by freezing, cauterization, surgery, laser vaporization, etc.

Herpes Viruses

Virus

Herpesviruses: Herpes Simplex (HSV-1, HSV-2) (Figure 46) Varicella-Zoster virus (VZV), Cytomegalovirus (CMV), Epstein-Barr virus (EBV)

General Concepts

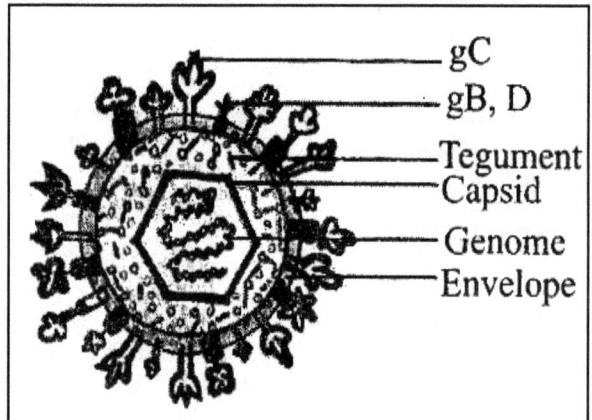

Figure 46. Structure of Herpes Simplex Virus

☆ The Herpesviruses are a large group containing more than 70 members that infect organisms from fungi to humans.

☆ The Herpesviruses contain a linear DNA genome and are enveloped (bilayered with surface projections derived from the host cell nuclear membrane). Replication occurs in the host cell nucleus.

☆ Herpesviruses cause acute infections but they are also capable of latency. This can lead to recurrent infections, which are important to the mechanism of host to host transmission. HSV-1, HSV-2 and VZV are termed "neurotropic" while CMV and EBV are "lymphotropic", referring to the cell type in which the latent infection is established.

☆ Some Herpesviruses have been associated with the production of cancers, providing the best evidence for a viral etiology for these diseases.

☆ Humans are the natural host for HSV, VZV, CMV and EBV.

Distinctive Properties

☆ The morphology of each the Herpesviruses is similar but unique.

☆ At the genome level, HSV-1 and HSV-2 share only about 50 per cent DNA homology; the genome molecular weights, however are similar (96×10^6 daltons). The genome molecular weights of the other viruses are somewhat larger; EBV is about 114×10^6; CMV is about 150×10^6.

☆ Except for HSV-1 and HSV-2, the proteins surrounding the genome are immunologically unrelated.

☆ Herpes DNA replication involves three coordinately regulated sequential events. Upon entering the host cell, the virus travels to the nucleus. There, a set of genes termed *immediate early* are transcribed by cellular transcription factors to produce a set of proteins, some of which are new transcription factors. This groups of factors recognize viral promoters preceding the *early* genes. Some of the resultant proteins are replication enzymes while others are a third set of transcription factors that synthesize mRNAs encoding the *late* genes. The late gene products, then, are structural proteins that encapsidate the newly formed DNA genome to produce new progeny virions.

Replication

☆ Virion binds to the extracellular protein through gB and gC receptors (Figure 47)

☆ Another viral protein gD interacts with a second cellular receptor.

☆ This interaction mediates fusion of virus with the host Plasma membrane.

☆ The virus is released into the cytoplasm.

☆ Viral nucleocapsid docks at the nuclear pore and releases the viral DNA into the nucleus, where the DNA circularizes

VP16 enhances transcription of viral genome and stimulates the transcription of immediate early genes by RNA polymerase H of host cells.

☆ Immediate early mRNAs are spliced and transported to the cytoplasm, where they are translated into proteins.

☆ Immediate early proteins (α proteins) are imported into the nucleus, where they activate the transcription of early genes.

☆ I protein genes are transported to the cytoplasm after transcription and they are translated proteins are imported to the nucleus where they induce DNA replication and synthesis of substrate for DNA synthesis.

☆ DNA replication produces long concatemeric DNA molecules, the templates for late gene expression.

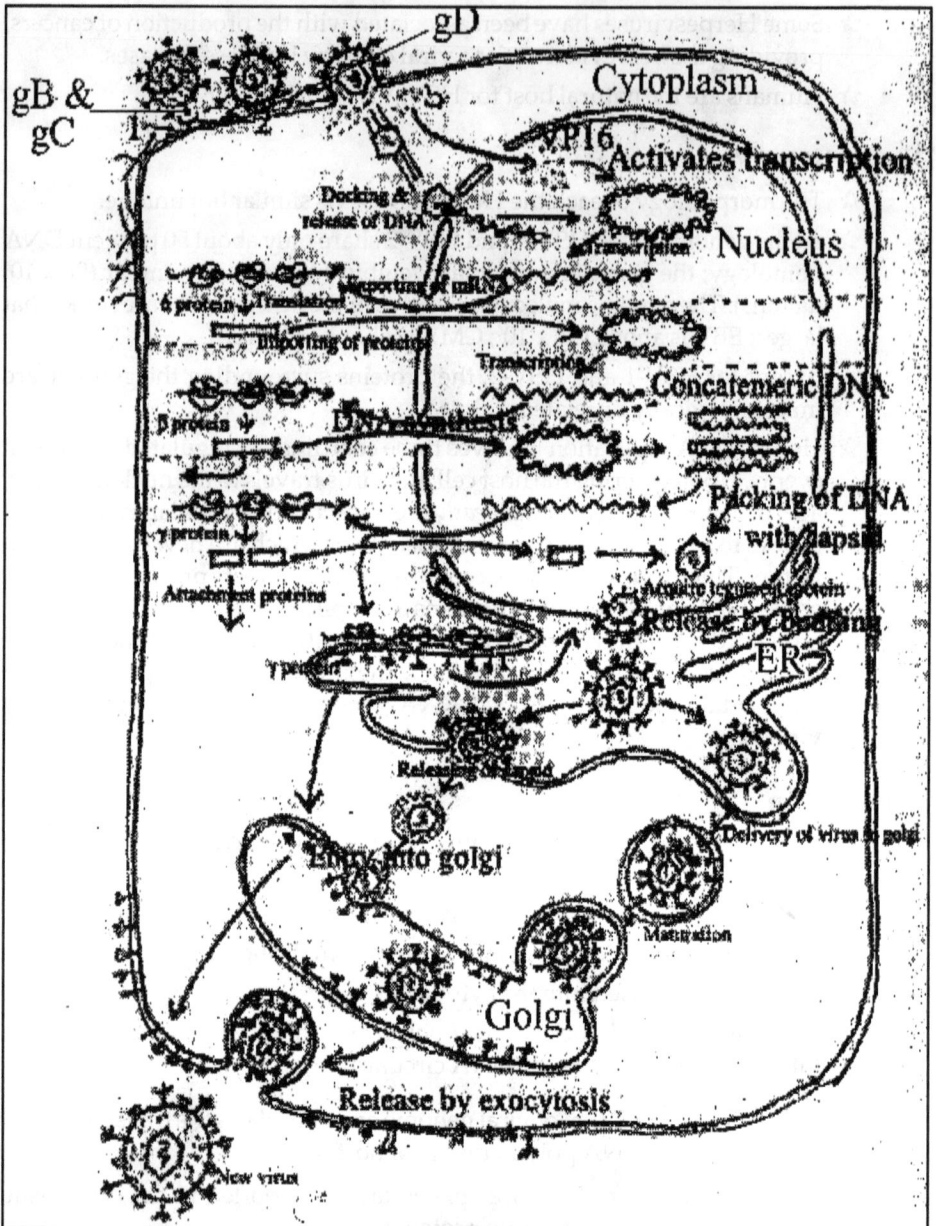

Figure 47. Replication of Herpes Simplex Virus

☆ Late mRNAs are transported to the cytoplasm for the synthesis of gamma protein. These proteins are structural proteins needed for the viral assembly.

☆ Some late proteins are inserted to ER and transported to Golgi apparatus for glycosylation.

☆ Mature glycoproteins are transported to plasma membrane of the infected cell.

☆ Some gamma proteins are transported o the nucleus for the assembly of viral capsid.

☆ Newly replicated viral DNA is packaged into preformed capsids.

☆ These capsids, together with some tegument proteins are budding off from the inner nuclear membrane into the lumen of Endoplasmic reticulum and acquire an envelope.

☆ The enveloped virus is then transported to the plasma membrane for release by exocytosis.

☆ Latent infection occurs primarily in neurons found sensory and autonomic ganglia. During this infection, Latency Associated Transcript (LAT) promoter is synthesized and involved in the protein synthesis.

Pathogenesis

☆ Herpesviruses infect a range of different cell types, causing different disease scenarios. The skin and mucus membranes are common sites of infection for HSV and VZV; CMV and EBV are more internal (EBV infects B lymphocytes).

☆ The viruses produce intranuclear inclusions and multinucleated giant cells

☆ With the exception of VZV, primary Herpes infections are often asymptomatic. The process of latency is poorly understood but, except for VZV, asymptomatic shedding is common. Recurrence of acute disease results because of emotional stress, surgery, trauma, cold, fever, immune suppression, etc. Recurrence of HSV is common, frequent and localized.

☆ More specifically:

HSV-1 is responsible for a variety of infections. Most commonly, HSV-1 produces the condition known as gingivostomatitis in which oral cavity vesicles or ulcers form. These lesions may recur frequently as "cold sores" (herpes labialis). Another condition produced by HSV-1 is herpetic keratitis, which may be serious if accompanied by conjunctivitis because this can lead to corneal scarring and blindness. Another condition known as "whitlows" appears as lesions on the fingers.

HSV-2 is commonly referred to as genital herpes. This virus produces lesions on the genitals, urethra and bladder. Recurrence may be frequent. In neonates, infection may be local or disseminated and has about 50 per cent mortality if untreated. HSV-2 may also cause meningitis or encephalitis.

VZV produces the disease varicella and zoster. Varicella is commonly known as chickenpox. This relatively mild infection in children can be more serious in adults, occasionally progressing to pneumonia. Varicella is characterized by a skin rash appearing first on the head and trunk, and later on the extremities. The skin lesions progress from macules to papules to vesicles to pustules to crusts. These lesions are not prone to scar. Zoster is commonly known as shingles and is a manifestation of varicella infection. Typically occurring in older individuals, the lesions are confined to skin areas innervated by sensory nerves of the dorsal ganglia, primarily in the thoracic and lumbar regions. These nervous tissues are thought to be the sight of latent infection by VZV.

CMV infections are often asymptomatic but when infection occurs in utero, cytomegalic inclusion disease may result. This condition is characterized by jaundice, hepatosplenomegaly and central nervous system disorder. CMV may also produce a form of mononucleosis characterized by fever, fatigue and atypical lymphocytes. This form of mononucleosis is different than that produced by EBV (below).

EBV is primarily responsible for infectious mononucleosis, characterized by fever, fatigue, malaise and pharyngitis. EBV latently infects B-cells, creating the potential for recurrence. In addition, a strong association between EBV and Burkitt's lymphoma in Africa and nasopharyngeal carcinoma in the Orient and Africa provides evidence for a viral etiology for some human cancers.

Host Defenses

Primary infection by Herpesviruses induces antibody, which is protective, but recurrent infections still occur.

The cell-mediated immune response is also important (viral antigens on the cell surface allow detection and killing of infected cells).

Epidemiology

In lower socioeconomic groups, most individuals are infected subclinically by HSV-1, CMV and EBV. In the higher socioeconomic groups, about half are infected. VZV infects both groups equally. HSV-2 is generally transmitted by sexual contact; the others more commonly by saliva. CMV can be spread transplacentally (0.5-2.5 per cent newborns have CMV in the urine).

Diagnosis

☆ Diagnosis may be made on clinical grounds. Patients with fever and vesicles may be considered as HSV infection.

☆ Vesicular and hepatic lesions of skin, cornea orb1 throat washings, CSF and stool are used as specimens for diagnosis.

☆ Virus isolation is one of the most reliable met1. for confirmation of the clinical diagnosis. The specimens are transported through viral transport medium and lated into tissue culture. As HSV has a wide host range, many cell culture system are susceptible. The appearance of typical cytopathic effects in cell culture in 2-3 days suggests the presence of HSV.

☆ Serological methods have been developed for rapid diagnosis. Antibodies of HSV are measured by neutralization test, C.1- ELISA, RIA and immuno fluorescence tests.

☆ HSV -DNA is detected by RT-PCR method.

Control

Sanitary
Avoidance of contacts reduces the incidence of disease but virus may be transmitted by asymptomatic individuals.

Immunological
Vaccines are in development but the problems associated with latency and possible cancers remains. A vaccine for VZV is available. Hyperimmune serum can be used for susceptible or high risk individuals.

Chemotherapeutic
Acyclovir (a nucleoside analog) is effective against HSV infections. Acyclovir and other analogs (iododeoxyuridine, trifluorothymidine, adenine arabinoside) have been used to treat other Herpes infections but often show little benefit.

Pox Viruses

Virus
- ☆ Poxviruses (Orthopoxviruses): Smallpox

General Concepts
- ☆ A passage from T. B. Macaulay's *The History of England from the Accession of James II*, volume IV, serves to preface this page describing the disease smallpox:

  "The smallpox was always present, filling the churchyards with corpses, tormenting with constant fears all whom it had stricken, leaving on those whose lives it spared the hideous traces of its power, turning the babe into a changeling at which the mother shuddered, and making the eyes and cheeks of the betrothed maiden objects of horror to the lover."

- ☆ The Poxviruses contain a dsDNA genome, have a complex morphology and replicate in the host cytoplasm.
- ☆ Poxviruses cause localized and generalized infections in humans and animals.
- ☆ Smallpox was the major human disease agent but it was officially declared eradicated by the World Health Organization (WHO) in 1980.
- ☆ Smallpox is transmitted via the respiratory route. In the body, it is spread by a transient viremia to internal sites, and then a second viremia to the skin where the characteristic lesions erupt.
- ☆ The lack of an animal reservoir made eradication of human disease possible.

Distinctive Properties
- ☆ Poxviruses are large and brick shaped having a complex symmetry (not icosahedral or helical) and small surface tubules.
- ☆ The genomic DNA has a molecular weight of 100-200 x 10^6 daltons.

☆ The virion contains a DNA-dependent RNA polymerase (DDRP, transcription enzyme) that is required for infection. Other enzymes in the virion complete the uncoating process and initiate early replication, which occurs in acidophilic intracytoplasmic inclusions.

☆ Poxviruses are antigenically complex but may share common internal antigens.

Pathogenesis

☆ Smallpox is a systemic infection with a very characteristic rash.

☆ Transmission of disease occurs via the respiratory tract. Following a 12 day incubation, lesions first appear in the mouth and throat and then on the skin. The rash develops in a synchronous and centrifugal manner with lesions first taking the macular form, then papular, vesicular, pustular and finally as crusts.

☆ Two forms of smallpox exist: the more serious form, variola major, has about a 50 per cent mortality rate; the less severe form, variola minor results in less than 1 per cent death.

Host Defenses

☆ Antiviral antibody is protective. Hyperimmune serum can prevent disease. The cell-mediated response is also important, however, since individuals with hypogammaglobulinemia still recover.

Epidemiology

☆ Before eradication, Asia and India were endemic for variola major; South America for variola minor.

Diagnosis

Clinical
The characteristic lesions with their centrifugal and synchronous development are diagnostic.

Laboratory
Electron microscopy of crusts can reveal the intracytoplasmic inclusions.

Control

Sanitary
Quarantine of diseased individuals.

Immunological
Vaccination of all individuals.

Chemotherapeutic
None.

Picorna Viruses

Picornaviruses are a family of icosahedral viruses containing small R.NA as the genome pico means small; rna means RNA virus). This family is popularly known as Picorriaviridae. It is the largest family of RNA viruses frequently infecting man and animals. Picornaviruses are non-enveloped viruses with the size of 20-30 nm diameter. The genome is a single stranded RNA With positive strand polarity. The icosahedral head consists of 60 subunits.

Structure

The picornaviruses are small (22 to 30 nm) non-enveloped, single-stranded RNA viruses with icosahedral symmetry (Figure 48). The virus capsid is composed of four viral VP_1 to VP_4. The genome is a single stranded RNA. The lar weight is approximately 2×10^6

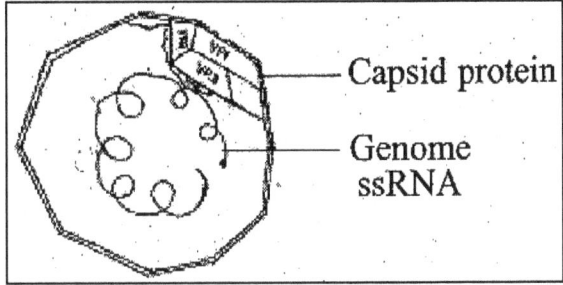

Figure 48. Structure of *Picornavirus*

to 3×10^6 daltons. The RNA is covalently bonded to a non-capsid viral protein (VPg). Picornaviruses can survive for long periods in L matter and are resistant to the low pH in the stomach (pH 3.0-5.0). Picornaviruses are inactivated by pasteurization, boiling, formalin and chlorine.They are ether resistant due to the absence of essential lipids. They replicate in the cytoplasm.

Replication

Replication is similar in all picornaviruses. It involves the following steps:

- ☆ The virus binds to a cellular receptor.
- ☆ The viral RNA is released into the cell. The mechanism of uncoating of RNA genome is not yet clear.
- ☆ Translation is initiated at an internal site of 741 nucleotides from the 5' end of the viral RNA and polyprotein precursors are synthesized.
- ☆ Polyproteins are cleaved into three individual proteins P_1, P_2 and P_3
- ☆ The P_1 protein units are assembled into viral structural proteins.
- ☆ The P_2 and P_3 proteins are responsible for the synthesis of proteases and RNA polymerase.
- ☆ The proteins that involve in RNA synthesis are transported to the membranous vesicles.

Positive sense RNA is also transported into the vesicles.

It is copied into minus sense RNA by RNA replication (Figure 49)

- ☆ Structural proteins are formed by partial cleavage of P_1 precursor proteins.
- ☆ These proteins are transported to vesicle.

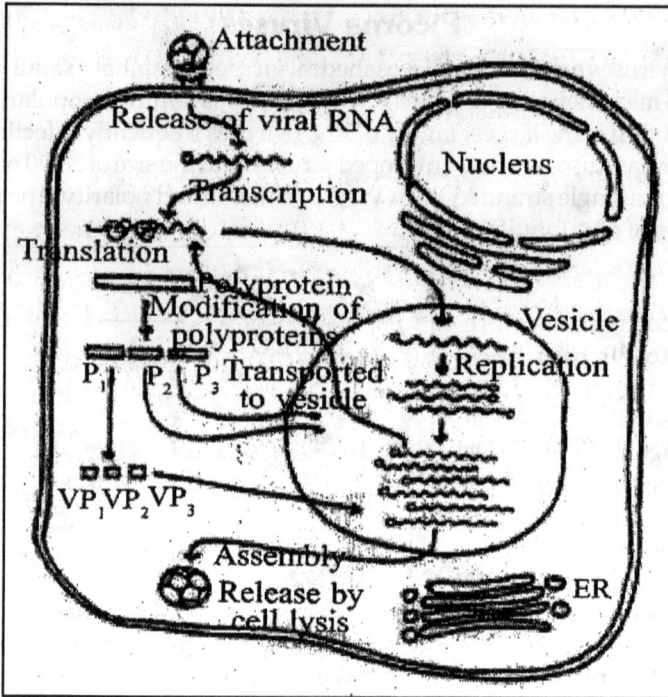

Figure 49. Replication of *Picornavirus*

★ Virus assembly takes place within the vesicle.

★ Mature virions are released after cell lysis

General Concepts

★ Picornaviruses are small, nonenveloped viruses containing a single positive strand RNA genome. They possess an icosahedral symmetry.

★ Picornaviruses are divided into two groups; the Enteroviruses (Poliovirus, Coxsackievirus and Echovirus) and the Rhinoviruses. There are about 63 serotypes of Enterovirus and more than 100 serotypes of Rhinovirus.

★ Picornaviruses commonly produce subclinical infections; acute disease may range from minor illness to paralytic disease.

★ The Enteroviruses enter via the intestinal tract and attach to receptors on intestinal epithelia. During this alimentary phase, the virus replicates in cytoplasm. They then spread into the lymphatic circulation (lymphatic phase) and then to the bloodstream (viremic phase). The viremic phase generally marks the end of the infection but an occasional neurologic phase can lead to more severe and permanent problems.

Distinctive Properties

Polioviruses

Poliovirus types 1, 2 and 3 are recognized. Their genome contains a 7000 base positive strand of RNA. These viruses adsorb only to intestinal epithelial cells and motor neuron cells of the central nervous system.

Coxsackie

These viruses are divided into two groups; A and B. There are 23 serotypes of A, 6 serotypes of B. In humans, Coxsackieviruses produce respiratory disease, herpangitis, "hand, foot and mouth" disease, febrile rashes, pleurodynia, pericarditis, myocarditis, aseptic meningitis and paralytic disease.

Echoviruses

An acronym for *Enteric Cytopathogenic Human Orphan viruses*, the Echoviruses contain 31 serotypes and produce respiratory disease, febrile illness (with or without a rash), aseptic meningitis and paralytic disease.

Rhinoviruses

This group of viruses are sensitive to acid pH and their optimal growth occurs at 33°. There are over 100 serotypes of Rhinoviruses and they produce the common cold.

Pathogenesis

Polioviruses

A number of syndromes can be observed, most (90-95 per cent) are subclinical. About 4-8 per cent present with a mild fever, sore throat and headache. Nonparalytic polio occurs in about 1 per cent of cases while paralytic disease affects only 0.1 per cent. Of the paralytic conditions, about 5-10 per cent affect the cranial nerve or medulla while the others are spinal and limited to muscle weakness rather than complete paralysis.

Coxsackievirus, Echovirus

Infection by Coxsackie- or Echoviruses resembles that produced by Poliovirus. A variety of syndromes are possible, ranging from trivial to severe.

Rhinovirus

Infection results via the nasopharynx by direct contact. Viral replication leads to inflammation and edema, symptoms of the common cold. Inapparent infection is common.

Host Defenses

Enteroviruses

Interferon is effective against these viruses. More specifically, IgA antibodies in the intestine and saliva are protective.

Rhino

Susceptibility to the Rhinoviruses is dependent on prior exposure. Antibody of the IgA and IgG classes is important. Non-specific defenses including interferon, gastric acidity and temperature may play a major role in controlling infection.

Epidemiology

Enteroviruses

Found worldwide, the enteroviruses are spread via the fecal-oral route. Most illnesses occur in the summer and fall. The virus may be carried in the throat for a week and shed in the feces for several weeks.

Rhinoviruses

These viruses are spread from person to person, usually by direct contact. Inapparent infections occur in about half. On average, most persons suffer with 2-4 colds per year during the fall and spring months and these represent different serotypes of the virus.

Diagnosis

Clinical

Diagnosis of enteroviral infections is usually not possible based on clinical presentation. However, some symptoms (pleurodynia, myocarditis) or conditions (aseptic meningitis) are suggestive. Diagnosis of rhinoviral infections, in contrast, is usually based on clinical presentation.

Laboratory

Recovery of Enterovirus from the throat or feces is diagnostic. Recovery of Rhinoviruses is simply not practical.

Control

Sanitary

Avoidance of contacts is the best means for preventing disease.

Immunological

Vaccines for Polio have been available for more than 40 years. The Salk vaccine is a trivalent (types 1, 2 and 3), formalin inactivated suspension that is given by injection (parenteral). The Sabin vaccine is an attenuated, live, trivalent, oral suspension that produces intestinal IgA but has rare vaccine-associated paralysis. Vaccines against the other picornaviruses are not practical (too many serotypes).

Chemotherapeutic

Supportive care is best.

Hepatitis Viruses

Virus

☆ Hepatitis A, B and C

General Concepts

☆ Type A hepatitis virus (HAV) is a member of the enterovirus group of Picornaviruses; Hepatitis B virus (HBV) is distinctly different and contains a DNA genome; Hepatitis C virus (HCV, also referred to as non-A, non-B or

NANBH) is not well characterized but appears to contain an RNA genome and may belong in the Flavivirus group.

☆ The disease caused by these agents is hepatitis, a generalized infection with inflammation and necrosis of the liver. The course of hepatitis may range from inapparent disease to chronic liver disease.

Distinctive Properties

Hepatitis A
As a member of the Enterovirus group, HAV contains a linear, positive strand RNA genome. It is also known as Enterovirus type 72 and is resistant to heat and acid.

Hepatitis B
HBV contains a circular, partly double-stranded DNA genome 3200 nucleotides in length. Interestingly, a 600-2100 single-stranded region is contained in the DNA molecule. The particle contains a DNA-dependent DNA polymerase and a reverse transcriptase and replicates via an RNA intermediate. HBV possesses several antigens; the "Australian antigen" is associated with the surface (HBsAg), the "core antigen" (HBcAg) is internal and the "e antigen" (HBeAg) is part of the same capsid polypeptide as the HBcAg. All of these antigens elicit specific antibodies.

Hepatitis C
HCV (or NANBH) contains a positive stranded RNA genome and is related to the Flaviviruses.

Pathogenesis

Hepatitis A
Clinical presentation of HAV infection varies from subclinical and mild in children to jaundice in adults. The virus usually enters by intestinal infection (fecal-oral transmission), spreads via the blood to the liver, which is its target organ. HAV is detectible in the feces during the incubation period (average 4 weeks), preceding a rise in serum levels of aminotransferase enzymes and the occurrence of pathologic changes in the liver. Most disease resolves within two weeks. Chronicity or fulminant hepatitis is rare.

Hepatitis B
HBV, in contrast, may produce a persistent carrier state in addition to liver damage. Infection early in life often produces a carrier state, but only about 5-10 per cent of cases if infection occurs later. Most disease is acquired via the parenteral route (blood transfusions. There are generally two patterns of HBV-associated disease:

☆ Chronic persistent: Infections are generally asymptomatic with a mild elevation of serum alanine transaminase (ALT) and little liver fibrosis.

☆ Chronic active: Infections produce jaundice with elevated ALT levels, liver damage and cirrhosis. Liver failure may predispose those affected to cancer.

Hepatitis C

HCV may also produce a persistent carrier state, a higher level of chronic disease and cirrhosis. At least 50 per cent of HCV infections result from blood transfusion (Table 9).

Table 9: Differences Between the Hepatitis Viruses

	Hepatitis A	Hepatitis B	Hepatitis C
Genome	+RNA	DNA	+RNA
Onset	Abrupt	Insidious	Insidious
Transmission	Fecal-Oral	Parenteral	Parenteral
Incubation (days)	15-40	60-180	28-112
Asymptomatic infection	usual	common	common
Carrier State	no	yes	yes
Chronicity	0 per cent	10 per cent	30-60 per cent
Sequelae	no	cirrhosis	cirrhosis

Host Defenses

Hepatitis A

Antibody usually develops late in the infection; IgM after a week, IgG later. These help to clear virus from the body via complement and antibody-dependent cell-mediated cytotoxicity, which may account for some of the liver damage.

Hepatitis B

Antibody and the cell-mediated responses are induced but do not seem to protect against infection and probably cause autoimmune responses (liver damage may relate to the host immune response). The possibility of immune complexes with surface antigen can also lead to immune complex disease.

Hepatitis C

Little is known about host defenses against this agent but immune serum does not seem to be an effective prophylactic remedy. Probably a combination of humoral and cell-mediated defenses are important.

Epidemiology

- Infectious hepatitis (HAV) is endemic throughout the world, but the incidence is difficult to estimate because many cases are subclinical. HAV has a low mortality but patients may be incapacitated for several weeks. It is spread via the fecal-oral route, person to person and reflects conditions of poor sanitation and overcrowding. The virus may be shed in the feces but there is no carrier state or progression to chronic liver disease.

- Serum hepatitis (HBV, HCV) is generally passed via blood, unsterile syringes, transfusions, etc. A carrier state can occur; in North America and Europe, the rate is about 0.1 per cent. The rate increases to about 25-30 per cent in Africa, Asia and the Pacific.

Diagnosis

Clinical
Clinical diagnosis depends on the symptomology, which may include fever, malaise, headache, dark-colored urine and jaundice.

Laboratory
Serology is the best method for determining infection.

Control

Sanitary
HAV infection may be prevented by ensuring a safe water supply; HBV and HCV may be prevented by testing the blood supply.

Immunological
Human immunoglobulin prevents or attenuates HAV and vaccines are under study. A vaccine for HBV is available and effective. A vaccine against HCV is under study.

Chemotherapeutic
None.

Toga Viruses

Virus
☆ Togaviruses: Alphavirus, Flavivirus

General Concepts
☆ The Togaviruses are divided into two groups; Alphaviruses and Flaviviruses.

☆ Togaviruses are enveloped and contain a positive stranded RNA genome. This RNA is 5'-capped and 3'-polyadenylated.

☆ This group of viruses are called "arboviruses" because their life cycle involves alternating between vertebrates and arthropod vectors, mostly mosquitoes.

☆ The Togavirus' natural cycle usually involves birds and mammals and rarely humans with the exception of those responsible for yellow fever and dengue fever.

☆ Overall, there are approximately 25 types that are pathogenic for humans.

Distinctive Properties

Alphaviruses
Members of this group contain a nucleoprotein capsid surrounded by a lipid bilayer envelope that is derived from the host cell membrane plus two 50 kD glycoproteins. Their RNA is 4×10^6 daltons and the capsid is composed of a single 30 kD capsid protein. Antibodies specific for the glycoproteins are neutralizing. The

viruses enter host cells by pinocytosis and they replicate in the cytoplasm. Translation of the genomic RNA gives a large polyprotein that is cleaved to yield an RNA polymerase and the structural proteins. Once assembled, the progeny virions bud from the host cell, picking up their envelope.

Flaviviruses

This group is very similar to the Alphavirus group with a few exceptions. First, the genomic RNA is not polyadenylated until it enters the host cell. Second, the single capsid protein is smaller (14 kD). Third, the completed virion forms in the cytoplasm in association with the endoplasmic reticulum instead of budding from the cell.

Pathogenesis

Alphaviruses

There are three agents that cause disease in humans;

Chikungunya

This disease is transmitted by a mosquito, resulting in viremia that presents as an acute, febrile illness with malaise, a rash and arthritis.

Eastern and Western Equine Encephalitis (EEE, WEE)

Following the bite of an infected mosquito, the resultant viremia is often asymptomatic. However, if the virus invades neural tissue, encephalitis may result. This condition is marked by high fever, delirium, coma and possibly death due to convulsions and paralysis.

Venezuelan Equine Encephalitis (VEE)

Similar to EEE and WEE, VEE has more systemic manifestations with less neural involvement.

Flaviviruses

There are four agents that cause disease in humans;

St. Louis Encephalitis (SLE)

Transmitted by mosquitos, a viremia occurs but in most cases no disease results. In some, however, central nervous system involvement gives inflammation and neuronal degeneration, clinically presenting with fever, headache, convulsions, coma and death.

West Nile Encephalitis (WNE)

WNE is very similar to SLE and it is also transmitted by mosquitos. In most cases no disease results. Mild, flu-like cases may be referred to as "West Nile fever". More severe cases of "West Nile encephalitis" or "West Nile meningitis" indicate central nervous system involvement that can lead to death.

Yellow Fever

This is a severe systemic disease (unlike SLE). The virus replicates in reticuloendothelial (RE) cells in many organs, producing liver damage and intestinal hemorrhages. Typically, phase 1 presents with a fever, headache, nausea and vomiting, while phase 2 shows toxicity, jaundice, shock and death.

Dengue Fever and Dengue Hemorrhagic Fever

The virus producing these diseases initially replicates in the skin at the site of the mosquito bite. Next, lymph nodes and the RE system become involved and viremia results. Fever and rash lasting 3-9 days defines the symptomology. Dengue fever is usually self-limiting but Dengue Hemorrhagic fever often involves additional processes (probably immunopathologic) that produce extreme vascular permeability, shock and death.

Host Defenses

Togaviruses induce interferon and are susceptible to its effects. Infections usually resolve once antiviral antibody is produced. The role of the cell-mediated response is less well known but is probably important.

Epidemiology

Alphaviruses are generally associated with birds and horses and their mosquitoes. Flaviviruses are generally associated with birds and bird-feeding mosquitoes for SLE and WNE. Yellow fever and Dengue fever are human in origin.

Diagnosis

Clinical

The epidemiology of the infection along with clinical suspicion and laboratory results are diagnostic.

Laboratory

One can isolate the virus from the blood during the viremic phase.

Control

Sanitary

Surveillance and vector control are excellent means for disease prevention.

Immunological

A live vaccine against Yellow fever and killed vaccines against EEE and WEE are available for those at high risk.

Chemotherapeutic

None available.

Rubella Virus

Virus

☆ Rubellavirus

General Concepts

☆ Rubella virus is a member of the Togaviridae family and produces the disease commonly known as German measles.

☆ Prior to the introduction of the vaccine, rubella was a common, mild disease with a rash. It was found worldwide and mostly affected children. It was uncommon in adults but had severe manifestations when affecting a pregnant woman during the first trimester because of the potential for birth defects.

Distinctive Properties

☆ The structure of Rubellavirus is similar to other Togaviruses.

☆ Humans are the only known reservoir for disease.

Pathogenesis

☆ Transmission of the virus is via the respiratory route.

☆ Initial multiplication occurs in the respiratory tract. One week after infection, viremia occurs and this is later followed by the skin rash. The highest concentration of virus in the respiratory tract begins 3 days prior to the rash and lasts until 3 days after.

☆ If the virus infects a woman during the first 3-4 months of pregnancy, it may infect the placenta or fetus, multiply in any fetal organ and cause damage or death of certain cell types. This congenital rubella syndrome affects the eyes, heart and brain. The virus may persist in the child's tissues for 3-4 years and be shed for up to a year after birth.

Host Defenses

☆ Humoral defenses soon after the viremia help to clear the virus from the blood and prevent continued spread. Cell-mediated defenses clear the virus from tissues. The immune responses generated by natural infections provide life-long immunity.

Epidemiology

☆ Before the vaccine, sporadic outbreaks and epidemics occurred every 6-9 years following a seasonal pattern (late winter, early spring, typical for respiratory disease).

☆ An epidemic in 1964-65 in the US produced 12.5 million cases of disease and resulted in 20,000 cases of congenital rubella syndrome.

Diagnosis

Clinical

Following a 14-20 day incubation period (the "prodrome"), a rash develops on the face and neck and then the trunk and extremities. Lymph nodes become enlarged about a week before the rash and persist for 1-2 weeks after. Congenital rubella symptoms vary depending on the organs affected but typically the eyes, ears, heart and brain are involved. Confirmation of disease is critical in pregnant women.

Laboratory

Illustrating a 4-fold increase in serum titer against Rubellavirus is diagnostic.

Control

Sanitary
Avoidance of contacts.

Immunological
A live vaccine is given in combination with measles and mumps (MMR) to all children over 15 months and selective young and adult women NOT pregnant or 3 months prior to becoming pregnant. Since its introduction in the late 1960's, the vaccine has greatly reduced the incidence of disease.

Chemotherapeutic
None available.

Rhabdo Viruses

Virus
☆ Rhabdovirus: Rabies

Structure
Rabies virus is a bullet shaped virus. Size of the virus is about 180 x 75 nm..Genome is negative sense single stranded RNA. It is nucleo capsid in nature. Matrix layer and outer envelope layer protects the genome. Matrix is made up of Mprotein. Outer envelope is made up of lipid bilayer as like the plasma membrane. The envçlope has spike like projections. It is madeup of glycoproteins. Spikes are responsible for pathogenic property of the virus. RNA dependent RNA polymerase is responsible for genome replication. L and P proteins control its activity. Rabies viruses of man and animals all over the world appears to be of a single antigenic type. Antigens of Rabies viruses are G protein, M protein, N protein and Haemagglutinin. Chemical compositions of the viruses are 4 per cent RNA,67 per cent protein, 26 per cent lipid and 3 per cent carbohydrate (Figure 49a).

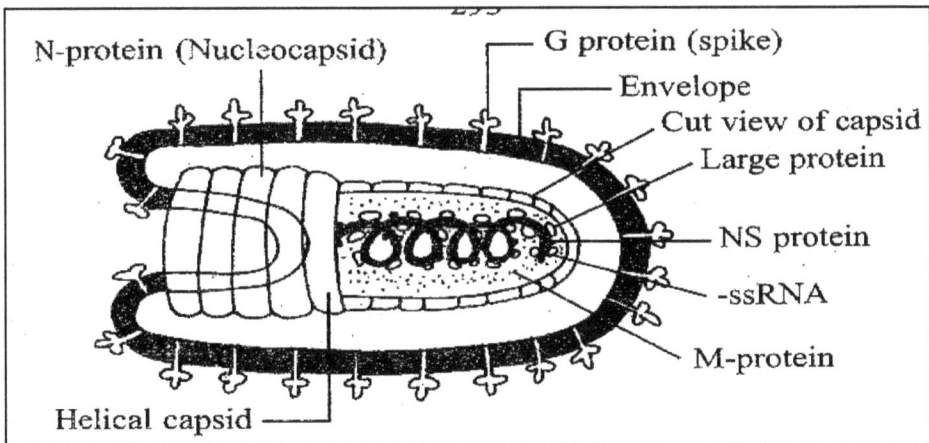

Figure 49a. Structure of *Rabies Virus*

Distinctive Properties

☆ Rhabdoviruses possess a lipid envelope displaying a surface glycoprotein. The RNA-binding nucleocapsid protein surrounds the RNA genome. The virion itself contains an RNA-dependent RNA polymerase (RDRP).

☆ During replication, the virus first makes short positive strand RNAs (mRNAs) which are translated to produce proteins. Later, a full length positive RNA is transcribed and this is used to produce the full length negative strand RNA that is packaged into progeny.

Replication

☆ Virus binds to the cellular receptor and enters the cell via receptor mediated endocytosis.

☆ The viral membrane fuses with the membrane of the vesicle, releasing viral nucleocapsid.

☆ The released virus has a negative sense RNA with nucleocapsid and a small number of L and P proteins which catalyze RNA replication. Negative sense RNA is copied into 5 subgenomic mRNA by L and P proteins.

☆ The N, P, M and L mRNAs are translated by free toplasmic mRNAs.

☆ G rnRNA is translated by ribosome's bound to the endoplasmic reticulum.

☆ Newly synthesized P, N and L proteins involve in RNA replication. This process begins with positive sense RNA synthesis.

☆ Positive sense RNA of the host serves as the template.

☆ Some of the negative sense RNA enters to viral protein synthesis.

☆ G mRNA transcribes and synthesizes glycoproteins.

☆ G proteins travel to the plasma membrane.

☆ Progeny nucleocapsid and M proteins are transported to the adjacent area of plasma membrane.

☆ Assembly takes place in the cytoplasm.

☆ New viruses are released through budding process (Figure 50).

Pathogenesis

☆ Rhabdoviruses generally enter via a bite or a wound infected with saliva.

☆ Initially, the virus replicates at the site and then infects central nervous system (CNS) tissue. Following a variable (6 days up to 1 year, average 30-70 days) incubation period, the virus spreads rapidly via the nerves. CNS damage produces the symptoms of disease.

☆ Neurons accumulate ribonucleoprotein as intracytoplasmic inclusions (Negri bodies). Infection of the thalamus, hypothalamus or pons may occur.

☆ Symptoms include fever, excitation, dilation of the pupils, excessive lacrimation, salivation, anxiety, hydrophobia due to spasms of the throat muscles and, eventually, death.

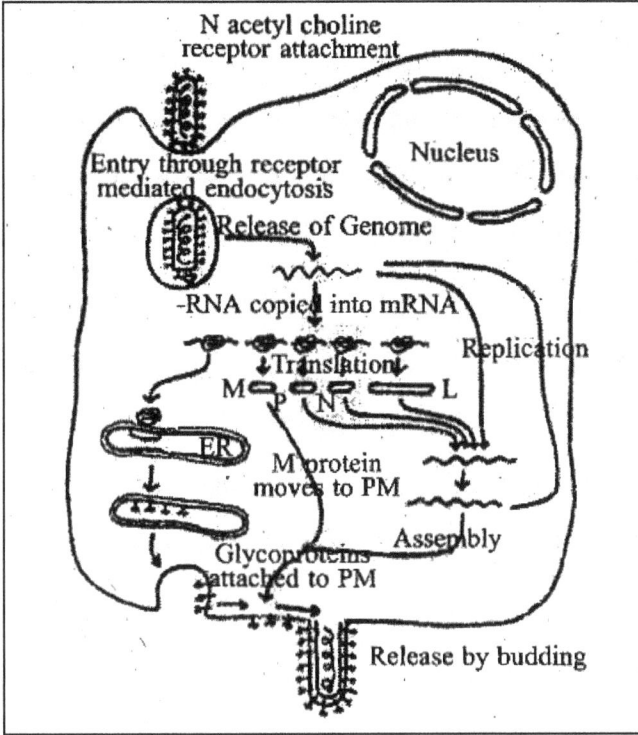

Figure 50. Replication of *Rabies Virus*

Host Defenses

☆ Many factors come into play. Interferon, humoral factors and the cell-mediated response are all important.

☆ Only about half of those bitten by an infected animal actually acquire disease; however, once symptomatic, death is certain.

Epidemiology

☆ Normally, rabies can be found in domestic and wild animals (dogs, cats, cattle, bats, foxes, skunks, raccoons). In the US, 8 per cent of animals screened tested positive. Approximately 10,000 humans become infected per year worldwide.

Diagnosis

Clinical

The patient's history along with symptoms of encephalitis are suggestive.

Laboratory

Direct immunofluorescence is a highly specific and sensitive method. Also, confirmation in the animal (if available) is important.

Control

Sanitary
Rigorous cleansing of a bite wound to reduce the number of viral particles can help to prevent disease. Use of the vaccine for all dogs (a modified live vaccine) and cats (a dead virus suspension) is important in preventing spread to humans.

Immunological
Both active and passive vaccination may be used to prevent human disease. The active vaccines are inactivated virus grown in human diploid cell cultures (HDCV) while the passive vaccine uses immunoglobulin.

Chemotherapeutic
None available.

Paramyxo Viruses

Virus
☆ Paramyxoviruses: Parainfluenza, Mumps, Measles, Respiratory Syncytial Virus

General Concepts
☆ The Paramyxoviruses are an important cause of respiratory disease in children. Illnesses include croup, bronchiolitis and pneumonitis.

☆ This group of viruses share similar features; they possess a double layered envelope with spikes, have a helical symmetry and contain a negative stranded RNA genome. An RNA-dependent RNA polymerase (RDRP) is contained within the virus particle (Figure 51).

☆ Paramyxoviruses replicate in the cytoplasm and are released by budding.

☆ Virus-specific antigens include those associated with the envelope and those contained within the nucleocapsid.

☆ The host range for the Paramyxoviruses includes humans and monkeys.

☆ These viruses produce multinucleated giant cells (syncytia) by production of a cell fusing factor, and then cause host cell lysis.

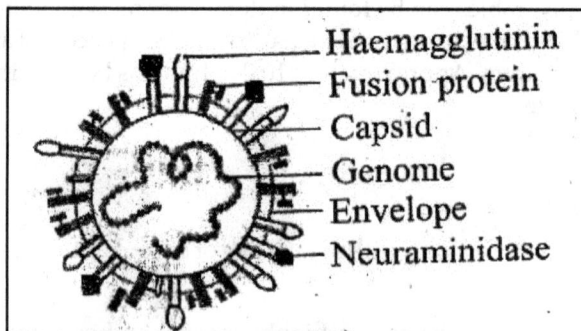

Haemagglutinin
Fusion protein
Capsid
Genome
Envelope
Neuraminidase

Figure 51. Structure of *Parainfluenza Virus*

Replication

☆ Parainfluenza viruses attach to the host cells by the haemagglutinins (Figure 52).

☆ It binds to the host cell's neuraminic acid receptor.

☆ Entry of virus to the cell by fusion with the cell membrane is mediated by the F_1 and F_2 glycopeptides.

☆ The viral particles contain a single-stranded negative sense RNA, which cannot serve as a messenger.

☆ The viral transcriptase initiates transcription into 5-8 positive sense RNA strands which act as mRNAs.

☆ They direct the viral protein synthesis and are copied into negative-sense RNA strands which are integrated in the new virions.

☆ Virus assembly occurs in the cytoplasm near plasma membrane.

☆ After assembly, progeny viruses are released budding.

☆ Thus the new virus particles acquire an envelope to become mature virions.

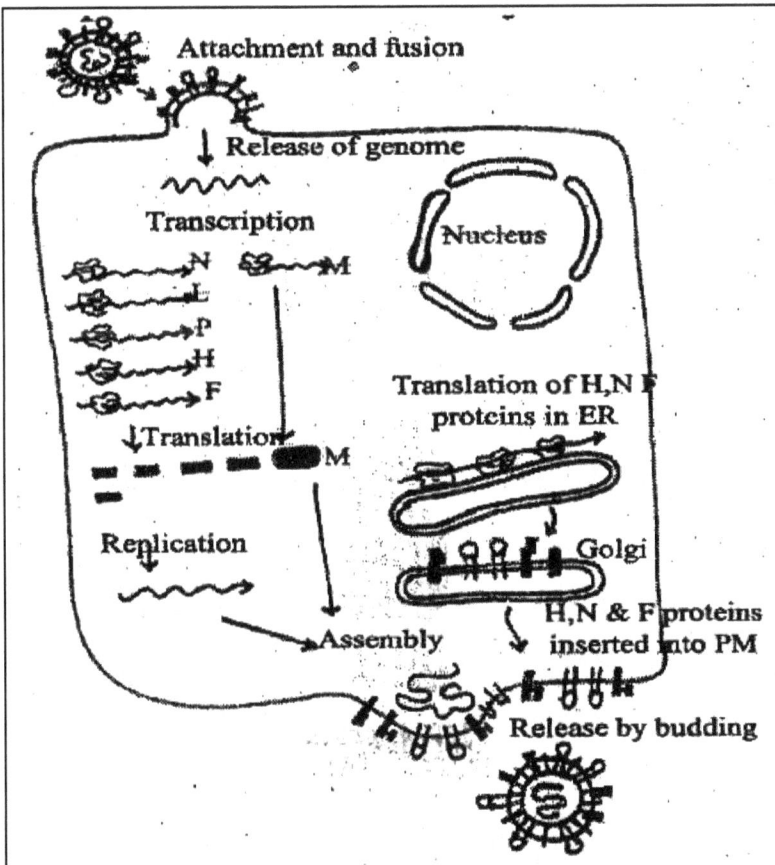

Figure 52. Replication of *Parainfluenza Virus*

Distinctive Properties

Parainfluenza

About 30-40 per cent of acute respiratory infections in infants and children can be attributed to the Parainfluenza virus. Disease ranges from mild and cold-like to life-threatening (croup, bronchiolitis and pneumonia). Parainfluenza viruses are the most common cause of croup and there are 4 serotypes, numbered 1-4.

Mumps

A common acute disease of children, the mumps virus produces inflammation of salivary glands leading to obvious enlargement. Some severe manifestations can result from infection but there is only one serotype.

Measles

Another acute childhood disease, measles virus commonly causes a fever with a rash, occasionally producing conjunctivitis and, sometimes, pneumonia. More serious complications include encephalitis and subacute sclerosing panencephalitis (SSPE). Only one serotype exists and disease is limited to humans and monkeys.

RSV

Respiratory Syncytial virus is a major cause of bronchiolitis and pneumonia in infants under 1 year. Reinfection in adults usually involves the upper respiratory tract. The viruses produce a characteristic syncytia formation, hence the name. There is only one serotype that affects humans.

Pathogenesis

Parainfluenza

These viruses generally produce local infections in the upper and lower respiratory tract. The viruses implant in ciliated epithelia of respiratory tract (nose and throat). The virus can be shed over 3-16 days and the main pathologic response is inflammation. The most important (*i.e.* serious) diseases are croup, bronchiolitis and pneumonia. The severe diseases occur most often with types 1 and 2.

Mumps

Mumps is a systemic infection spread by viremia. The major targets include glandular and nervous tissue. The virus enters via the pharynx or conjunctiva, there is local multiplication followed by viremia. A secondary viremia disseminates the virus to salivary glands, testes, ovaries, pancreas and the brain. The incubation period is 18-21 days and disease is asymptomatic in about 35 per cent of those infected. The most characteristic response is painful enlargement of the parotid glands. Severe cases may progress to include epididymoorchitis in prepubescent males, which can cause atrophy of the testes, but rarely sterility. Mumps can also produce a transient high frequency deafness.

Measles

Measles is also a systemic infection spread by viremia. Acute disease affects the lymphatic and respiratory systems and occasionally the brain. Persistence of the virus can produce SSPE. Measles virus generally enters via the oropharynx. Local multiplication precedes the viremia and spread to the reticuloendothelial system. Extensive multiplication produces a secondary viremia 5-7 days later with spread to the mucosa of the respiratory, urogenital or gastrointestinal tracts or the central nervous system. Clinically, there are respiratory symptoms during the prodromal stage (malaise, fever, cough) and a rash during the eruptive phase (Koplik's spots, rash on head then body). Complications are mainly bacterial superinfection but other rare complications (*e.g.* SSPE) can occur.

RSV

The RS virus initiates a local infection in the upper or lower respiratory tract but illness varies with age and previous experience. The virus infects ciliated epithelia of the nose, eye and mouth and remains generally confined. Virus spreads extracellularly and by fusion. Severe disease may present as bronchiolitis, pneumonia or croup, particularly in infants. Some evidence suggests that there are possible immunopathologic mechanisms involved.

Host Defenses

Parainfluenza
Interferon and neutralizing IgA are important but IgA is often short lived.

Mumps
Interferon limits viral spread while IgM and IgG are protective. Long lasting IgG affords life-long immunity.

Measles
Interferon affects viral spread but the cell-mediated response is associated with recovery (disease is more severe in persons with a T-cell immunodeficiency). Life long persistence of IgG affords protection.

RSV
Interferon, age, immune competence and neutralizing IgA (short lived) all contribute to host defense. Breast milk may contain neutralizing factors as well.

Epidemiology

Parainfluenza
Found worldwide, Parainfluenza affects mostly children. The virus is endemic in some areas and epidemic at times. The source for infections is the respiratory tract of humans. Disease is contagious for 3-16 days, transmission is person to person or by droplets. At 5 years of age, about 90-100 per cent have antibodies against type 3, about 74 per cent for type 1 and about 58 per cent for type 2.

Mumps
Also found worldwide, mumps is endemic with peaks of acute disease appearing from January through May. Epidemics occur every 2-7 years and humans are the only host. Transmission is via salivary secretions and disease is contagious just before and after the symptoms. It is less contagious than Parainfluenza so intimate contact is usually required. In unvaccinated populations, about 45 per cent of people

are infected by age 5 and about 95 per cent by 15 years of age. About 35 per cent of those infected are subclinical, however.

Measles

Endemic worldwide, epidemics occur every 2-4 years in developed countries when the population susceptibility reaches about 40 per cent. The source of the virus is human (respiratory) secretions. Disease generally affects those in the 4-7 year age group. The incidence of SSPE is about 6-20 per million.

RSV

Also endemic worldwide, epidemics occur yearly, usually between January and March. Reinfection is common. The source for infection is the human respiratory tract and disease is usually contagious about 4-5 days after the symptoms. Most infants experience upper respiratory infections but between 15 per cent and 50 per cent experience lower (more serious) respiratory infections.

Diagnosis

Parainfluenza

Clinically not possible. Isolation of virus or serotests are required.

Mumps

Typical symptoms on clinical presentation make diagnosis relatively easy.

Measles

Typical symptoms on clinical presentation make diagnosis relatively easy.

RSV

Strongly suspect in infants with lower respiratory tract infection.

Control

Parainfluenza

Supportive care and isolation.

Mumps

Immunize at 15 months (measles, mumps, rubella live vaccine).

Measles

Immunize at 15 months (measles, mumps, rubella live vaccine).

RSV

Supportive care and isolation.

Orthomyxo Viruses

Virus

☆ Orthomyxoviruses: Influenza

General Concepts

☆ The Orthomyxoviruses are composed of one genus and 3 types; A, B and C.

☆ The disease caused by these viruses, influenza, is an acute respiratory disease with prominent systemic symptoms despite the fact that the infection rarely extends beyond the respiratory tract mucosa.

☆ Type A is responsible for periodic worldwide epidemics; types A and B cause regional epidemics during the winter.

☆ The recurring pattern of the influenza viruses is due to their ability to exhibit variation in surface antigens. Two phenomenon account for this variability:

Antigenic drift is due to mutations in the RNA that leads to changes in the antigenic character of the H and N molecules. Antigenic drift involves subtle changes that may cause epidemics but not pandemics.

Antigenic shift is due to rearrangement of different segments of the viral genome that produces major changes in the antigenic character of the H and N molecules. Antigenic shift usually occurs in animal hosts and is responsible for producing both epidemics and pandemics.

☆ Influenza epidemics have been documented since 1173 AD; a pandemic in 1918 was responsible for 20 million deaths worldwide.

☆ The table outlines some of the differences between the three types of influenza virus (Table 10).

Table 10: Differences Between the Three Types of Influenza Virus

	A	B	C
Severity	++++	++	+
Animal Reservoir	Yes	No	No
Population Spread	Pandemic, epidemic	Epidemic	Sporadic
Antigenic Changes	shift, drift	drift	drift

Distinctive Properties

☆ Orthomyxoviruses contain a single stranded, negative RNA genome divided into 8 segments.

☆ The viruses have a lipid bilayer envelope with surface glycoproteins (hemagglutinin and neuraminidase)

☆ There are 3 viral antigens of importance: the nucleoprotein antigen that determines the virus type (A, B or C), the hemagglutinin (H) antigen, and the neuraminidase (N) antigen. The H and N antigens are variable. There are about 13 different H antigens and 9 different N antigens found in birds. This provides a total of 117 (13 x 9) possible combinations, 71 of which have been observed. There are only about 3 combinations that affect humans, however.

☆ Viral attachment is mediated by the hemagglutinin. The virus enters host cells by pinocytosis and uncoating occurs by fusion of the viral envelope with the membrane of the vacuole. The RNA is capped and replication proceeds in the nucleus. The progeny are released by budding and cell death ensues (Figure 53).

☆ The segmented genome of the influenza virus allows rearrangements to occur in simultaneously infected cells. This accounts for the periodic appearance of new variants. The new variants are responsible for the process of antigenic shift.

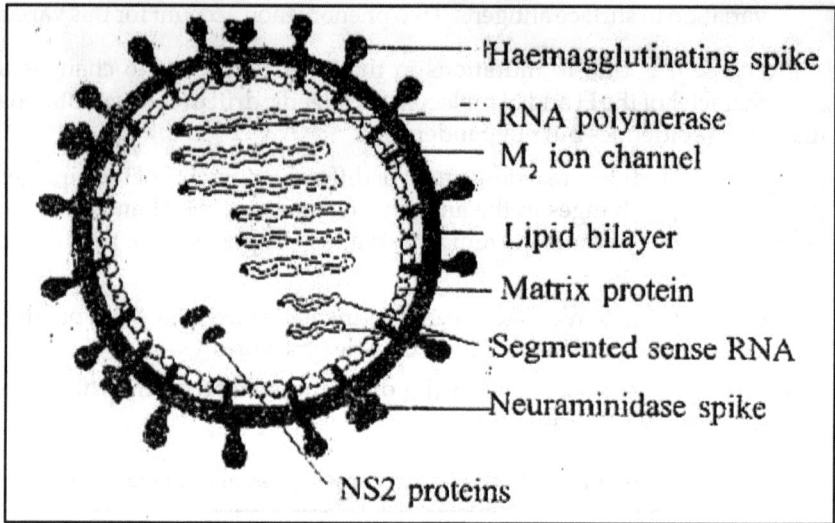

Figure 53. Structure of *Influenza Virus*

Replication

☆ Virus replication requires about 6 hours to kill the host cell.

☆ The virus is attached to sialic acid receptor on the cell via the haemagglutinin subunit (Figure 54).

☆ The virus is taken into the cell by pinocytosis it is sent to endosome (vacuole). The acid environment of the endosome, uncoats the nucleocapsid and releases the RNA into the cytoplasm.

☆ A transrnembrane protein derived from the matrix ion channel for the release of RNP-RNA complex into the cytoplasm. The latter is transported to the nucleus for transcription and replication.

☆ Influenza virus replication depends on the presence of active host cell DNA.

☆ The synthesized viral mRNA is transported to the cytoplasm, where it is translated by host ribosome.

☆ mRNAs specifying viral membrane proteins (*HA,NA,M*) are translated by ribosome bound to endoplasmic reticulum and they undergo glycosylation.

☆ The nucleocapsid is assembled at the inner surface of host cell membrane.

☆ After the attachment of.M 1 protein to newly synthesized RNA, viral RNA synthesis is stopped.

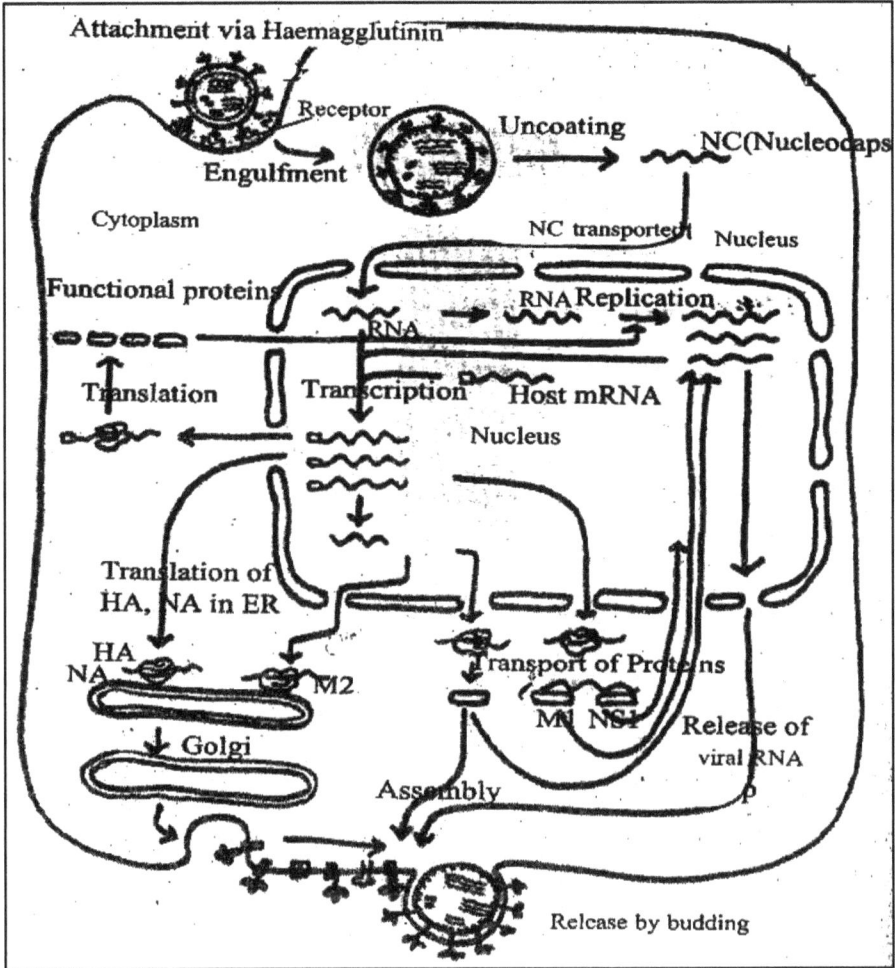

Figure 54. Replication of *Influenza Virus*

☆ HA and NA proteins are transported to the cell surface and incorporated into the plasma membrane.

☆ Virion nucleocapsids along with NS2 associated with regions of plasmamembrane containing HA and NA proteins.

☆ The virion is budding off to acquire the envelope.

☆ The released virus undergoes maturation to become an infective particle.

Pathogenesis

☆ Transmission of disease is airborne. The viruses deposit in lower respiratory tract, their primary site is the tracheobronchial mucosa.

☆ Neuraminidase produces liquefaction, which leads to viral spread.

☆ Respiratory symptoms include a cough, sore throat and nasal discharge. There is no viremia but systemic symptoms such as fever and muscle aches do occur.

☆ The extent of respiratory tract cell destruction is a probable factor in the disease.

☆ Severe complications include pneumonia (viral or bacterial).

Host Defenses

☆ Interferon is one non-specific defense but antibody is the prime defense. IgA is produced in the upper respiratory tract and IgG is produced in the lower respiratory tract. These antibodies are directed primarily against the hemagglutinin and neuraminidase. Cell-mediated (*i.e.* CTL) defenses are important in recovery.

Epidemiology

☆ Influenza displays a typical pattern: school children bring the disease home and infect siblings and parents.

☆ Epidemics usually last from 3-6 weeks and the highest attack rates are for 5-19 year olds (generally Type A).

☆ Every winter, the recurring population susceptibility is due to changes in the surface antigens; major changes are referred to as antigenic shift. These changes are responsible for pandemics and they result from rearrangement of the viral genome segments. Minor changes are called antigenic drift and these are responsible for many epidemics. They result from mutation in the viral RNA. Antigenic drift occurs every 2-3 years while antigenic shift only occurs every 10 years.

Diagnosis

Clinical

Influenza usually displays a sudden onset with fever, malaise, headache, muscle aches, sore throat, cough and rhinorrhea, generally in winter. The presence of disease in the community (*i.e.* epidemiology) is helpful in diagnosis.

Laboratory

Serology on the patient's serum can be performed or the viruses may be isolated in chick embryos if necessary.

Control

Sanitary

Avoid contacts.

Immunological

Every year, inactivated vaccines are prepared using the most likely types and antigenic characters expected for any particular season. These vaccines are given parenterally in the fall, primarily for those at risk (older persons or those with chronic disease). Protection against disease is variable (50-90 per cent).

Chemotherapeutic

The drug amantadine HCl can be used for influenza type A but not type B in patients with other disease conditions. Generally, however, pain relievers (*e.g.* acetaminophen) are more generally employed.

Retro Viruses

Virus

☆ Retroviruses: HTLV, HIV

General Concepts

☆ The Retroviruses are composed of three subfamilies, two that infect humans. They are:

- Oncornaviruses: HTLV 1, HTLV 2, HTLV 5
- Lentiviruses: HIV 1, HIV 2

☆ The HTLV or Human T Cell Lymphotrophic Viruses are divided into three types based on the type of diseases they produce:

- HTLV-I produces cutaneous T-cell lymphomas,
- HTLV-II produces hairy T-cell leukemias,
- HTLV-V produces T-cell lymphomas and leukemias.

☆ The HIV or Human Immunodeficiency Viruses are divided into two types based on the type of diseases they produce:

- HIV-1 produces Acquired Immunodeficiency Syndrome (AIDS),
- HIV-2 produces a related disease syndrome, restricted to W. Africa.

☆ The Retroviruses are RNA viruses and their name is derived from the viral enzyme Reverse Transcriptase, which makes circular DNA from linear RNA. The viral DNA has the capacity to integrate into the host cell genome.

☆ In general, the viruses are not cytopathic. The Oncornaviruses generally transform cells, causing leukemias, sarcomas and carcinomas. The Lentiviruses attack T-cells, all but abolishing the host immune response.

Distinctive Properties

☆ Most of the information about Retroviruses comes from studies of HIV.

☆ HIV is enveloped and displays a viral glycoprotein (gp120) that recognizes and binds to the CD4 receptor on T-helper cells. This glycoprotein is antigenically variable.

☆ The viral genome is composed of 2 positive strands of RNA that are 5'-capped and 3'-polyadenylated. A cellularly derived tRNA binds the 3'-end and serves as a primer for DNA synthesis.

☆ The order of the genes encoded by the viral RNA is as follows:

> 5'-   LTR   *gag*   *pol*   *env*   *tax/rex*   LTR   -3'

☆ These genes correspond to the following viral products or functions:

- *LTR:* The Long Terminal Repeats are used for integration of the virus into the host genome and also contain promoter and enhancer sequences.
- *gag:* The Group-specific Antigen corresponds to the core and capsid proteins.
- *pol:* This gene encodes the reverse transcriptase, which actually has several functions.
- *env:* This Envelope gene encodes the gp120 glycoprotein.
- *tax/rex:* This region encodes factors involved in transactivation and other regulatory functions.

☆ Upon entering a host cell, translation of the RNA produces a polyprotein that is cleaved to give the individual components. Cleavage of the polyprotein requires a specific protease that is the target for new anti-HIV drugs.

☆ During replication, the reverse transcriptase (RT) uses the tRNA as a primer for DNA synthesis, creating a DNA/RNA hybrid duplex. Next, RT degrades the RNA strand using its RNAseH function and synthesizes a new DNA strand to produce a DNA duplex that circularizes. The circular DNA then integrates into the host genome, where it remains to be transcribed to produce new progeny RNA molecules. Following replication, the viruses escape by budding.

Pathogenesis

☆ HIV is transmitted via body fluids. Blood is the best and persons can acquire through sex, parenteral drug usage or transfusions.

☆ The virus has a specific trophism for $CD4^+$ (T-helper) cells. Following infection, it may remain latent for many years. Eventually, the infected T-cells lose their ability to function, resulting in a loss of both humoral and cell-mediated immunity.

☆ Patients generally die of secondary manifestations including Kaposi's sarcoma or opportunistic infections.

☆ Several specific syndromes are associated with infection by HIV. These include:

1. *Lymphadenopathy and fever* has an insidious onset and is characterized by weight loss and malaise.
2. *Opportunistic infections:* Many diseases that rarely affect normal individuals may occur in persons infected with HIV. These include: *Pneumocystis carinii* pneumonia, Candidiasis, severe Herpesvirus infections and frequent diarrhoea caused by *Salmonella, Shigella* and *Campylobacter.*

3. *Malignancies:* Kaposi's sarcoma is a rare type of cancer that occurs in HIV-infected persons. These normally benign lesions become malignant and disseminate to involve visceral organs.

4. *Wasting:* Also known as "slim disease", this syndrome is common in Africa.

5. *AIDS dementia:* This condition mimics Alzheimer's disease and may involve HIV infection of the brain.

Host Defenses

☆ Host defenses are essentially eliminated, but there is some evidence suggesting a possible genetic component in some individuals that causes suppression of the virus.

☆ In addition, the ability of the virus to remain latent and frequent antigenic changes in the gp120 protein (antigenic drift) diminish host defense capabilities.

Epidemiology

☆ The first case of AIDS was described on June 5, 1981. Now, there are estimated to be more than 30 million total HIV infections worldwide. About half of these are in sub-Saharan Africa and about one-tenth of them occurred in 1996 alone, giving daily infection rates of about 8500 per day, approximately 1000 in children.

☆ Current global trends include the following:

● The majority of new adult HIV infections involves persons 15-24 years old,

● Between 75 per cent and 85 per cent of HIV-positive adults have been infected through unprotected sexual intercourse, with heterosexual (male-female) intercourse accounting for more than 70 per cent and male-to-male intercourse accounting for approximately 5-10 per cent,

● Transfusion of HIV-infected blood and the sharing of HIV-infected injection equipment by drug users account for 3-5 per cent and 5-10 per cent of all global adult infections, respectively.

☆ In the US, approximately 750,000 persons are infected with HIV and about 580,000 have AIDS. The following graphic illustrates the distribution of AIDS throughout the US.

☆ While the focus on HIV/AIDS has always been homosexual men, a much larger percentage of infections now occur in women and children. Men who have sex with men continue to provide about 40-50 per cent of new HIV infections, while about 20 per cent of new infections involve women and about 10 per cent involve children. Many (92 per cent) of these children are infected by vertical transmission (*in utero* or during birth).

In terms of race/ethnicity, overall AIDS infections in the US predominate in non-Hispanic whites, but the number of new cases in blacks has now exceeded those in whites. However, dividing the percent of cases by the percent of the population represented by each group reveals that a disproportionate number of cases occur in African-Americans and Hispanics.

Diagnosis

Clinical

Diagnosis is often helped by the occurrence of rare diseases or infections such as Kaposi's sarcoma or *Pneumocystis* pneumonia or recurrent or serious opportunistic infections. The patient's history (life-style, drug use, etc.) is also informative.

Laboratory

Laboratory diagnosis is based on measuring HIV antibodies using the ELISA (Enzyme-Linked Immuno-Sorbent Assay) test. Positive results are confirmed with another test known as a Western Blot. Together, the two tests are more than 99.9 per cent accurate.

Control

Sanitary

The use of condoms during sexual intercourse can greatly reduce the chance of infections. Some spermicidal creams may also have anti-HIV properties. Non-use of intravenous drugs or sharing of syringes/needles prevents direct inoculation or the virus into the blood. Testing of the blood supply reduces transmission by transfusion. Education is perhaps the best means of preventing disease.

Immunological

No vaccines are available but some possibilities do exist.

Chemotherapeutic

Anti-HIV drugs fall into three categories: the nucleosides, the non-nucleosides, and the protease inhibitors. Nucleosides and non-nucleosides are both known as reverse transcriptase inhibitors.

- ☆ Nucleoside reverse transcriptase inhibitors include: Retrovir (zidovudine, AZT), Videx (didanosine, ddI), Hivid (zalcitabine, ddC), Zerit (stavudine, d4T) and Epivir (lamivudine, 3TC).

- ☆ Non-nucleoside reverse transcriptase inhibitors include: Viramune (nevirapine) and Rescriptor (delavirdine).

- ☆ Protease Inhibitors include: Invirase (saquinavir), Norvir (ritonavir), Crixivan (indinavir) and Viracept (nelfinavir).

These drugs are often given in combinations of two or three in order to attack the HIV virus in different ways.

Adeno Virus

Adenovirus was first isolated in 1953 from *adenoids* and *tonsils*. The name adenovirus is given to it because it was first isolated from adenoids. Adenoviruses are especially valuable systems for the molecular biology and biochemical studies of eukaryotic cell process. Adenovirus infections are common in young childrens. This virus is associated with respiratory tract, conjunctiva and digestive tract. Adenoviruses are widespread in nature and have been isolated from animal species ranging from frogs to human.

The family Adenoviridae is subdivided into two genera, *Mastadeno virus* and *Aviadeno virus*. The human adenoviruses comprise of 49 distinct serotypes that are grouped into 6 serogroups based on various immunologic, biologic and biochemical characteristics.

Structure

☆ The adenovirus particle consists of an icosahedral protein shell surrounding a protein core that contains the linear, double-stranded DNA genome.

☆ The shell is 70 to 100 nm in diameter and is made up of 252 structural *capsomeres*.

☆ The 12 vertices of the icosahedron are occupied by units called *pentons*, each of which has a slender projection called a *fibre*.

☆ The 240 capsomeres that make up the 20 faces and the edges of the icosahedral shape are called *hexons* because they form hexagonal arrays.

☆ The shell also contains some additional, minor polypeptide elements. The core particle is made up of two major proteins (polypeptide V and polypeptide VII) and a minor arginine rich protein (μ).

☆ Genome of adenovirus has different types of genes, which are expressed in a sequence. The order of expressions is as follows-E_1A, E_1B, E_2A, E_2B, E_3, E_4, L_1, L_2, L_3, L_4 and L_5, $_{VA}RNA$, IVA_2.(Figure 55).

Replication

☆ Completion of adenovirus replication cycle requires 24-30 hours.

☆ Adenovirus fibres attach to a specific receptor on the cell surface (Epidermal growth factor). The virion is then taken into endocytic pits, where it enters the cytoplasm.

☆ Uncoating takes place and the viral genome associated with core protein VII is imported into the nucleus.

☆ Host cell RNA polymerase II transcribes the immediate early E_1A gene.

☆ E_1A mRNA is transported to the cytoplasm.

☆ E_1A protein is synthesized by the cellular translation machinery.

☆ The protein is extensively modified by phosphorylation and sent to the nucleus

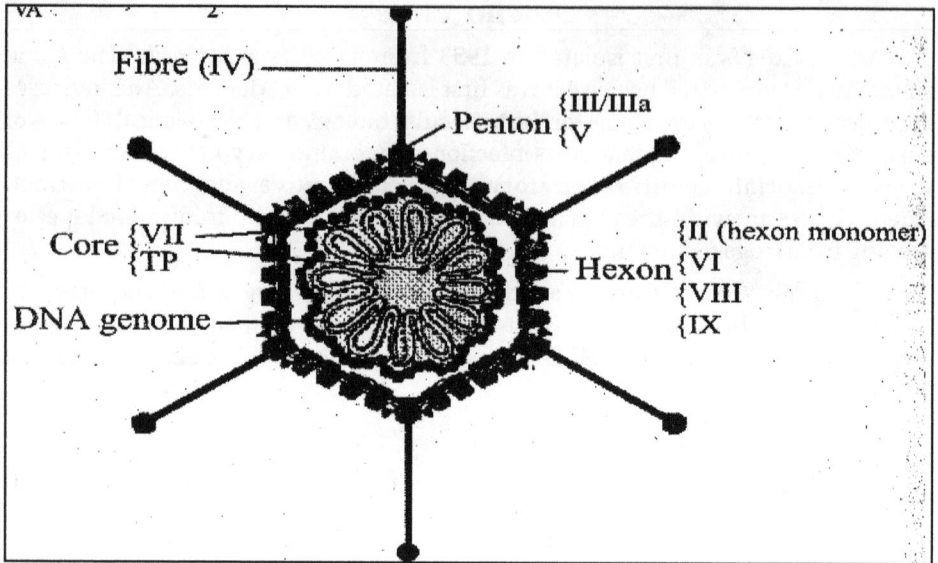

Figure 55. Structure of *Adeno Virus*

☆ The larger E_1A protein stimulates the transcription of cellular early viral genes by cellular RNA polymerase.

☆ Early mRNAs are translated into the early proteins.

☆ These early proteins include DNA polymerase, DNA binding proteins and viral DNA replication initiation proteins.

☆ Early proteins co-operate in the viral DNA synthesis.

☆ Replicated DNA serves as a template for further rounds of replication.

☆ DNA helps in transcription of late genes.

☆ Processed mRNAs are sent to the cytoplasm.

☆ Translation of late mRNA needs $_{VA}$RNA $_{VA}$RNA$_1$

☆ Processed viral proteins are transported to the nucleus for virus assembly.

☆ Within nucleus capsids are assembled to form non infectious immature virions.

☆ By the action of protease on these virions mature virions are formed.

☆ Progeny vinons are released usually upon the lysis of the host cell (Figure 56)

Pathogenesis

Adenoviruses infect and replicate in epithelial cells of the respiratory tract, eye, gastro-intestinal tract, urinary bladder and liver. They usually do not spread beyond the regional lymph node. Some of the viruses persist as latent infection for years in adenoids, tonsils and shed in the faeces. Infection may be productive, abortive or

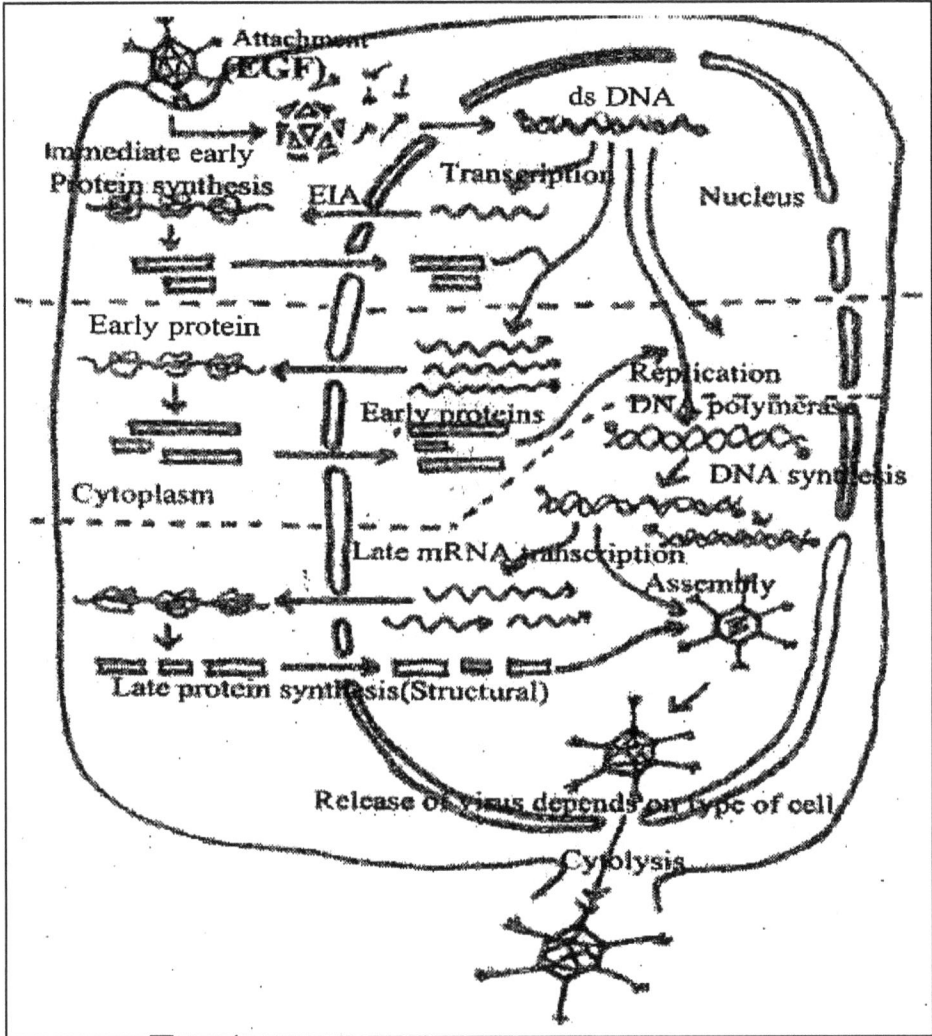

Figure 56. Replication of *Adeno Virus*

latent. In productive infections, the viral genorne is transcribed in the nucleus; mRNA is translated in the cytoplasm and virions are self-assembled in the nucleus. Latent infections, occur in transformed and tumour cells; the viral DNA is integrated into the host genome. Virus-host DNA recombinants are also found in productive infections.

Adenovirus-host interaction may lead to the following types of infections in man:

Productive Infection

Complete replication of infectious virion and release of virus.

Abortive Infection

Synthesis of viral protein without production of infectious virion.

Semipermissive Infection

Complete replication with a low yield of infectious virion.

Malignant Transformation

Associated with integration of viral DNA and differential viral and cellular gene expression.

Symptoms

Acute febrile pharyngitis, Pneumonia, Pharyngo conjunctival fever, Conjunctivitis, Gastroenteritis, Fever, Acute haemorrhagic cystitis and *cough.*

Lab Diagnosis

☆ Virus is isolated from stools and nasal discharges arid cultivated in primary human embryonic kidney cells.

☆ Viral antigens are demonstrated with immunofluorescence technique using MCAs.

☆ Viral antigens are detected by using serological markers in ELISA.

☆ Viral DNA is detected by RT-PCR and nucleic acid hybridization

Control

There is no treatment. Whole-virus vaccines are not because of the potential risk of oncogenesis. Other vaccine including recombinant vaccines, are under development, but adenoviruses do not represent a serious health hazard.

Rota Virus

It belongs to the family *Reoviridae* and the genus *Rotavirus.* In 1973, rotaviruses were discovered in duodenal biopsies obtained from the acute gastroenteritis patients.

Structure

Rotaviruses have a distinctive wheel like shape. They are 70 nm in diameter. It consists of an envelope, a capsid and a nucleic acid. The capsid is three layered and icosahedral. The capsid is surrounded by a lipid envelope. The capsid contains all enzymes for mRNA producion. Antigens are found in the capsid. The inner capsid

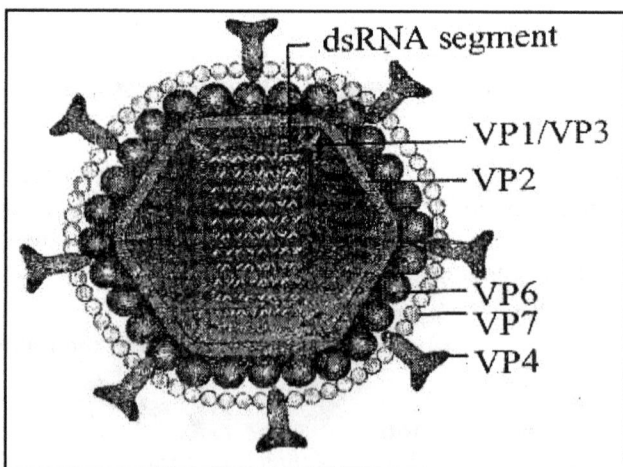

Figure 57. Structure of *Rota Virus*

contains the major antigen VP6, which determines the groups and sub-groups of this virus. The outer capsid contains VP7 and VP4 which determine some serotypes of this virus, 'Within the inner capsid, a core is present, which contain the RNA genome. The rotavirus genome contains 11 segments of double stranded RNA. The segmented genome of rotavirus readily under goes genetic reassortment during coinfection, Rotavirus RNA segments 1, 2, 3 and 6 encode for inner capsid polypeptides VP1, VP2, VP3 and VP6, RNA segments 4 and 7, 8 or 9 encode the major outer capsid polypeptides VP4 and VP7 (Figure 57).

Replication

The virus enters the cell by endocytosis or by direct membrane penetration. Replication occurs in the cytoplasm. Removal of the outer shell of the capsid leads to activation of viral RNA polymerase by lysosomes. Newly assembled subviral particles acquire the envelope by budding through the endoplasmic reticulum and are released by cell lysis (Figure 58).

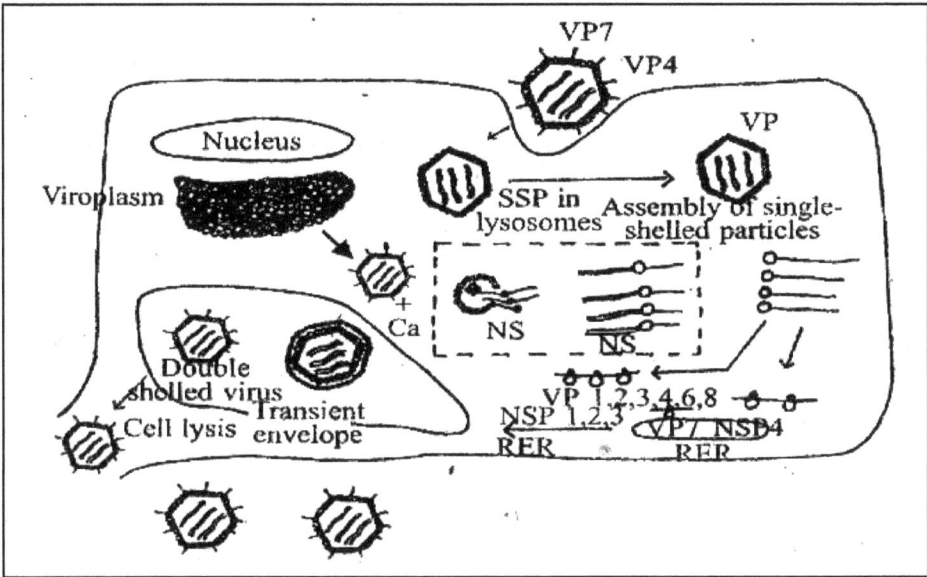

Figure 58. Replication of *Rota Virus*

Pathogenesis

Rotaviruses are transmitted by faecal-oral route, Incubation period of rotavirus diarrhoeal illness is 48 hours. Large numbers of virus particles are sheded in the stool following multiplication in epithelial cells of the small intestine.Histopathologic studies of tissue from the small intestine of infants and young children with rotavirus infection show shortening and blunting of villi on a patch, irregular, but predominantly intact mucosa and mononuclear cell infiltration of the lamina propria.

Diagnosis

Clinical findings are non-specific. Diagnosis depend on detecting virus in faeces (*e.g.* by immunoassay) or on demonstrating a serum antibody response.

Control

Dehydration is treated by fluid and electrolyte replacement. Vaccine development is under way and appears promising. Oral bismuth subsalicylate (BSS) as an adjunct hydration therapy is reported to shorten the course of rotavirus diarrhoea in a pediatric study.

Simian Virus (SV40)

Group	–	Group I (dsDNA)
Family	–	*Polyomaviridae*
Genus	–	*Polyomavirus*
Species	–	*Simian virus 40*

SV40 is an abbreviation for *Simian vacuolating virus 40* or *Simian virus 40*, a polyomavirus that is found in both monkeys and humans. Like other polyomaviruses, SV40 is a DNA virus that has the potential to cause tumors, but most often persists as a latent infection. The virion is a naked icosahedral virion that consists of a closed circular dsDNA of 5kb.

Architecture and Replication

As with polyomavirus genomes, the SV40 genome consists of two sets of expressed genes: the early expressed genes, which encode proteins required for viral replication, including the small and large T antigens (small-t Ag, large-T Ag) and the 17KT protein; and the late expressed genes, which encode structural proteins necessary for viral assembly, including three coat proteins (VP1, VP2, VP3) and a maturation protein (agnoprotein). Different SV40 strains show variability within the variable domain of the T antigen proteins, as well as within the non-translated regulatory region that contains the origin of replication (ori) and promoters and enhancers of replication (Figure 59).

Alternative splicing produces two early proteins, large-T Ag and small-t Ag. The large-T Ag functions as a replicative helicase, binding to specific sites in the ori to promote the local unwinding of DNA, as well as recruiting cellular DNA replication proteins to the site, including topoisomerase I, replication protein A, and DNA polymerases. The large-T Ag autoregulates the early promoter, and can indirectly contribute to the activation of late transcription. The large-T Ag also plays a role in viral maturation by influencing the phosphorylation of capsid proteins. The small-t Ag plays a key role in infections by increasing the level of virus during the lytic cycle, and by enhancing cell transformation by the large-T Ag.

The DNA-binding maturation protein, agnoprotein, is thought to be involved in viral assembly or release, and acts to increase the efficiency of plaque formation. The agnoprotein from JCV, which shows 50-60 per cent homology to SV40 agnoprotein,

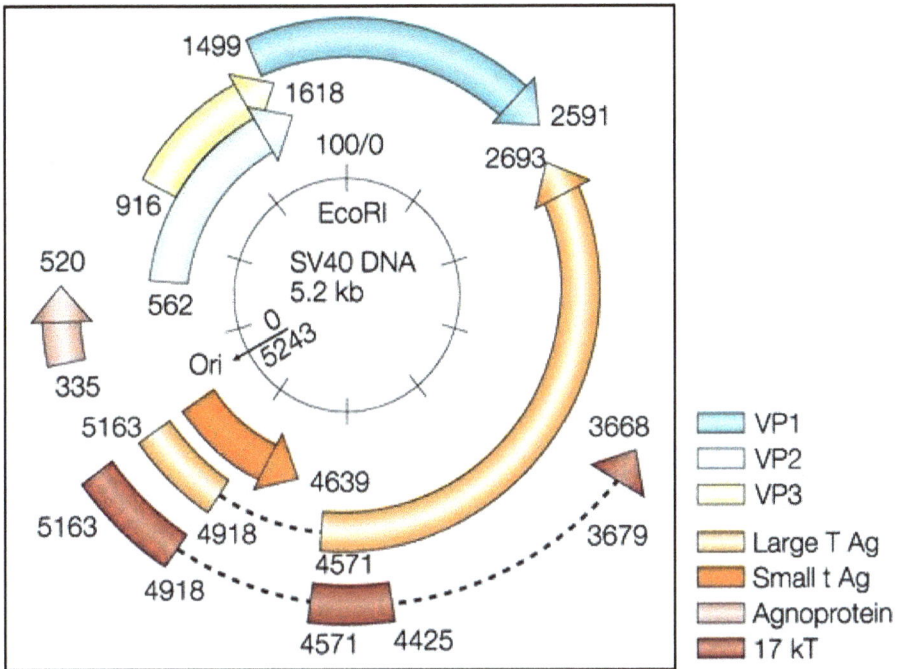

Figure 59. Genome of SV 40

co-localizes with cytoskeletal tubulin, suggesting a role in the stabilization of microtubules during infection.

The virion adheres to cell surface receptors of MHC class 1 by the virion glycoprotein VP1. Penetration into the cell is through a caveolin vesicle. Inside the cell nucleus the cellular RNA polymerase II acts to promote early gene expression. This results in an mRNA that is spliced into two segments. The small and large T antigens result from this. The large T antigen has two functions: 5 per cent will go to the plasma membrane of the cell and 95 per cent will go back to the nucleus, driven by the amino acid sequence lys lys lys arg lys val glu. Once in the nucleus the large T binds three viral DNA sites, I, II, and III. Binding of sites I, and II autoregulates early RNA synthesis. Binding site II happens for each round of replication. Binding site I initiates DNA replication at the ORI. Early transcription gives two spliced RNAs that are both 19s. Late transcription gives both a longer 16s, which synthesises the major viral capsid protein VP1, and the smaller 19s, which gives Vp2, and Vp3 through leaky scanning. All of the proteins, besides the 5 per cent of large T go back to the nucleus because assembly of the viral particle happens in the nucleus. Eventual release of the viral partical is cytolytic and results in cell death.

Pathogenesis

The viral capsid is composed of pentamers of VP1. Two minor structural proteins, VP2 and VP3, produced by alternative splicing, bridge VP1 capsid to the SV40 genome

(Figure 60). VP3 is essential for formation of infection particles, and may be involved in virus-cell interactions during post-packaging steps.

SV40 and Cancer

Several viruses have been implicated in the pathogenesis of different cancers, including papillomavirus, Epstein-Barr virus, hepatitis B virus and HIV-1. The study of tumour-promoting viruses has greatly increased our understanding of cancer pathology. SV40 has been found in various tumour types, including brain tumours, bone tumours, mesotheliomas and non-Hodgkin's lymphomas, suggesting that SV40 maybe a transforming virus under certain circumstances. All these tumour-types can be induced with SV40 in laboratory animals.

The SV40 small-t Ag and large-T Ag are considered to be oncoproteins under certain conditions, and together they can transform cells in culture. The transformation of cultured cells requires both the stimulation of cell division and the blocking cell apoptosis; the large-T Ag can bring both of these functions about.

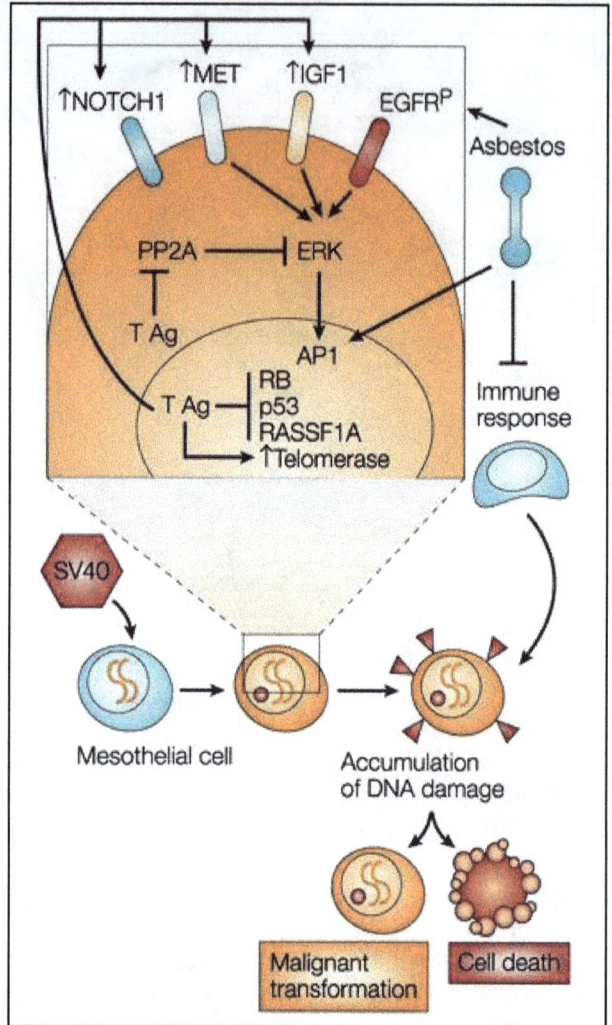

Figure 60. SV40 and the Pathogenesis of Mesothelioma Diagram Shows an *in vitro* Transformation Model, where Asbestos and SV40 Large-T Ag are Acting as Co-carcinogens. T-Ag is shown to block p53, RB and RASSF1A, and upregulate Notch1 and MET oncogenes.

Reprinted by permission from Nature Reviews Cancer 2:957-964, A. Gazdar, J. S. Butel and M. Carbone.

"SV40 and tumours: myth, association or causality?" copyright (2002) Macmillan Magazines Ltd.

PMID: 12459734

The large-T Ag is a multifunctional protein concerned with a wide range of cellular processes, including transcriptional activation and repression, blocking of differentiation, stimulation of the cell cycle, repression of apoptosis and cell transformation. The large-T Ag has three domains: a DnaJ domain (acts as a molecular chaperone), a RB-binding domain and a p53-binding domain, while the small-t Ag has the DnaJ domain and a serine/threonine phosphatase PP2A-binding domain. The large-T Ag gains control of the cell through its interactions with cellular proteins such as p53, RB (retinoblastoma), p107, p130, CBP/p300, and RASSF1A. The large-T Ag is able to disrupt both the RB and the p53 tumour suppressor pathways by binding and inactivating the cell cycle control proteins RB and p53, which stimulates the host cell to enter S phase and undergo DNA synthesis. RB acts to arrest cells in the G1 phase of cell division by repressing the transcription of genes required for entry into S phase, while p53 controls an apoptosis pathway; by inactivating both these proteins, the large-T Ag is able to stimulate cell division and block apoptosis.

The small-t Ag acts to transform cell by binding to PP2A, an abundant family of serine-threonine phosphatases. The loss of PP2A activity is thought to cause defects in the biogenesis and properties of tight junctions, leading to the disorganisation of the actin cytoskeleton. PP2A loss also causes the deregulation of Rho GTPases, F-actin and intercellular adhesion. Defects in the actin cytoskeleton and the disruption of tight junctions have been linked to tumour invasiveness.

Laboratory Diagnosis

☆ Viruses are detected by electron microscopy of cell cultures and fine sections of tumourous growths.

☆ Vacuolation in cells can be viewed under a light microscope.

☆ Viral antigen can be detected in the urine by ELISA

☆ Nucleic acid of SV4O is detected with PCR amplification followed by nucleic acid hybridization.

☆ Virus may be isolated from urine of diseased monkeys

Arbo Virus (Arthropod-borne virus)

Arboviruses (Arthropod-borne viruses) are viruses of vertebrates biologically transmitted by insect vectors. Mosquitoes, ticks, flea and other insects transmit these viruses. Arbo viruses and rodent-borne viruses are placed among the Toga, Flavi, Bunya, Rhabdo, Arena and Filovirus groups. They are more than 450 arboviruses and rodent-borne viruses.Of these about 100 are known to be pathogens for human.

The arthropod-borne viruses (arboviruses) multiply in both vertebrates and arthropods. The former serve as reservoirs in the cycle of transmission and the latter mostly as vectors, acquiring infection with a blood meal. After the virus is propagated in the arthropod's gut and attains a high titer in its salivary glands, virus is transmitted when a fresh host is bitten. The viruses often cause disease in the vertebrate hosts, but none is evident in the arthropods.

Yellow fever virus was the first arthropod borne virus to be discovered, through the work of Major Walter Reed. He headed the U.S. Army Yellow Fever Commission, established in 1901, to try to overcome the disastrous effect of yellow fever on American troops in Cuba during the Spanish-American War.

Classification

The presently known viruses isolated from arthropods may be divided into atleast 21 anrigenically distinct groups, many viruses are still ungrouped. Table shows the principal viruses that infect man, the groups to which they have been assigned and some of their clinical and epidemiological characteristics.

Family	Group	Sub Group	Viral Species	Vector	Clinical Disease of Man
Toga virus	A	I	Eastern equine encephalitis (EEE)	Mosquito	Encephalitis
			Venezuelan equine encephalitis (VEE)	Mosquito	Encephalitis
			Western equine encephalitis (WEE)	Mosquito	Encephalitis
		II	Sindbis	Mosquito	Subclinical
			Chikungunya	Mosquito	Headache, fever, rash, joint and muscle pains
			Semliki Forest	Mosquito	Fever or none
			Mayora	Mosquito	Headache, fever, joint and muscle pains
Toga virus	B	I	St.Louis encephalitis	Mosquito	Encephalitis
			Japanese B Encephalitis	Mosquito	Encephalitis
			Murray Valley Encephalitis	Mosquito	Encephalitis
			LIheus	Mosquito	Encephalitis
			West Nile	Mosquito	Headache, fever, myalgia, rash, lymphadenopathy
		II	Dengue (4 types)	Mosquito	Headache, fever, myalgia, prostration
		III	Yellow fever	Mosquito	Fever, prostration, hepatitis, nephritis
		IV	Tick-borne group (Rus-spring-summer encephalitis group) 9 viruses	Tick	Encephalitis; meningosian encephalitis, hemorrhagic fever

In general arboviruses belong to three families:

1. *Togaviruses*–genera alphaviruses *e.g.* EEE, WEE, VEE.
2. *Bunyaviruses*–*e.g.* Sicilian Sandfly Fever, Rift Valley Fever, Crimean-Congo Haemorrhagic Fever.
3. *Flaviviruses*–*e.g.* St Louis encephalitis, Japanese Encephalitis, Yellow Fever, Dengue.

Alphaviruses

The *Togaviridae* comprises of 4 genera: Alphavirus, Rubivirus, Pestivirus, and Arterivirus. The only member of the rubivirus genus is rubella. The Pestivirus genus contains 3 viruses of veterinary importance; bovine diarrhoea virus, hog cholera virus, and border disease virus. The arterivirus genus contains a single member; equine arteritis virus. The alphavirus genus has 25 members of which 11 are recognized to be pathogenic for man. All alphaviruses are mosquitoe-borne and are geographically distributed mainly in the new world.

Properties

Belong to the family of *Togaviridae* ssRNA enveloped viruses,icosahedral symmetry virions 60-65nm in diameter positive stranded RNA genome. A 26S subgenomic mRNA which represents the 3' one third of the genome codes for structural proteins. Translation of the whole genome from the 5' end produces non-structural proteins alphaviruses are classified into 6 antigenic groups on the basis of neutralization tests and molecular biology

The American Equine Encephalitides

The alphavirus genus contains 4 viruses which produce encephalitis; EEE, WEE, VEE and Everglades. EEE, WEE and VEE were first isolated from the brains of dead horses in the 1930s. Focal epidemics of EEE have occurred from time to time in the Eastern USA, EEE and WEE are maintained in nature between mosquitoes and birds, VEE between mosquitoes and rodents. The strain of VEE pathogenic for man are mainly amplified in horses, producing equine disease prior to the beginning of human disease. This is in contrast to the EEE and WEE viruses, where horses appear to be a dead end of infection. (Figure 61).

The clinical manifestations may vary from an inapparent or a mild influenza like illness to the syndrome of encephalitis. The probability of developing encephalitis vary widely between the different viruses; it is highest for the EEE and lowest for the VEE. The mortality rate of encephalitis vary from 5 per cent for WEE, 35 per cent for VEE to 50 per cent for EEE. Laboratory diagnosis may be made by the isolation of the virus from the brain, blood, CSF or throat washings. The viruses can be isolated using suckling mice or a variety of vertebrate cell lines such as Vero, where they produce a CPE, or using mosquito cell lines, where no CPE is seen. Alternatively, a serological diagnosis may be made for which a variety of tests are available.

No specific therapy is available against these viruses.

Alphaviruses Producing Fever, Polyarthritis, and Rash Chikungunya Virus

The best example of a old world alphavirus is Chikungunya virus. It is found in areas of Africa and Asia, especially jungle areas. There are 3 possible modes of transmission, the first mode is sporadic cases arising when man come in contact with the jungle areas where the reservoir is in monkeys. The third type of Chikungunya exists in India, when during the rainy season, the population of mosquitoes increases greatly and epidemics can occur.

The clinical disease is a flu-like fever of acute onset, followed by a pharyngitis, maculopapular rash and arthritis. Minor haemorrhagic manifestations are occasionally seen. The diagnosis of infection is usually made by serology.

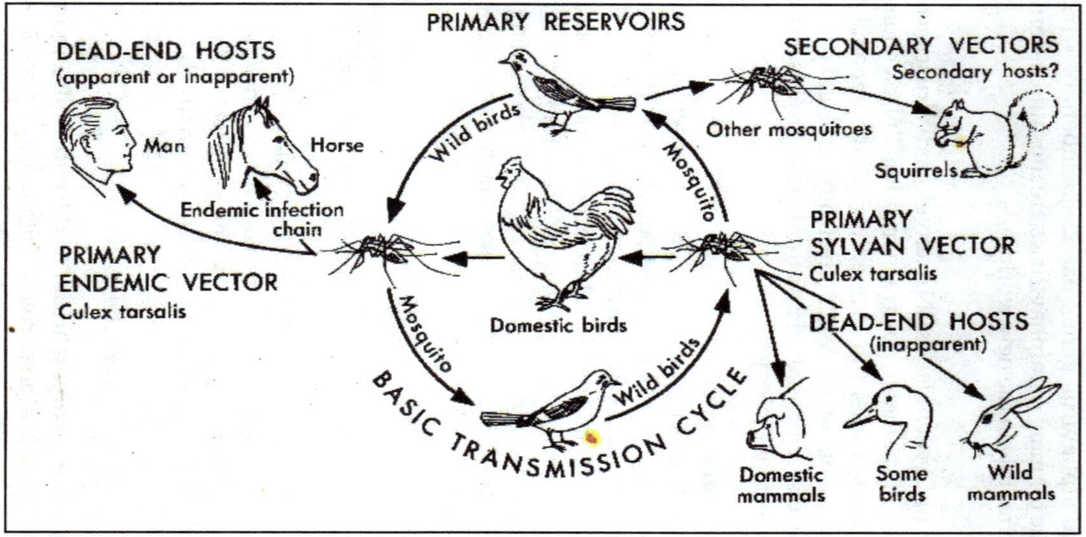

Figure 61. Epidemiological Pattern for WEE Virus Infections.

The chains for rural St. Louis encephalitis are similar, except that horses are inapparent rather than apparent hosts. EEE infections also have a similar summer infection chain, but a few significant differences exist: 1) the identity of the vector infecting man is unknown, 2) domestic birds do not appear to be a significant link in the chain, and 3) it has a bird-to-bird secondary cycle in pheasants whose role is unclear.

Bunyavirus

The Bunyaviridae family comprises more than 200 named viruses. Membership is usually based on antigenic interrelatedness or morphological similarity.

Bunyaviruses, with the exception of Hantaviruses, are thought to be transmitted in nature by arthropods; most frequently mosquitoes but occasionally phlebotomine sandflies, midges and ticks. Vertebrate reservoirs have been demonstrated for some viruses. In others, tranovarial transmission is thought to play a dominant role in virus maintenance. Man is not known to be a natural or reservoir for any of these viruses, with the probable exception of sandfly fever.

Properties

Enveloped ssRNA viruses, virions 100 nm in diameter 5-10 nm projections visible on the surface genome consists of 3 pieces of negative stranded RNA (the small RNA of phlebovirus is ambisense) virion has 2 surface glycoproteins G1 and G2, with HA and virus neutralization epitopes uncertain whether reassortment takes place uncommon event.

Naples and Sicilian Sandfly Fever Viruses

They are the best known and the most widely distributed phleboviruses and are thought to be the cause of most clinically described sandfly fever in the Mediterranean, SW Asia, India and perhaps China. It causes a febrile illness of acute onset, with severe headache, general malaise and arthralgia. The disease is self-limiting with complete recovery. In areas where sandflies are endemic, most of the population is thought to be infected during childhood.

Rift Valley Fever Virus

This virus is able to cause severe disease in domestic animals and man. The disease is found in all parts of Africa, where epidemics have occurred from time to time with significant morbidity and mortality. It was originally isolated in Kenya during an epizootic of fatal hepatic necrosis and abortion in sheep. Since that time, it has repeatedly infected herds of sheep, cattle, and goats in which it produced 10 to 30 per cent mortality. The impact of the virus on domestic animal raisers in the Rift valley and southern Africa is substantial. Rift valley fever virus extended into Egypt in 1977 causing a widespread epidemic with at least 600 deaths. The virus subsequently disappeared from Egypt. The virus is thought to transmitted mainly by mosquitoes although it can be transmitted by sandflies. Aerosol transmission had also been documented where man had become infected after coming into contact with animal carcasses. Most infections are symptomatic and usually present as a mild non-specific febrile illness, a small proportion (~1 per cent) develop haemorrhagic fever, retinal vasculitis and encephalitis. Treatment is supportive, although ribavirin, interferon and passive immunization have been shown to be useful in animal models. Certain ribavirin should be considered as part of the management. A formalin-inactivated cell culture vaccine is available and is thought to be effective and safe, although it is very expensive and thus its use should be confined to susceptible laboratory and veterinary workers.

Laboratory Diagnosis Bunyavirus Infections

Virus Isolation

Intracranial inoculation of suckling mice is thought to be the most sensitive system available for virus isolation. However, several sensitive cell culture systems are available such as vero. LLC-MC2 and mosquito cells. Once isolated the virus can be types by neutralizing tests.

Rapid Diagnosis

Antigen detection systems and the detection of specific IgM antibodies are becoming available as means of rapid diagnosis.

Serology

A wide variety of serological techniques are available such as HI, CFT, IFA, neutralization tests and ELISAs.

Flaviviruses

The viruses of the family Flaviviridae are important arthropod-borne viruses in both human and veterinary medicine. They are transmitted by mosquito and ticks and usually are maintained in a transmission cycle in nature. They are widely distributed throughout the world with the exception of the polar region, although a specific flavivirus may be geographically restricted to a continent or a particular part. They produce a broad spectrum of clinical responses in humans ranging from asymptomatic infection to fulminant encephalitis or haemorrhagic fever. Nearly 60 flaviviruses are known to exist but many are yet to be shown to cause disease in humans.

Properties

ssRNA enveloped viruses, 40-50 nm in diameter positive RNA genome of around 10,000 nucleotides 3 structural proteins (an envelope glycoprotein, a nucleoprotein, and a small membrane protein), and a number of non-structural proteins, one single reading frame, no subgenomic mRNA classified in terms of cross-reactivity and the host.

The flavivirus family is divided into 7 subgroup of viruses. The antigenic determinants are carried by the envelope glycoprotein which is recognized by both HI and neutralization tests. 3 types of antigenic determinants are found for flaviviruses;

1. Type-specific determinants
2. Complex-reactive determinants shared by closely related viruses
3. Group-reactive determinants shared by all serologically related flaviviruses.

The envelope glycoprotein contain mainly type-specific determinants and to a lesser extent complex-reactive and group-reactive determinants. The nucleocapsid proteins had only been shown to contain group-reactive determinants.

Flaviruses Producing Encephalitis

The flavivirus family contains many viral agents which produces encephalitis. Flavivirus encephalitides are either mosquito- borne, tick-borne, or have an unknown vector.

Mosquito-Borne Flaviviruses
St Louis Encephalitis

St. Louis encephalitis occurs in endemic and epidemic form throughout the Americas and is the most important arboviral disease of North America. It is closely related to Japanese encephalitis and the Murray Valley encephalitis viruses.The virus is maintained in nature by a bird-mosquito-bird cycle.The incubation period is 21 days. The ratio of inapparent to apparent infection ranges from 16:1 to 425:1. Children are much more likely to have inapparent infection than adults. The morbidity and mortality rate increases with age. Patients who are symptomatic will usually present with or progress to one of three syndromes (1) febrile headache (2) aseptic meningitis (3) encephalitis. The laboratory diagnosis is usually made by serology. Treatment is supportive and no vaccine is available.

Japanese Encephalitis

Japanese encephalitis is a major public health problem in Asia, SE Asia, and the Indian subcontinent. Prior to 1967, thousands of cases with several hundred deaths were reported each year. In endemic areas where vector control and vaccination had been undertaken, the incidence had dropped dramatically. Epidemics had been reported from Japan, China, Korea, Taiwan, USSR, Vietnam, Philippines, ASEAN countries, India and Bangladesh. The transmission cycle in nature involve the Culex and Aedes mosquitoes and domestic animals, birds, bats, and reptiles. Man is not a preferred host for Culex mosquitoes and transmission of JE virus does not usually occur until mosquito populations are large.

Some patients will only show an undifferentiated febrile illness or have mild respiratory tract complaints. The diagnosis is usually made serologically as virus isolation is not usually successful. No specific treatment is available. An inactivated suckling mouse brain vaccine had been available since the early 1960s which had been extensively used throughout Asia. The efficacy rate ranges from 60 to 90 per cent. Despite the vaccine being a mouse brain preparation, no postvaccination demyelinating allergic encephalitis had been reported. Mild symptoms occur in 1 per cent of vaccines and thus the vaccine is generally to be considered as safe.

Murray Valley Encephalitis

This virus is closely related to Japanese encephalitis and resembles JE clinically. It is confined to the Australia and New Guinea, where it is an important cause of epidemic encephalitis periodically. The diagnosis is made by serology and no specific treatment or vaccine is available.

West Nile Fever

West Nile fever is a dengue-like illness that occurs in both epidemic and endemic forms in Africa, Asia, and the Mediterranean countries. Areas of high endemicity include Egypt and Iran where most of the adult population will have antibodies. West Nile virus is a member of the St Louis encephalitis complex and is transmitted by Culex mosquitoes. The virus is maintained in nature through a transmission cycle involving mosquitoes and birds. Children will usually experience an inapparent or a mild febrile illness. Adults may experience a dengue-like illness whilst the elderly may develop an encephalitis which is sometimes fatal. The diagnosis is usually made by serology although the virus can be isolated from the blood in tissue culture. No vaccine fro the virus is available and there is no specific therapy. Among the arboviruses, it is difficult to distinguish clinically between West Nile, dengue and Chikungunya. In the absence of a rash, a number of toga and bunyaviruses should also be considered in the differential diagnosis.

Ilheus Virus

This virus is found in Latin America where it causes a febrile illness with arthralgia. Occasionally a mild encephalitis is seen. The virus can often be confused with dengue, St Louis encephalitis, yellow fever and influenza viruses.

Flaviviruses Producing Haemorrhagic Fever

Yellow Fever

Epidemiology

Yellow fever occurs in 2 major forms: urban and jungle (sylvatic) yellow fever. Jungle YF is the natural reservoir of the disease in a cycle involving nonhuman primates and forest mosquitoes. Man may become incidentally infected on venturing into jungle areas. The S American monkeys are more prone to mortality once infected with YF than the old world monkeys, suggesting that American YF probably originated from the old world as a result of sailing ships. Jungle yellow fever transmitted in a monkey-mosquito-monkey cycle. In these areas, YF is reintroduced into urban populations from time to time as a result of contact with jungle areas. YF cases occur more frequently at times of the year when there are high temperatures at high rainfall, conditions which are most conducive to mosquito reproduction.

Once infected, the mosquito vector remains infectious for life, transovarial transmission of Aedes aegypti had been demonstrated and may provide a mechanism for the continuation of the jungle or urban cycle. Once the virus is inoculated into human skin, local replication occurs with eventual spread to the local lymph nodes and viraemia occurs. The target organs are the lymph nodes, liver, spleen, heart, kidney and foregut.

Clinical Features

The incubation period varies from 3 to 6 days, following which there is an abrupt onset of chills, fever, and headache. Generalized myalgias and GI complaints (N+V) follows and signs may include facial flushing, red tongue and conjunctival injection. Some patients may experience an asymptomatic infection or a mild undifferentiated febrile illness. After a period of 3 to 4 days, improvement should be seen in most patients. The moderately ill should begin to recover, however, the more severely ill patients with a classical YF course will see a return of fever, bradycardia (Faget's sign), jaundice, and haemorrhagic manifestations. The haemorrhagic manifestations may vary from petechial lesions to epitaxis, bleeding gums, GI haemorrhage (black vomit of YF). 50 per cent of patients with frank YF will develop fatal disease characterized by severe haemorrhagic manifestations, oliguria and hypotension. Frank renal failure is rare. Rarely, other clinical findings such as meningoencephalitis in the absence of other findings have been described.

Laboratory Diagnosis

The differential diagnosis of YF include typhoid, leptospirosis, tick-borne relapsing fever, typhus, Q fever, malaria, severe viral hepatitis, Rift valley fever, Crimean-Congo haemorrhagic fever, Lassa, Marburg and Ebola fever. Yellow fever can be diagnosed serologically or by virus isolation. The serological diagnosis can be made by HI, CF and PRN tests. Virus isolation can be attempted from the blood which should be obtained within the first 4 days of illness. A variety of techniques are available for virus isolation, such as intracerebral inoculation of newborn Swiss mice or inoculation into Vero, LLC MK-2, BHK, or arthropod cell lines.

Treatment and Prevention

No specific antiviral therapy is available and treatment is supportive. Intensive medical treatment may be required but this is difficult to provide as many epidemics occur in remote areas. A live attenuated vaccine known as the 17-D had been available since 1937.

Kyasanur Forest Disease

This is a tick borne disease closely related to the tick-borne encephalitis complex and is geographically restricted to Karnataka State in India. Haemorrhagic fever and meningoencephalitis may be seen. The case-fatality rate is 5 per cent.

Dengue

Hundreds of thousands of cases of dengue occur every year in endemic and epidemic forms in tropical and subtropical areas of the world. The attack rates during epidemics can reach as high as 50 per cent. Dengue is a prevalent public health problem in SE Asia, the Caribbean, Central America, Northern South America and Africa. In hyperendemic areas, most cases occur in young children as the majority of the population had already been infected with multiple serotypes. In other areas, older children and adults are more likely to be affected. Maximum number of cases occur during the months of the year with the highest rainfall and temperatures, when Aedes aegypti populations are at their highest. *A. aegypti* mosquitoes deposit their eggs in waterfilled containers and thus reproduction is highest during periods of high rainfall.

Clinical Manifestations

The clinical presentation of dengue in children is varied. The disease may be manifested as an undifferentiated febrile illness, an acute respiratory illness, or as a GI illness: atypical presentations which may not be recognized by clinicians as dengue. Older children and adults infected with dengue the first time will display more classical symptoms: sudden onset of fever, severe muscle aches, bone and joint pains, chills, frontal headache and retroorbital pain, altered taste sensation, lymphadenopathy, and a skin rash which appears 3 days after the onset of fever. The rash may be maculopapular, petechial or purpuric and is often preceded by flushing of the skin. Other haemorrhagic manifestations may be seen such as epitaxis, gingival bleeding, ecchymoses, GI bleeding, vaginal bleeding and haematuria. Severe cases of bleeding should not be diagnosed as Dengue haemorrhagic fever (DHF) or Dengue shock syndrome (DSS) unless they meet the criteria below.

DHF or DSS are usually seen in children and usually occurs in 2 stages. The first milder stage resembles that of classical dengue and consists of a fever of acute onset, general malaise, headache, anorexia and vomiting. A cough is frequently present. After 2 to 5 days, the patient's condition rapidly worsens as shock begins to appear. Haemorrhagic manifestations ranging from petechie and bleeding form the gums to GI bleeding may be seen.

Laboratory Diagnosis

Serology

HI, CF and PRN tests are commonly used. The high degree of cross-reactivity between flaviviruses can make the interpretation of serological results very difficult.

Virus Isolation

This can be accomplished by the intracerebral inoculation of sera from patients into suckling mice. Sera can also be inoculated intrathoracically into Aedes mosquitoes. Head squash preparations are examined for the presence of antigen by the FA technique. Cell cultures such as LLC MK-2 and several mosquito-derived cell lines can be used.

Direct Detection

RT-PCR may detect virus in serum and can reliably identify the infecting dengue type.

Treatment and Prevention

There is no specific antiviral treatment available. Management is supportive and intensive medical management is required for cases of severe DHF and DSS. No vaccine for dengue is available but a tetravalent live-attenuated vaccine has been evaluated in Thailand with favourable results.

Oncogenic Viruses

Cancers are the result of a disruption of the normal restraints on cellular proliferation. It is apparent that the number of ways in which such disruption can occur is strictly limited and there may be as few as forty cellular genes in which mutation or some other disruption of their expression leads to unrestrained cell growth.

There are two classes of these genes in which altered expression can lead to loss of growth control: (a) Those genes that are stimulatory for growth and which cause cancer when hyperactive. Mutations in these genes will be dominant. These genes are called oncogenes. (b) Those genes that inhibit cell growth and which cause cancer when they are turned off. Mutations in these genes will be recessive. These are the anti-oncogenes or tumor-suppressor genes.

Viruses are involved in cancers because they can either carry a copy of one of these genes or can alter expression of the cell's copy of one of these genes.

Classes of Tumor Viruses

There are two classes of tumor viruses:

1. DNA tumor viruses
2. RNA tumor viruses, the latter also being referred to as RETROVIRUSES.

DNA Tumor Viruses

DNA tumor viruses have two life-styles:

In permissive cells, all parts of the viral genome are expressed. This leads to viral replication, cell lysis and cell death

In cells that are non-permissive for replication, viral DNA is usually, but not always, integrated into the cell chromosomes at random sites. Only part of the viral genome is expressed. This is the early, control functions (*e.g.* T antigens) of the virus. Viral structural proteins are not made and no progeny virus is released.

DNA Tumor Viruses Involved in Human Cancers

The first DNA tumor viruses to be discovered were rabbit fibroma virus and Shope papilloma virus, both discovered by Richard Shope in the 1930s. Papillomas are benign growths, such as warts, of epithelial cells. They were discovered by making a filtered extract of a tumor from a wild rabbit and injecting the filtrate into another rabbit in which a benign papilloma grew. However, when the filtrate was injected into a domestic rabbit, the result was a carcinoma, that is a malignant growth. A seminal observation was that it was no longer possible to isolate infections virus from the malignant growth. This was because the virus had become integrated into the chromosomes of the malignant cells.

Papilloma Viruses

Papilloma viruses have a genome size about 8 kilobases. They cause warts and also *human and animal cancers*. Warts are usually benign but can convert to malignant carcinomas. This occurs in patients with *epidermodysplasia verruciformis*.

Epidermodysplasia verruciformis is also known as *Lewandowsky-Lutz dysplasia* or *Lutz-Lewandowsky epidermodysplasia verruciformis* and is very rare. It is an autosomal recessive mutation that leads to abnormal, uncontrolled papilloma virus replication. This results in the growth of scaly macules and papules on many parts of the body but especially on the hands and feet. Epidermodysplasia verruciformis, which is associated with a high risk of skin carcinoma, is typically associated with HPV types 5 and 8 (but other types may also be involved). These infect most people (up to 80 per cent of the population) and are usually asymptomatic.

Papilloma viruses are also found associated with human penile, uterine, cervical and anal carcinomas and are very likely to be their cause; moreover, genital warts can convert to carcinomas. Squamous cell carcinomas of larynx, esophagus and lung appear very like cervical carcinoma histologically and these may also involve papilloma viruses. Recently, a strong causal link between certain oral-pharyngeal cancers and HPV16 has been demontsrated. Penile and cervical cancers associated with type 16 and type 18 papilloma viruses (and others) but the most common genital human papilloma viruses (HPV) are types 6 and 11. As might be expected if they are indeed the causes of certain cancers, types 16 and 18 cause transformation of human keratinocytes

Adenoviruses

These viruses are somewhat larger than polyoma and papilloma viruses with a genome size of about 35 kilobases. They were originally isolated from human tonsils and adenoids, are highly oncogenic in animals and only a portion of the virus is integrated into the host genome. This portion codes several T antigens that carry out early functions. Tumor-bearing animals make antibodies against the T antigens.

No humans cancers have been unequivocally associated with adenoviruses.

Tumor Antigens are Oncogenes

Tumors caused by papilloma virus, adenovirus or polyoma virus contain viral DNA but do not produce infectious virus. The presence of the virus, however, elicits the formation of antibodies against the tumor antigens. In the case of adenoviruses, only part of the viral genome is found in the host cell chromosomes whereas SV40 may integrate part or all of its genome. Whether or not the whole SV40 genome is integrated, only a part of the genome is transcribed into mRNA and protein and this is the region that encodes the early functions of the virus replication cycle.

Many DNA viruses have early and late functions. Early functions are the result of the expression of proteins that prime the cell for virus production and are involved in viral DNA replication. These proteins are expressed before genome replication and do not usually end up in the mature virus particle. Late functions are the results of the expression of viral structural proteins that combine to form the mature virus. They are expressed during and after the process of DNA replication. Since early functions are involved in the replication of the viral genome, it is not surprising that they can also alter the replication of host cell DNA.

RNA Tumor Viruses (Retroviruses)

Retroviruses are different from DNA tumor viruses in that their genome is RNA but they are similar to many DNA tumor viruses in that the genome is integrated into host genome.

Since RNA makes up the genome of the mature virus particle, it must be copied to DNA prior to integration into the host cell chromosome. This life style goes against the central dogma of molecular biology in which that DNA is copied into RNA.

Complex Tumor Viruses

Herpesviruses

Herpesviruses are much larger than the DNA viruses described above and have a genome size of 100 to 200 kilobases. Because of their large size, a lot remains to be discovered concerning the way in which these viruses transform cells.

There is considerable circumstantial evidence that implicates these large enveloped viruses in human cancers and they are highly tumorigenic in animals. The herpes virus genome integrates into the host cell at specific sites and may cause chromosomal breakage or other damage. Herpesviruses are often co-carcinogens. They may have a hit and run mechanism of oncogenesis, perhaps by expressing proteins early in infection that lead to chromosomal breakage or other damage.

Herpesviruses have over 100 genes. When these viruses infect cells which are non-permissive for virus production but which are transformed, only a subset (about 9) of viral genes are expressed. These genes code of nuclear antigens or membrane proteins. Not all nine transformation-associated genes are expressed in all herpes transformed cells.

Epstein-Barr Virus (Human Herpes Virus 4)

EBV is the herpes virus that is most strongly associated with cancer. It infects primarily lymphocytes and epithelial cells. In lymphocytes, the infection is usually non-productive, while virus is shed (productive infection) from infected epithelial cells.

EBV is causally associated with:

Burkitt's lymphoma in the tropics, where it is more common in malaria-endemic regions

Nasopharyngeal cancer, particularly in China and SE Asia, where certain diets may act as co-carcinogens

B cell lymphomas in immune suppressed individuals (such as in organ transplantation or HIV)

Hodgkin's lymphoma in which it has been detected in a high percentage of cases (about 40 per cent of affected patients)

X-linked lymphoproliferative Disease (Duncan's syndrome)

EBV can cause lymphoma in Marmosets and transform human B lymphocytes *in vitro*.

EBV also causes infectious mononucleosis, otherwise known as glandular fever. This is a self-resolving infection of B-lymphocytes which proliferate benignly. Often infection goes unnoticed (it is sub-clinical) and about half of the population in western countries has been infected by the time they reach 20 years of age. Why this virus causes a benign disease in some populations but malignant disease in others is unknown.

Hepatitis B Virus

Hepatitis B virus is very different from the other DNA tumor viruses. Indeed, even though it is a DNA virus, it is much more similar to the oncornaviruses (RNA tumor viruses) in its mode of replication. The DNA is transcribed into RNA not only for the manufacture of viral proteins but for genome replication. Genomic RNA is transcribed back into genomic DNA. This is called reverse transcription. The latter is not typical of most DNA tumor viruses but reverse transcription is a very important factor in the life cycles of RNA-tumor viruses.

Human Oncogenic Viruses

Viruses generally do not cause cancer in humans, however, there are a couple that do.

Epstein-Barr Virus (EBV)

Burkitt's lymphoma: Cancer of lymphoid tissue in the jaw. Rare except in Central Africa where malaria is common. first thought that Burkitt's was transmitted by mosquito like malaria, but now we know that malaria just wipes out resistance and induces virus to cause cancer.

Adult T-cell Leukemia

HTLV-1 retrovirus (human T lymphocyte virus1). Endemic in Japan, The Caribbean, South and Central America in U.S. about 1 million people were infected with T- cell Leukemia cancer

Liver Cancer

Hepatitis B chronic infections have a 200 times more likelihood to get liver cancer.

Both very common in developing countries. in U.S. chronic hepatitis B only 0.1 per cent. Liver cancer also rare.

Liver cancer is the MOST COMMON cancer in the world.

Human Immunodeficiency Virus (HIV):

It has been known for a while that HIV weakens the immune system (kills T-helper cells) and also allows the development of tumors and cancers. eg: Kaposi's sarcoma. HIV induces mixed cell lymphoma (B cells, T cells and macrophages all divide out of control).

Chapter 13
Interferon

Introduction

Interferons play an important role in the first line of defense against viral infections. They are part of the *non-specific immune system* and are induced at an early stage in viral infection–before the specific immune system has had time to respond. Interferons are made by cells in response to an appropriate stimulus, and are released into the surrounding medium; they then bind to receptors on target cells and induce transcription of approximately 20-30 genes in the target cells, and this results in an anti-viral state in the target cells.

Types of Interferon

There are three major classes of interferons that have been described for humans according to the type of receptor through which they signal:

☆ *Interferon type I*: All type I IFNs bind to a specific cell surface receptor complex known as the IFN-α receptor (IFNAR) that consists of IFNAR1 and IFNAR2 chains. The type I interferons present in humans are IFN-α, IFN-β and IFN-ω.[1]

☆ *Interferon type II*: Binds to IFNGR. In humans this is IFN-γ.

☆ *Interferon type III*: Signal through a receptor complex consisting of IL10R2 (also called CRF2-4) and IFNLR1 (also called CRF2-12)

Type I Interferon

Interferon-alpha (leukocyte interferon) is produced by virus-infected leukocytes, etc

Interferon-beta (fibroblast interferon) is produced by virus-infected fibroblasts, or virus-infected epithelial cells, etc

Interferon-a (a family of about 20 related proteins) and interferon-b are particularly potent as antiviral agents. They are not expressed in normal cells, but viral infection of a cell causes interferons to be made and released from the cell (that cell will often eventually die as a result of the infection). The interferon binds to target cells and induces an antiviral state. Both DNA and RNA viruses induce interferon but RNA viruses tend to induce higher levels. Double-stranded RNA produced during viral infection may be an important inducing agent. Other stimuli will also cause these interferons to be made: *e.g.* exogenous double-stranded RNA, lipopolysaccharide, other components of certain bacteria.

Type II Inteferon

Interferon-gamma (immune interferon) is produced by certain activated T-cells and NK cells. Interferon-gamma is made in response to antigen (including viral antigens) or mitogen stimulation of lymphocytes.

Interferon-alpha and Interferon-beta (Type I Interferons)

These interferons induce about 20-30 proteins, and the function of many of these is not fully understood. However, three of the proteins that appear to play an important role in the induction of the anti-viral state have been intensively studied. Expression of one of these proteins (2'5' oligo A synthase) results in activation of the second of these proteins (a ribonuclease) which can break down mRNA, and expression of the third protein (a protein kinase) results in inhibition of the initiation step of protein synthesis. These activities target viral protein synthesis, but also result in inhibition of host protein synthesis. *Thus it is important that these proteins are only made and activated when needed.*

Interferon treatment induces the synthesis of the inactive form of these proteins in the target cell. Double-stranded RNA is needed for activation of these proteins. It directly activates 2'5' oligo A synthase and protein kinase R, and indirectly activates ribonuclease L (since this needs 2'5'oligo A, the product of 2'5' oligo A synthase, for activation). Thus, these potentially toxic pathways are only activated in the interferon-treated cell if double-stranded RNA is made, this will usually only happen if virus infection actually occurs. The activation of these proteins may sometimes result in the death of the cell, but at least the progress of the infection is prevented.

Other Effects of Interferons

The pathway described above is by no means the only way that interferons protect cells against viruses and other pathogens. All three interferons increase expression of class I MHC molecules and thus promote recognition by cytotoxic T cells. All three interferons can activate NK cells which can then kill virus-infected cells.

Interferon-gamma increases expression of class II MHC molecules on antigen-presenting cells and thus promotes presentation of antigens to helper T cells. Interferon-gamma can also activate the ability of macrophages to resist viral infection (intrinsic antiviral activity) and to kill other cells if they are infected (extrinsic antiviral activity). Interferons have many other effects on gene expression, not all of which are understood.

Therapeutic Uses of Interferons

Interferons-alpha and -beta have been used to treat various viral infections. One currently approved use for various types of interferon-a is in the treatment of certain cases of acute and chronic hepatitis C and chronic hepatitis B.

Interferon-gamma has been used to treat a variety of disease in which macrophage activation might play an important role in recovery, *e.g.* lepromatous leprosy, leishmaniasis, toxoplasmosis. Since interferons have anti-proliferative effects, they have also been used to treat certain tumors such as melanoma and Kaposi's sarcoma.

Side Effects of Interferons

Common side effects of interferons: Fever, malaise, fatigue, muscle pains High levels of interferons can cause kidney, liver, bone marrow and heart toxicity.

Viral Defences Against the Non-Specific and Specific Immune Systems

Not surprisingly, some viruses have developed defenses against the interferon-induced antiviral response and other aspects of the immune defense system. For example, viruses may code for proteins which block interferon binding to cells, inhibit the action of the interferon-induced protein kinase, inhibit NK function, interfere with cell surface expression of MHC, block complement activation, prevent the host cell committing apoptosis, etc.

Pharmaceutical Uses

Interferon therapy is used (in combination with chemotherapy and radiation) as a treatment for many cancers. More than half of hepatitis C patients treated with interferon respond with viral elimination (sustained virological response), better blood tests and better liver histology (detected on biopsy). There is some evidence that giving interferon immediately following infection can prevent chronic hepatitis C. However, people infected by HCV often do not display symptoms of HCV infection until months or years later making early treatment difficult.

Interferons (interferon beta-1a and interferon beta-1b) are also used in the treatment and control of multiple sclerosis, an autoimmune disorder.

Administered intranasally in very low doses, interferon is extensively used in Eastern Europe and Russia as a method to prevent and treat viral respiratory diseases such as cold and flu. However, mechanisms of such action of interferon are not well understood; it is thought that doses must be larger by several orders of magnitude to have any effect on the virus. Consequently, most Western scientists are skeptical of any claims of good efficacy.

Route of Administration

When used in the systemic therapy, IFN-α and IFN-γ are mostly administered by an intramuscular injection. The injection of interferons in the muscle, in the vein, or under skin is generally well tolerated. Interferon alpha can also be induced with small imidazoquinoline molecules by activation of TLR7 receptor. Aldara (Imiquimod) cream works with this mechanism to induce IFN alpha and IL12 and approved by FDA to treat Actinic keratosis, Superficial Basal Cell Carcinoma, and External Genital Warts.

Adverse Effects

The most frequent adverse effects are flu-like symptoms: increased body temperature, feeling ill, fatigue, headache, muscle pain, convulsion, dizziness, hair thinning, and depression. Erythema, pain and hardness on the spot of injection are also frequently observed. Interferon therapy causes immunosuppression, in particular through neutropenia and can result in some infections manifesting in unusual ways. All known adverse effects are usually reversible and disappear a few days after the therapy has been finished.

Types

Several different types of interferon are now approved for use in humans.

More recently, the FDA approved pegylated interferon-alpha, in which polyethylene glycol is added to make the interferon last longer in the body. (Pegylated interferon-alpha-2b was approved in January 2001; pegylated interferon-alpha-2a was approved in October 2002.) The pegylated form is injected once weekly, rather than three times per week for conventional interferon-alpha. Used in combination with the antiviral drug ribavirin, pegylated interferon produces sustained cure rates of 75 per cent or better in people with genotype 2 or 3 hepatitis C (which is easier to treat) but still less than 50 per cent in people with genotype 1 (which is most common in the U.S. and Western Europe). Interferon-beta (Interferon beta-1a and Interferon beta-1b) is used in the treatment and control of multiple sclerosis. By an as-yet-unknown mechanism, interferon-beta inhibits the production of Th1 cytokines and the activation of monocytes (Table 11).

Table 11: Pharmaceutical Forms of Interferons in the Market

Generic Name	Trade Name
Interferon alpha 2a	Roferon A
Interferon alpha 2b	Intron A
Human leukocyte Interferon-alpha (HuIFN-alpha-Le)	Multiferon
Interferon beta 1a, liquid form	Rebif
Interferon beta 1a, lyophilized	Avonex
Interferon beta 1a, biogeneric (Iran)	Cinnovex
Interferon beta 1b	Betaseron/ Betaferon
Pegylated interferon alpha 2a	Pegasys
Pegylated interferon alpha 2a (Egypt)	Reiferon Retard
Pegylated interferon alpha 2b	PegIntron
Pegylated interferon alpha 2b plus ribavirin (Canada)	Pegetron

Chapter 14
Antiviral Drugs

Antiviral drugs are a class of medication used specifically for treating viral infections. Like antibiotics for bacteria, specific antivirals are used for specific viruses. Antiviral drugs are one class of antimicrobials, a larger group which also includes antibiotic, antifungal and antiparasitic drugs. They are relatively harmless to the host, and therefore can be used to treat infections. They should be distinguished from viricides, which actively destroy virus particles outside the body.

Most of the antivirals now available are designed to help deal with HIV, herpes viruses (best known for causing cold sores and genital herpes, but actually causing a wide range of diseases), the hepatitis B and C viruses, which can cause liver cancer, and influenza A and B viruses. Researchers are now working to extend the range of antivirals to other families of pathogens.

Designing safe and effective antiviral drugs is difficult, because viruses use the host's cells to replicate. This makes it difficult to find targets for the drug that would interfere with the virus without harming the host organism's cells.

The emergence of antivirals is the product of a greatly expanded knowledge of the genetic and molecular function of organisms, allowing biomedical researchers to understand the structure and function of viruses, major advances in the techniques for finding new drugs, and the intense pressure placed on the medical profession to deal with the human immunodeficiency virus (HIV), the cause of the deadly acquired immunodeficiency syndrome (AIDS) pandemic.

Almost all anti-microbials, including anti-virals, are subject to drug resistance as the pathogens evolve to survive exposure to the treatment.

History

Modern medical science and practice has an array of effective tools, ranging from antiseptics to vaccines and antibiotics. One field in which medicine has historically been weak, however, is in finding drugs to deal with viral infections. Highly effective vaccines have been recently developed to prevent such diseases, but formerly, when someone contracted a virus, there was little that could be done but to recommend rest and plenty of fluids until the disease ran its course.

The first experimental antivirals were developed in the 1960s, mostly to deal with herpes viruses, and were found using traditional trial-and-error drug discovery methods. Researchers grew cultures of cells and infected them with the target virus. They then introduced chemicals into the cultures they thought were likely to inhibit viral activity, and observed whether the level of virus in the cultures rose or fell. Chemicals that seemed to have an effect were selected for closer study.

This was a very time-consuming, hit-or-miss procedure, and in the absence of a good knowledge of how the target virus worked, it was not efficient in discovering antivirals that were effective and had few side effects. It was not until the 1980s, when the full genetic sequences of viruses began to be unraveled, that researchers began to learn how viruses worked in detail, and exactly what chemicals were needed to thwart their reproductive cycle (Table 12). Dozens of antiviral treatments are now available, and medical research is rapidly exploiting new knowledge and technology to develop more.

Table 12: Antiviral Treatment Drugs

Drug	Viruses	Chemical Type	Target
Vidarabine	Herpesviruses	Nucleoside analogue	Virus polymerase
Acyclovir	Herpes simplex (HSV)	Nucleoside analogue	Virus polymerase
Gancyclovir and Valcyte ™ (valganciclovir)	Cytomegalovirus (CMV)	Nucleoside analogue	Virus polymerase (needs virus UL98 kinase for activation)
Nucleoside-analog reverse transcriptase inhibitors (NRTI): AZT (Zidovudine), ddl (Didanosine), ddC (Zalcitabine), d4T (Stavudine), 3TC (Lamivudine)	Retroviruses (HIV)	Nucleoside analogue	Reverse transcriptase
Non-nucleoside reverse transcriptase inhibitors (NNRTI): Nevirapine, Delavirdine	Retroviruses (HIV)	Nucleoside analogue	Reverse transcriptase
Protease Inhibitors: Saquinavir, Ritonavir, Indinavir, Nelfinavir	HIV	Peptide analogue	HIV protease

Contd...

Table 12–Contd...

Drug	Viruses	Chemical Type	Target
Ribavirin	Broad spectrum: HCV, HSV, measles, mumps, Lassa fever	Triazole carboxamide	RNA mutagen
Amantadine/ Rimantadine	Influenza A strains	Tricyclic amine	Matrix protein/ haemagglutinin
Relenza and Tamiflu	Influenza strains A and B	Neuraminic acid mimetic	Neuraminidase Inhibitor
Pleconaril	Picornaviruses	Small cyclic	Blocks attachment and uncoating
Interferons	Hepatitis B and C	Protein	Cell defense proteins activated

Chemotherapeutic Index

Dose of drug which inhibits virus replication / Dose of drug which is toxic to host

The smaller this value of this number the better, *i.e.* several orders of magnitude difference is required for a really safe drug.

Modern technology allows deliberate design of drugs, but to do this, need to "know your enemy":

☆ Molecular biology–understanding viral replication and producing specific targets for inhibition

☆ Computer aided design (C.A.D.)

Antiviral Chemotherapy

Vaccines have, to date, occupied the central position in attempts to control virus infections. Vaccines are relatively cheap and safe and the immunity is often lifelong. However, some viruses, for some reasons, are not fully amenable to this approach, such as influenza, retroviruses, herpesviruses, the slow viruses, rhinoviruses and arboviruses. Obstacles to the use of vaccines include (1) multiplicity of serotypes *e.g..* rhinoviruses, togaviruses (2) antigenic change *e.g..* influenza, retroviruses and (3) Latent infections. Only relatively recently have notable successes on a large scale been achieved with antiviral drugs such as acyclovir and AZT, in situations where no vaccine is available. However. acyclovir and AZT do not approach penicillin in their spectra of activity or degree of inhibition. They are more analogous to some of the first antibacterial agents such as salvarsan. No antiviral compound tested has been able to inhibit completely the replication of any virus and a proportion of viral particles always seems to be able to circumvent the drug-induced blockade.

A. The Chemistry of Antiviral Compounds

There are few restrictions on the types of molecules that inhibit virus replication, at least in the laboratory. They vary greatly in complexity and include natural products of plants, synthetic oligonucleotides, oligosaccharides, simple inorganic and organic compounds and nucleoside analogues. Examples of antiviral compounds in current use include:

1. Nucleoside Analogues

Thousands of analogues of naturally occurring nucleosides have now been synthesized and tested in the laboratory, initially as herpesvirus inhibitors and many now are retested as anti-HIV agents. In addition to purine and pyrimidine nucleosides, ara-, amino, aza-nucleosides or nucleotides have been synthesized. Even single atomic substitution may change an active to an inactive molecule.

2. Pyrophosphate Analogues

Forscarnet is an example of a pyrophosphate analogue. This specifically inhibits herpesvirus DNA polymerase at the pyrophosphate binding sites and it also has anti-HIV activity.

3. Amantidine Molecules

Amantidine is licensed for the treatment of influenza A infection. Addition of a methyl grouping (rimantidine) alters the pharmacological distribution of the drug and prevents entry to the brain, thus reducing the side effect described as "jitteriness".

The Search for New Antiviral Compounds

All the antiviral drugs now known were discovered by random search in the laboratory. With developments in molecular biology and the advent of HIV, more attention is now being paid to the development of drugs designed to act on specific targets on the virus itself or in its replication. Attempts are being made to exploit data on viral nucleic acid sequences, X-ray crystallographic studies of viral proteins and enzymes. A recent example of the success of this new rational approach has been the discovery of potent neuraminidase inhibitors of influenza viruses.

Resistance of Viruses to Inhibitors

A disappointing feature of antiviral chemotherapy has been the failure so far of any antiviral molecule to inhibit virus replication completely. Antiviral activity tends to produce a 100 to 1000 fold reduction in virus titre which, although significant, still allows some infective particles to survive. This may have important consequences in immunocompromized patients who may be unable to eradicate any residual virus. It is not known for certain whether these virions are drug-resistant mutants or with different biologically or genetically from the major portions of the virus population.

B. Points of Action of Antiviral Molecules

Thousands of compounds can inhibit viral replication in cell culture. In general, the more complex the regulatory mechanisms of a virus, the easier it is to find molecules that can inhibit it. It is often very hard to decide which of the compounds should be investigated further. A broad estimate of the ratios of the activity of antiviral compounds in cell culture, animal models and humans is 1000:10:1.

1. Cell-free Virus

Few antiviral compounds inhibit or inactivate extracellular virus in vivo. An exception is the series of WIN compounds which bind to the external proteins of picornaviruses. These compounds bind to and fit within the canyons which exists on the surface of picornavirus virions and thereby stabilizing the particle and prevent uncoating.

2. Virus Adsorption

There is considerable theoretical interest in compounds able to block the adsorption of virus to susceptible cells. In the case of HIV, which binds specifically to CD4 receptors, short peptides have been synthesized to correspond with the sequence of the receptor binding site of CD4 molecule and with the binding protein gp120. These peptides should block the interaction of the receptor region and gp120 without interrupting other receptor functions of the CD4.

3. Virus Entry and Uncoating

Viruses such as influenza and certain flaviviruses enter by viropexis or engulfment. Immediately afterwards, while in a cytoplasmic endosome (vacuole), the virus catalyses fusion between the viral lipid-containing membrane and the membrane of the intracellular vacuole. The fusion is mediated by a sequence of hydrophobic amino acids or one of the glycoproteins of the virus. A compound that interrupts fusion would block virus replication at this early stage. In the case of influenza A, the fusion sequence on the HA molecule can only act after a structural 3- dimensional rearrangement of the HA molecule. This major change, whereby the HA trimer opens out like the petals of a flower, probably occurs only at a low pH5.5 found in lysosomal vacuoles. Amantidine appear to inhibit influenza A replication in part by raising the pH of the cytoplasmic vacuole, thus preventing virus-induced fusion and hence virus uncoating. Other enveloped viruses such as paramyxoviruses and HIV, enter cells by virus-induced fusion with the plasma membrane of the cell. This "fusion from without" may be susceptible to short peptides which may act on the fusion sequence extracellularly.

4. Transcription and Translation of Viral Nucleic Acids and Release of Virus

Most of the antiviral drugs now known act by inhibiting the replication or transcription of viral nucleic acids.

(a) Inhibitors of Herpes DNA Polymerase

By far the most amenable target for antiviral drugs is the herpes simplex DNA polymerase. The most successful antiviral compound yet developed is acyclovir inhibits the function of this enzyme. The ideal antiviral drug should (1) be taken up only into infected cells (2) the actual inhibitory molecule should be generated inside the infected cell by enzymatic activity (3) the inhibitor should have a selective effect on a virus enzyme. Acyclovir demonstrates all of the above characteristics.

(b) Inhibitors of Viral Reverse Transcriptase

AZT and the majority of other compounds act as chain terminators. AZT triphosphate binds to and inhibit virus RT more effectively than normal cellular DNA polymerases and so some antiviral specificity is achieved. However, the compound is certainly not comparable to acyclovir in terms of antiviral specificity. This is reflected in the toxicity of AZT in clinical practice. This cellular toxicity may be partly explained by the fact that normal cellular enzymes phosphorylate AZT and is thus activated in both infected and uninfected cells.

5. Translation

It may be possible to interfere with the viral mRNA itself. Small anti-sense oligonucleotides can be constructed which are complementary to specific genes, such as the rev gene. Fomivirsen (Vitravene) is a 21-base anti-sense oligonucleotide complementary to the early region 2 mRNA of CMV. It is approved for the local treatment of CMV retinitis in AIDS patients.

6. Assembly

HIV protease is required for the cleavage of the gag-pol fusion protein. Inhibitors of this enzyme may therefore block the assembly of HIV.

C. Some Commonly Used Antiviral Agents

1. Acyclovir

Acyclovir is a synthetic guanine nucleoside analogue. The initial step of phoshorylation to ACV monophosphate is preferentially carried out by viral thymidine kinase rather than cellular kinases. The monophosphate cannot leave infected cells so that more non-phosphorylated compound enters to make up for the depleted intracellular concentration, only to be converted to the monophosphate. In this manner, the drug accumulates in the herpes-infected cells rather than in the uninfected counterparts. The monophosphate is then phosphorylated to the di and tri-phosphate forms by cellular enzymes. ACV triphosphate is the pharmacologically active form of the drug. It inhibits herpes DNA polymerase with little effect on the host cell DNA polymerase. It also has some chain termination activity and thereby it behaves as a "suicide inhibitor" (Figure 62)

Acyclovir resistant strains of HSV have mutations in either the viral thymidine kinase gene or the viral DNA polymerase. Acyclovir also has antiviral activity against other herpesviruses such as VZV, CMV and EBV, although the mechanism is not so well understood in these cases. Forscarnet is the preferred drug in the treatment of acyclovir-resistant strains.

2. Valacyclovir

Valacylovir is an ester of acyclovir that is well-absorbed. Its bioavailibility is 2-5* greater than acyclovir. It is used for the treatment and suppression of genital herpes infection.

3. Famciclovir

Famiciclovir is the prodrug of penciclovir which is the active form and a guanosine analog. It has a very high bioavailability of 77 per cent. It is converted into penciclovir by a two step process. The first step occurs in the gut and the second step in the liver. It has a long half life in the gut. It has a higher affinity for HSV thymidine kinase than acyclovir but a lower affinity for HSV DNA polymerase than acyclovir. It acts as an inhibitor of viral DNA polymerase and also as a chain terminator. At present famciclovir is licensed for the treatment of shingles and the dosage is 250mg tds. It is also used for the treatment and suppression of genital herpes infection.

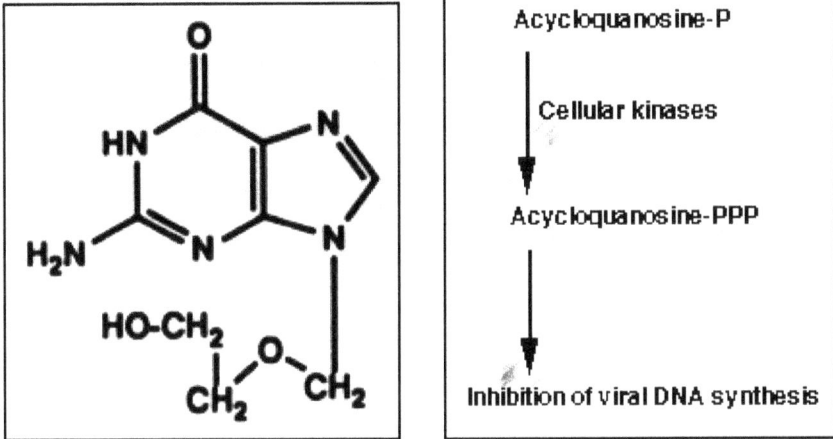

Figure 62. Acyclovir

4. Ganciclovir

Ganciclovir is a guanine nucleoside chemically related to acyclovir. It acts as a chain terminator and subsequent termination of viral DNA replication. The active form is thought to be the triphosphate. CMV does not specify TK and the initial phosphorylation of ganciclovir is thought to be mediated by other cellular enzymes. Ganciclovir has potent in vitro activity against all herpesviruses, including CMV. It has some activity against other DNA viruses such as vaccinia and adenovirus. Ganciclovir is more active against CMV than acyclovir. Ganciclovir has been shown to be of value in treating severe CMV infections in the immunocompromized, especially in conjunction with hyperimmune immunoglobulin. Reversible neutropenia is the most frequent adverse reaction. Ganciclovir resistance has been reported in immunocompromised patients being treated for CMV disease and is thought to be due to lack of phosphorylation of the drug by CMV infected cells. A recent prospective study estimated that 8 per cent of patients receiving ganciclovir for more than 3 months developed resistant CMV (Figure 63)

5. Ribavirin

Ribavirin is a synthetic triazole nucleoside and the active form is ribavirin triphosphate. It is not incorporated into the primary structure of DNA or RNA during cellular synthesis of nucleic acids. In the case of influenza viruses, it inhibits the 5' capping of viral mRNAs. It has also been shown to inhibit influenza viral RNA polymerase complex. It has further been postulated that ribavirin triphosphate inhibits several steps in viral replication and this phenomenon may explain the failure to detect viral isolates that are resistant to ribavirin. Ribavirin has been shown to possess activity against both DNA and RNA viruses in infected cells. It has been found to have activity against adenoviruses, herpesviruses, CMV. vaccinia. influenza A and B, parainfluenza 1, 2, 3, measles, mumps, RSV, rhinovirus. Ribavirin has made a major contribution to the therapy of children infected with RSV where is given as an aerosol in hospital. It has also been shown to be effective against influenza A and B. It has also been reported to be of value in the treatment of Lassa fever, hantavirus disease and hepatitis C.

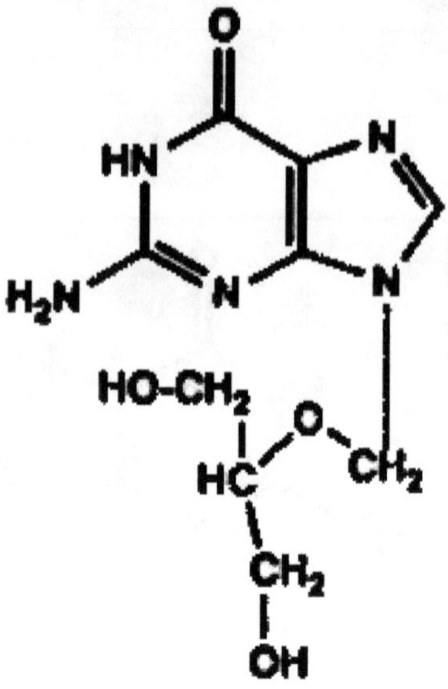

6. Zidovudine (AZT)

AZT is a synthetic analogue of thymidine. It requires conversion to the triphosphate form by cellular enzymes. It inhibits viral reverse transcriptase by acting as a chain terminator. Viral reverse transcriptase is 100 times more susceptible to inhibition by zidovudine triphosphate than host cellular DNA polymerase. Once incorporated into the viral DNA chain, viral DNA synthesis is terminated as no more phosphodiester bonds could be formed. AZT is active in vitro against many human retroviruses, including HTLV-I and HIV. AZT is currently indicated for the management of patients with HIV infection who have impaired immunity. (T4 cell count of 400-500 or less) AZT has been clearly shown to prolong the life of individuals infected with HIV. It has also been shown to be of benefit for the treatment of symptomless individuals although this is controversial (Figure 64).

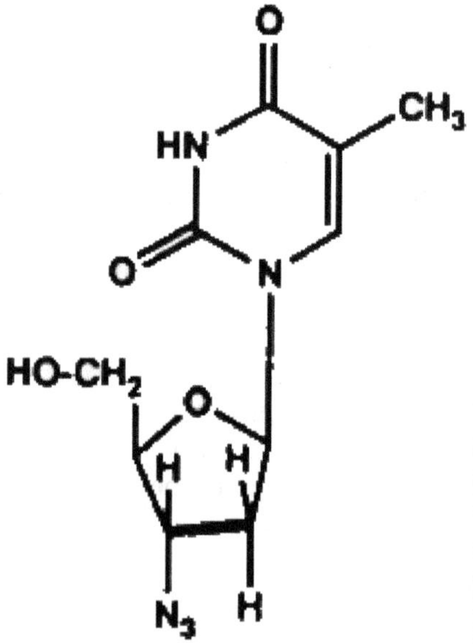

7. Lamivudine

Lamivudine is a potent reverse transcriptase inhibitor. It is generally well tolerated by patients. It now usually forms

Figure 63. Ganciclovir

Figure 64. Zidovudine

an essential component in the combination therapy of HIV patients. Recently, it had been approved for the treatment of chronic hepatitis B.

8. Forscarnet

Forscarnet is a pyrophosphate analog and unlike nucleoside analogues, forscarnet does not need to be activated by cellular or viral kinases. Forscarnet binds directly to the pyrophoshate- binding sites of RNA or DNA polymerases. Forscarnet is difficult to use as it must be given continuously intravenously via an infusion pump. It is used for the treatment of CMV retinitis in AIDS patients receiving AZT therapy, as it does not have overlapping toxicity with AZT. It is also used in the treatment of AZT resistant HSV infections. Its major adverse effect is on renal function.

9. Amantidine

This compound inhibit the growth of influenza viruses in cell culture and in experimental animals. Amantidine is only effective against influenza A, and some naturally occurring strains of influenza A are resistant to it. The mechanism of action of amantadine is not known. It is thought to act at the level of virus uncoating. The compound has been shown to have both therapeutic and prophylactic effects. Amantidine significantly reduced the duration of fever (51 hours as opposed to 74 hours) and illness. The compound also conferred 70 per cent protection against influenza A when given prophylactically. Amantidine can occasionally induce mild neurological symptoms such as insomnia, loss of concentration and mental disorientation. However, these symptoms quickly developed in susceptible individuals and cease when treatment is stopped. The therapeutic and prophylactic activity of amantidine is now generally accepted and numerous analogues of this compound have been prepared. Rimantadine is not as effective as amantadine but is less toxic. One factor that limits the usefulness of amantidine and rimantidine is the rapid development of resistance of these molecules in 30 per cent of patients. These resistant mutants have been reported to be as capable of being transmitted and causing disease as the wild virus.

10. Zanamivir

The rational approach to drug design has led to the design of several potent inhibitors of influenza neuraminidase of which two, zanamivir and oseltamivir are licensed for the treatment of influenza A and B infections. In clinical trials, both agents have demonstrated efficacy with minimal side effects. Because of its poor bioavailibility, Zanamivir must be given by inhalation whilst oseltamivir can be given orally. Because selection of drug-resistant mutants characterized by changes in NA requires prolonged passage in tissue culture, development of zanamivir-resistant viruses is not expected to occur readily in patients. The available information suggests that mutants may be less stable in vivo. The significance of changes in hemagglutinin remains to be evaluated. Overall the NA family of anti-influenza drugs is showing considerable promise; resistant variants do not occur readily and may be biological cripples.

11. Immunoglobulins

Immunoglobulins are available in three different formulations; intramuscular form, IVIG, and hyperimmune globulins against individual viruses. Immunoglobulins are more effective when used prophylactically rather than therapeutically. Currently, HNIG is used primarily for the prevention of hepatitis A. HNIG can also be given to non- immunized contacts of measles. Hyperimmune globulins are used for the postexposure prevention of hepatitis B, chickenpox and rabies. It has also been used in the treatment of arenavirus infections, Crimean-Congo haemorrhagic fever and Rift valley fever. CMV Ig is given prophylactically to seronegative recipients of kidneys from seropositive donors. The use of prophylactic CMV Ig in BMT patients is controversial. CMV IVIG is used in conjunction with ganciclovir in the treatment of CMV pneumonitis. IVIG is also used in the treatment of chronic enteroviral meningoencephalitis in children with agammablobinaemia.

D. Anti-HIV Therapy

The huge resources that had gone into HIV research had resulted in the development of a large number of anti-HIV agents. The speed of advance in this area is unprecedented in the history of medicine. Therapy of HIV is complicated by the fact that the HIV genome is incorporated into the host cell genome and can remain there in a dormant state for prolonged periods until it is reactivated. However although it may or not be possible to actually eradicate the virus completely, it seems possible that the infection can be indefinitely contained so that an infected will die with HIV infection rather than from it.

Zidovudine (AZT) was the first anti-viral agent used for the treatment of HIV and was introduced in 1987. However, it became clear with Concorde study in 1994 that monotherapy with AZT did not provide durable efficiency and hardly made any dent in the mortality rate. In 1995, results of the European DELTA and the American ACTG 175 studies became available and showed that combination therapy with two nucleoside analogues were better than monotherapy with one alone. A further breakthrough occurred with the introduction of HIV protease inhibitors which were specifically designed against HIV protease and were shown to be the most potent anti-HIV effect to date. An early clinical trial reported that the use of oral ritonavir decreased HIV mortality from 38 per cent to 22 per cent. Combination therapy, otherwise known as HAART (highly active antiretroviral therapy) using two or three agents became available. The rationale for this approach is that by combining drugs that are synergistic, non-cross-resistant and no overlapping toxicity, it may be possible to reduce toxicity, improve efficacy and prevent resistance from arising. The final breakthrough occurred when David HO (Time Magazine Man of the Year 1996) finally elucidated the pathogenesis of HIV infection. He showed that far from being latent during the "latent phase" as previously thought, there is actually massive replication during this period. David Ho had coined the slogan "hit hard and early". The results of the new approach was seen quickly where within four years, between 1994 and 1998, the incidence of AIDS in Europe sank from 30.7 to 2.5/100 patient years *i.e.* to less than a tenth.

What is less hopeful is the possibility of ever eradicating HIV from the body *i.e.* complete cure. In the beginning, it is thought that continuous treatment for 3 years would be sufficient to eradicate all the remaining latently infected cells. However, the period of treatment required kept on being revised upwards as new data became available. The most recent estimate of eradication of all latently infected cells is 73.3 years. Therefore, it is clear it will not be possible to achieve a complete cure in the short term. Compliance is a major issue when therapy is expected to be life-long. There is clearly a great need to have formulations whereby the number of tablets to be taken per day is kept to a minimum. The development of side-effects with long term use is another issue.

As knowledge builds up on the risks and efficacy on various agents and regimens, recommendations on HIV are being continually revised. So now, instead of "hit hard and early", there is a shift towards "hit hard, but only when necessary". There is still a lot of debate on when to actually commence therapy. Two measures are used for determining whether to start HIV therapy: CD4 counts and viral loads. It is generally agreed that HIV therapy should be given when the CD4 count is below 200. Some experts would recommend treatment for any patient whose CD4 count is below 350. What is less clear are patients with CD4 counts of 300-500 and a modest viral loads. A decision to start therapy must be taken on an individual basis with the patient after thorough discussion and counseling.

1. Anti-Retroviral Agents
A. Nucleoside Reverse Transcriptase Inhibitor
1. Zidovudine (AZT)
2. Didanosine (ddI)
3. Zalcitabine (ddC)
4. Emtricitabine (FTC)
5. Lamivudine (3TC)
6. Tenofovir (TDF)
7. Stavudine (d4T)
8. Abacavir (ABC)

B. Non-Nucleoside Reverse Transcriptase Inhibitor
1. Nevirapine (NVP)
2. Delaviridine (DLV)
3. Efavirenz (EFV)

C. HIV Protease Inhibitors
1. Tipranavi (TPV)
2. Amprenavir (APV)
3. Indinavir (IDV)
4. Saquinavir (SQV)
5. Ritonavir (RTV)
6. Atazanavir (ATV)
7. Fosamprenavir (FPV)
8. Nelfinavir (NFV)

D. HIV Fusion Inhibitors
1. Enfuvirtid (T-20)
2. Maraviroc (MVC)

There are a number of combination preparations on the market *e.g.* CBV (AZT+3TC), TZV (AZT+3TC+ABC), TVD (FTC+TDF), Kaletra (Lopinavir/ritonavir). The use of combination preparations will reduce the numbed of tablets that need to be taken each time.

2. Monitoring Anti-HIV Therapy
(a) Viral Load
(*i*) *Initiation*: Viral load is now the preferred method of monitoring therapy. There should be >= 1 log reduction in viral load, preferably to less than 10,000 copies/ml HIV-RNA within 2-4 weeks after the commencement of treatment. If <0.5 log reduction in viral, or HIV-RNA stays above 100,000, then the treatment should be adjusted by either adding or switching drugs.

(*ii*) *Monitoring*: Viral load measurement should be repeated every 4-6 months if patient is clinically stable. If viral load returns to 0.3-0.5 log of pre-treatment levels, then the therapy is no longer working and should be changed.

(b) CD4 count
(*i*) *Initiation*: Within 2-4 weeks of starting treatment, CD4 count should be increased by at least 30 cells/mm^3. If this is not achieved, then the therapy should be changed.

(*ii*) *Monitoring*: CD4 counts should be obtained every 3-6 months during periods of clinical stability, and more frequently should symptomatic disease occurs. If CD4 count drops to baseline (or below 50 per cent of increase from pre-treatment), then the therapy should be changed.

(c) Anti-HIV Drug Resistance Testing
Anti-retroviral drug resistance testing has become part and parcel of patient management in N. America and W. Europe. Many studies in treatment experienced patients have shown strong associations between the presence of drug resistance and failure of the antiretroviral treatment regimen to suppress HIV replication.

(*i*) *Genotypic Assays*: Genotypic assays detect drug resistance mutations that are present in the relevant viral genes (*i.e.* RT and protease). Some genotyping assays involve sequencing of the entire RT and protease genes, while others utilize oligonucleotide probes to detect selected mutations that are known to confer drug resistance. Genotyping assays can be performed relatively rapidly, such that results can be reported within 1-2 weeks of sample collection. Interpretation of test results requires an appreciation of the range of mutations that are selected for by various antiretroviral drugs, as well as the potential for cross-resistance to other drugs conferred by some of these mutations.

(*ii*) *Phenotypic Assays*: Phenotypic assays measure the ability of viruses to grow in various concentrations of antiretroviral drugs. Automated, recombinant phenotyping assays have recently become commercially available with turn-around times of 2-3 weeks; however, phenotyping assays are generally more costly to perform compared with genotypic assays. Recombinant phenotyping assays involve insertion of the RT and protease gene sequences derived from patient plasma HIV RNA into a laboratory clone of HIV. Replication of the recombinant virus at various drug concentrations is monitored by expression of a reporter gene and is compared with replication of a reference strain of HIV. The concentrations of drugs that inhibit 50 per cent and 90 per cent of viral replication (*i.e.* the IC50 and IC90) are calculated, and the ratio of the IC50s of the test and reference viruses is reported as the fold increase in IC50, or fold resistance. Interpretation of phenotyping assay results is complicated by the paucity of data on the specific level of resistance (fold increase in IC50) that is associated with failure of different drugs.

(*iii*) *Use in Clinical Setting*: Resistance assays may be useful in the setting of virological failure on antiretroviral therapy. Recent prospective data supporting the use of resistance testing in clinical practice come from trials in which the utility of resistance tests were assessed in the setting of virological failure. The VIRADAPT and GART studies compared virological responses to antiretroviral treatment regimens when genotyping resistance tests were available to help guide therapy with those observed when changes in therapy were guided solely by clinical judgment. The results of both studies indicated that the short-term virological response to therapy was significantly greater when results of resistance testing were available. Similarly, a recent prospective, randomized, multicenter trial has shown that therapy selected on the basis of phenotypic resistance testing significantly improves the virological response to antiretroviral therapy, compared with therapy selected without the aid of phenotypic testing. Thus, resistance testing appears to be a useful tool in selecting active drugs when changing antiretroviral regimens in the setting of virological failure.

The rapidity of developments in anti-HIV therapy makes it virtually impossible for this site to keep up. For the latest information on HIV and anti-retroviral therapy, I recommend the HIV page at Medscape.com

E. Interferons

There are 3 classes of interferons: alpha, beta and gamma. Interferon-a exists as at least 15 subtypes, the genes for which shows 85 per cent homology. IFN b1 shows 30 per cent homology with IFNa. IFNb2 is now known as IL-6 and shows no homology with alpha or b1 types. IFN gamma is also a lymphokine and shows no homology with the other types. IFNs mediate their actions through specific receptors at hormone like concentrations. Interferon inducible response elements in the cellular genome are activated. There are 2 main types of IFN receptors, one for alpha and beta1 and the other for gamma.

IFNs are released form many cell types in response to virus infection, dsRNA, endotoxin, mitogenic and antigenic stimuli. DsRNA appears to be a particularly important inducer. Usually, good IFN inducers are viruses that multiply slowly and do not block the synthesis of host protein early or markedly damage the cells. IFN is usually assayed by determining its effect on the multiplication of a test virus, usually vesicular stomatitis virus, a rhabdovirus. Viral strains capable of high IFN production give rise to autointerference in endpoint assays. In general IFN gamma differs from the others in that it is released as a lymphokine from activated T-cells and occasionally from macrophages.

1. Mechanism of Action

The antiviral effects of IFNs are exerted through several pathways;-

(1) *Increased expression of Class I and Class II MHC glycoproteins,* thereby facilitating the recognition of viral antigens by the immune system.

(2) *Immunoregulatory effects*–activation of cells with the ability to destroy virus-infected targets; these include NK cells and macrophages. IFNs appear to drive a shift from humoral to cellular immunity.

(3) *Direct inhibition of viral replication:* several mechanisms contribute to the third pathway.

   (a) Production of specific inhibitory proteins *e.g.* the Mx protein which has specific anti-influenza action. It is likely that more specific inhibitory proteins will be identified.

   (b) Inhibition of viral processes such as penetration, uncoating and budding from infected cells have been reported.

   (c) *In vitro* studies with extracts of IFN-treated cells show that the main target of IFN action is translation, which is blocked by 2 mechanisms, both requiring the presence of minute amounts of dsRNA:

   (i) Activation of a dsRNA dependent protein kinase–this phosphorylates and inactivates the translation initiation factor eIF-2. The phosphorylation freezes the initiation complex formed by eIF-2, GTP and met-tRNAf with the small ribosomal subunit and mRNA. Because eIF-2 cannot be recycled, protein synthesis is inhibited or stopped.

   (ii) Activation of 2-5 oligo A synthetases synthesis of 2-5A activates endonuclease L (itself induced by IFN) degradation of mRNA inhibition of protein synthesis.

The combination of cell growth inhibition and enhancement of CMI accounts for the antitumour effect of IFN.

2. Protective Role in Virus Infections

A protective role of IFN in animals is suggested by many observations:

1. In mice recovering from influenza virus infection, the titre of IFN is maximal at the time when the virus titre begins to decrease and before a rise in Abs

can be detected. At this stage, the IFN titre in the animal is sufficient to protect them against the lethal action of a togavirus.

2. Administration of a potent antiserum to IFN markedly increases the lethality of mouse hepatitis virus infection.

3. Suckling mice, which are susceptible to coxsackievirus B1, produces little IFN in response to this virus, whereas adult mice, which are resistant, produces large amounts.

These studies suggest that IFN has a major protective role in at least some viral infections. Much depends on the dynamics of the disease.

3. Possible Therapeutic Use

Clinically, effective prophylaxis was demonstrated against rhinovirus infection of human volunteers, with decreased incidence of infection and reduction of symptoms. Contacts of infected patients can be protected by intranasal spray of large doses of IFN. Also reduced is CMV reactivation in seropositive patients undergoing kidney transplants. IFNs could theoretically be ideal antiviral agents, since they act on many different viruses and have high activity. However, their therapeutic value is limited by various factors: IFNs are effective only during relatively short periods and have no effect on viral synthesis that is already initiated in a cell. Moreover, at high doses they have serious toxic effects on the host.

Attempts to use exogenous IFN for the treatment of human viral diseases had been met with limited success. IFN had prophylactic effect against influenza infection during epidemics, and local administration lessens the severity of respiratory diseases, IFN had also reported to be successful in treating genital warts and juvenile laryngeal papillomatosis. More recently, synthetic alpha-interferon was licensed for use in the treatment of hepatitis B carriers and it is also being used for the treatment of hepatitis C carriers with chronic active hepatitis.

4. Interferon Therapy for Chronic HBV Carriers

It has been postulated that chronic carriage of HBV is due mainly to inadequate production of interferon and the failure of the body to respond to interferon in the presence of acute HBV infection. Two preparations of interferons are currently available: Alpha-Interfron (Intron A) and Peginterferon (Pegasys). In In early clinical trials, interferon therapy is associated with HBeAg loss in 30-40 per cent of patients, and in approximately 10 per cent lost HbsAg altogether. If a patient loses HBeAg loss during interferon therapy, HBsAg loss follows therapy in approximately 80 per cent of patients followed for a decade. In addition, improved survival, complication-free survival, and a reduction in the frequency of hepatocellular carcinoma have been reported in those who responded to interferon. Interferon therapy is more effective in patients with low-level HBV DNA 100,000–40 million copies per mL, elevated ALT (esp if >200 IU/mL), immunocompetence, normal liver function (albumin, bilirubin and coagulation), and acquisition of infection in adulthood. Early studies suggested that the efficacy of interferon was low in patients with pre-core-mutant HBV infection (HBeAg negative strains), but recent observations have renewed interest in interferon for this indication. Emerging data on PEG interferon may result in the first line use of

PEG products alone or in combination with oral agents. However, interferon requires inconvenient injection therapy, is associated with a lot of side effects, and is no better than lamivudine in terms eAg seroconversion. Morover, it isof limited value in certain subgroups although it is the only medication that offers a chance at a complete cure.

Interferon-alpha (Intron A) is given by injection several times a week for six months to a year, or sometimes longer. The drug can cause side effects such as flu-like symptoms, depression, and headaches. Approved in 1991 and available for both children and adults.

Pegylated Interferon (Pegasys)–peginterferon is is modified form of interferon that has been approved for the treatment of HBV and HCV. It has a similar but larger chemical structure than interferon-alpha. This improves the efficiency of the drug so that it only needs to be injected once weekly, usually for six months to a year. The drug can cause side effects such as flu-like symptoms, depression and other mental health problems. Approved May 2005 for adults.

5. Interferon Therapy of Hepatitis C Carriers

Early studies indicate that interferon and ribavirin are effective of cases of acute and chronic hepatitis C. A combination of interferon and ribavirin may be useful. There is more experience in the use of interferon for the treatment of hepatitis C. The current recommendation is that interferon treatment may be considered in those with chronic active hepatitis who are at risk of progression to cirrhosis and HCC. The recommended regimen is 3 MU tds sc or im for 6 months. The response rate is around 50 per cent. However, approximately 50 per cent of responders relapse upon cessation of treatment. At present, it is not clear what factors predict response to interferon therapy. There is some data to suggest that older patients and those with established cirrhosis respond less well. There is growing evidence that the genotype of the infecting HCV determines the response to IFN. Most responders will have significant reduction of SGPT level within 2 months of interferon therapy. One may try a higher dose such as 5 or 10 MU in non-responders although it is not certain whether the higher doses work. At present, it is not clear what factors predict relapse after treatment. For those who relapse after treatment, they may be offered a second course and then put on maintenance therapy for 6 to 12 months. There is data to suggest that combination therapy with interferon and ribavirin is more effective than interferon alone. In fact, a pharmaceutical preparation of both these agents together is available for this purpose. It is now also routine to test for the HCV genotype before the commencement of Interferon/Ribavirin therapy. Genotypes 1 and 4 carry a poorer prognosis and response and typically these patients are treated for 48 weeks instead of 24 weeks for other genotypes.

Chapter 15

Prions and Viroids

Prions

Structure

Prions are infectious agents composed exclusively of a single sialoglycoprotein called PrP 27-30. They contain no nucleic acid. PrP 27-30 has a mass of 27,000–30,000 daltons and is composed of 145 amino acids with glycosylation at or near amino acids 181 and 197. The carboxy terminus contains a phosphatidylinositol glycolipid whose components are ethanolamine, phosphate, myo-inositol and stearic acid. This protein polymerizes into rods possessing the ultrastructural and histochemical characteristics of amyloid. *Amyloid* is a generic term referring to any optically homogenous, waxy, translucent glycoprotein; it is deposited intercellularly and/or intracellularly in many human diseases such as:

☆ Alzheimer's disease

☆ Creutzfeldt-Jakob disease

☆ Down's syndrome

☆ Fatal familial insomnia

☆ Gerstmann-Straussler syndrome

☆ Kuru Leprosy

Replication

The prion is a product of a human gene, termed the PrP gene, found on chromosome 20. This gene contains two exons separated by a single intron. Exon I and Exon II are transcribed and the two RNAs ligated into a single mRNA (Figure 65a). This mRNA contains an open reading frame (ORF) or protein coding region

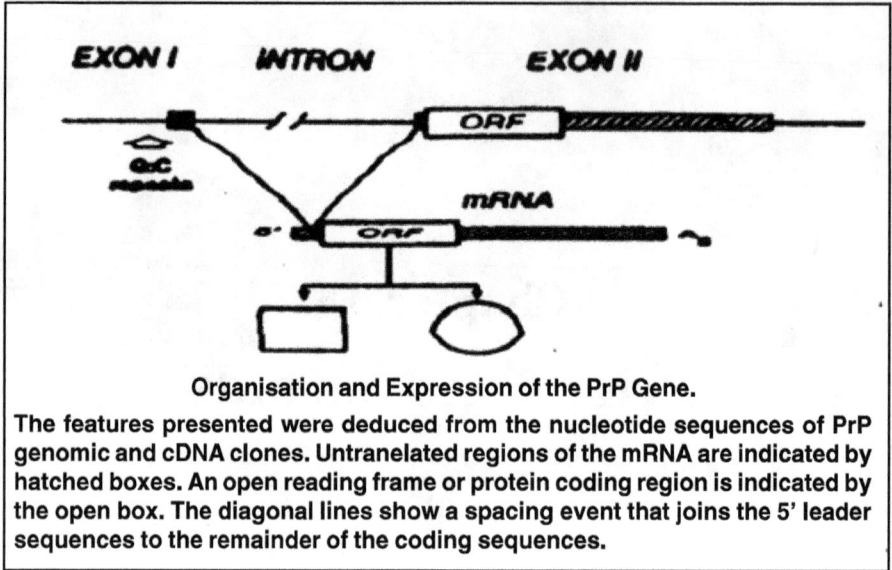

Organisation and Expression of the PrP Gene.

The features presented were deduced from the nucleotide sequences of PrP genomic and cDNA clones. Untranelated regions of the mRNA are indicated by hatched boxes. An open reading frame or protein coding region is indicated by the open box. The diagonal lines show a spacing event that joins the 5' leader sequences to the remainder of the coding sequences.

Figure 65a. Replication of Prions

which is translated into the PrP protein. The PrP protein is a precursor of the prion protein. It is termed PrP 33-35.

The PrP 33-35 undergoes several post-translational events to become the prion protein (PrP 27-30):

1. Glycosylation–at two sites. 2. Formation of a disulfide bond between two cysteine residues. 3. Removal of the N-terminal signal peptide. 4. Removal of the C-terminal hydrophobic segment. 5. Addition of a phosphatidylinositol glycolipid at the C-terminal. 6. Removal of the N-terminal first 57 amino acids.

In normal cells only the PrP 33-35 protein is synthesized. It is found in the neural cell membrane where it's function is to sequester Cu^{++} ions. In abnormal ("infected") cells, the PrP 27-30 is produced from the PrP 33-35 protein. The PrP 27-30 triggers a series of reactions that produce more PrP 27-30 proteins, *i.e.*, PrP 27-30 induces its own synthesis. In addition to the post translational modifications, the PrP 27-30 protein differs from the PrP 33-35 protein in a single amino acid residue. Residue 178 in the PrP 27-30 contains an asparagine residue whereas the PrP 33-35 protein has an aspartate residue at this position (Figure 65b).This causes a conformational change in the PrP 27-30 protein from an a-helix to a b-sheet. This conformational change in the PrP 27-30 protein has three effects:

1. It imparts to the PrP 27-30 protein the ability to induce the same a-helix to b-sheet conformation in the PrP 33-35 protein. This is a permanent conformational change. It thus induces its own "replication."
2. The b-sheet-forming peptides aggregate to form amyloid fibrils.
3. The amyloid fibrils kill thalamus neurons through apoptosis, a programmed series of events that leads to cell death.

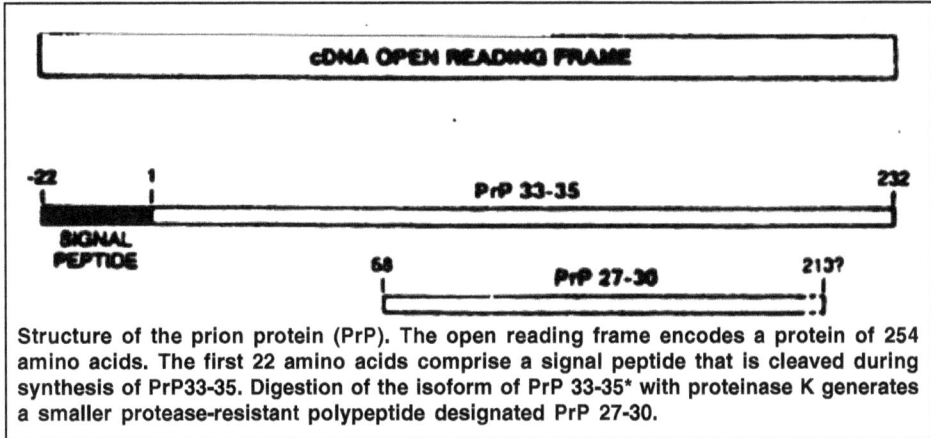

Structure of the prion protein (PrP). The open reading frame encodes a protein of 254 amino acids. The first 22 amino acids comprise a signal peptide that is cleaved during synthesis of PrP33-35. Digestion of the isoform of PrP 33-35* with proteinase K generates a smaller protease-resistant polypeptide designated PrP 27-30.

Figure 65b. Post Translational Modifications of Prions

Pathologies Induced by Prions

All diseases known to be of prion etiology, in animals and humans, are neurodegenerative diseases. In the human this includes:

Creutzfeldt-Jakob disease (CJD)

It is a neurodegenerative disease. Transmission through abrasions or cuts in the skin, via infected tissue, pituitary hormone and gonadotrophin acquired from cadavers with prion diseases and use of contaminated medical device. Symptoms of the CJD is Loss of motor nerve control, dementia, paralysis, pneumonia, muscle spasms and death in about 6 months to 2 years.

Fatal Familial Insomnia (FFI)

It is rare disease of prion in man. It is aslow virus disease. It is characterized by inflammatory lesions, amyloid plaque formation and atrophy of thalamous.

Gerstmann-Straussler- Scheinker Syndrome (GSSD)

These disease are transmitted through cuts in skin or tissues. Use of contaminated medical devise. Ingestion of contaminated food. This disease shows non inflammatory lesions with vacuoles and motor nerve problems.

Kuru

This is a neurological disease. It is reported in the tribes of New Guinea and has been linked with canabolistic behaviour. It is caused by a prion. It is a slow viral disease. This disease is transmitted by ingestion of brain tissue of dead realtives by tribes. Primary stage of the symptom is cerebellar Ataxia, unsteadiness in walking, tremor with change in face expression. Secondary stage of the symptom is patient cannot walk without support and the tertiary stage paralysis and death.

Transmission

Spread of the disease is via *horizontal transmission*, *i.e.*, transmission from one person to another, either directly or by fomites or by ingestion of contaminated meat.

Diagnosis

In the past, diagnosis of prion disease was made through examination of brain biopsies taken from patients in advanced stages of the disease or, more commonly, after they had died. In January of 1999 it was found that the prion protein accumulated in the tonsils and could be detected by an immunofluorescence test on tonsilar biopsies. A second test was simultaneously developed which was based on a Western blot. Later that year a third test was developed that had the high sensitivity necessary to detect the prion protein in blood. This test is based on capillary electrophoresis with laser-induced fluorescence. It detects as little as 10^{-18} mole.

Viroids

Viroids are naked circular single stranded RNAs that are not at all associated with a capsid or helper virus. They infect host cells and cause diseases in them as viruses do. They are also called virusoids.

Structure

Viroids are infectious agents composed exclusively of a single piece of circular single stranded RNA which has some double-stranded regions (Figure 66)

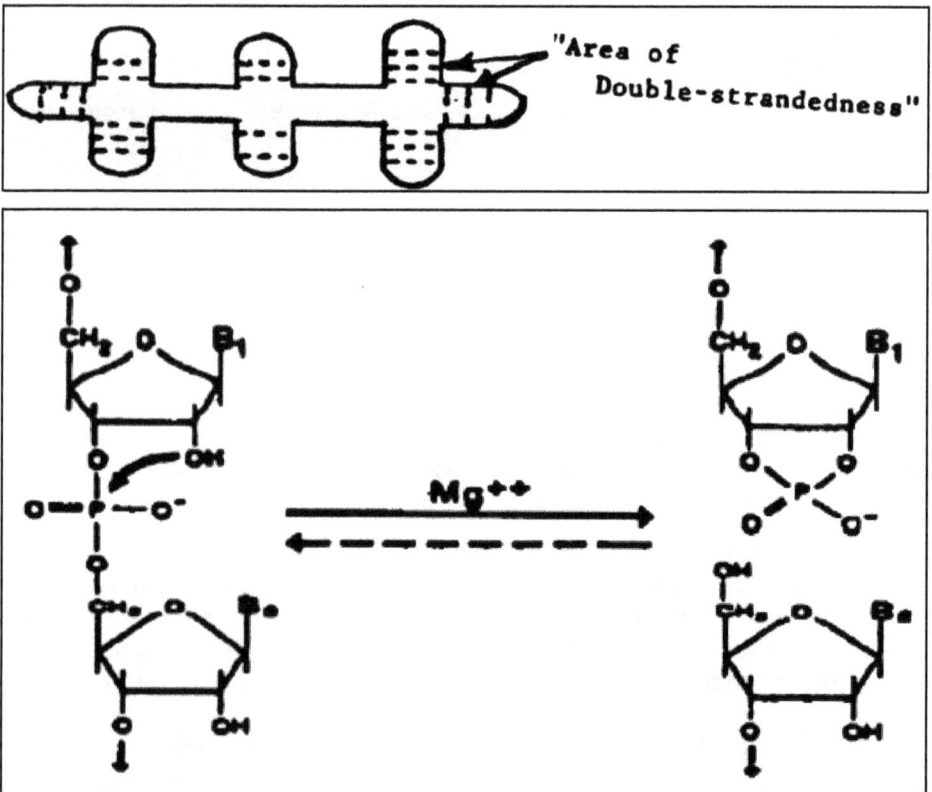

Figure 66. Structure of Viroids

Because of their simplified structures both prions and viroids are sometimes called *subviral particles*. Viroids mainly cause plant diseases but have recently been reported to cause a human disease.

Replication

Circular, pathogenic RNAs are replicated by a rolling circle mechanism *in vivo*. There are two variations of this rolling circle mechanism (Figure 67).

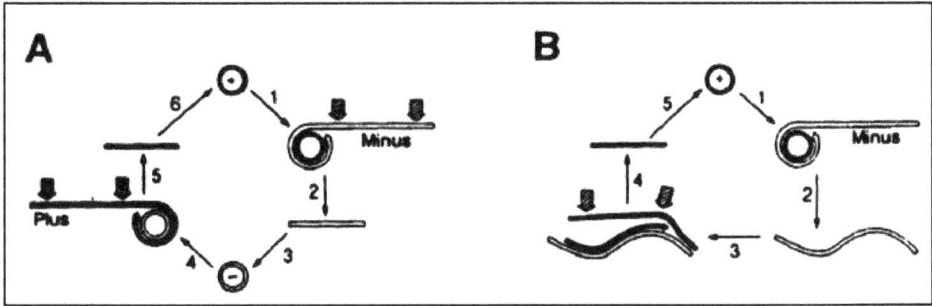

Figure 67. Replication of Prions

In the first variation (A), the circular plus strand is copied by viroid RNA-dependent RNA polymerase to form a concatameric minus strand (step 2). Site-specific cleavage (arrows) of this strand produces a monomer that is circularized by a host RNA ligase (step 3) and then copied by the RNA polymerase to produce a concatameric plus strand. Cleavage of this strand (step 5) produces monomers which, on circularization, produces the progeny circular, plus RNA, the dominant form *in vivo*.

In the other variation (B), the concatameric minus strand of step 1 is not cleaved but is copied directly to give a concatameric plus strand (step 3), which is cleared specifically to monomers for ligation to the circular progeny. Those RNAs that self-cleave only in the plus strand *in vitro* are considered to follow this route.

The hepatitis D viroid genome is a minus strand that gives rise to two RNA species. One of these is a mRNA for the delta antigen and the other is a complete complimentary copy (plus strand or anti-genome). The anti-genome acts as a template to make more minus strands. The minus strand self-cleaves and self-ligates. HDV replication takes place in the nucleus but delta antigen is made in the cytoplasm. The delta antigen is the only protein made by the HDV mRNA. It has a +12 charge at physiologic pH, accumulates in the nucleus and binds to minus strand RNA as a dimer. The delta antigen is necessary for viroid assembly but its exact mode of action is unknown.

Human Pathologies Induced by Viroids

The only human disease known to be caused by a viroid is hepatitis D. This disease was previously ascribed to a defective virus called the *delta agent*. However, it now is known that the delta agent is a viroid enclosed in a hepatitis B virus capsid. For hepatitis D to occur there must be simultaneous infection of a cell with both the hepatitis B virus and the hepatitis D viroid. There is extensive sequence complementarity between the hepatitis D viroid RNA and human liver cell 7S RNA, a small cytoplasmic RNA that is a component of the signal recognition particle, the structure involved in the translocation of secretory and membrane-associated particles. The hepatitis D viroid causes liver cell death via sequestering this 7S RNA and/or cleaving it.

Transmission

The hepatitis D viroid can only enter a human liver cell if it is enclosed in a capsid that contains a binding protein. It obtains this from the hepatitis B virus. The delta agent then enters the blood stream and can be transmitted via blood or serum transfusions.

Chapter 16
Virus Genomes

The Structure and Complexity of Virus Genomes

The composition and structure of virus genomes (*i.e.* the nucleic acid which encodes the genetic information of the virus) is more varied than any of those seen in the entire bacterial, plant or animal kingdoms. The nucleic acid comprising the genome may be single-stranded or double-stranded, and in a linear, circular or segmented configuration. Single-stranded virus genomes may be:

☆ Positive (+)sense, *i.e.* of the same polarity (nucleotide sequence) as mRNA

☆ Negative (-)sense

☆ Ambisense–a mixture of the two.

Virus genomes range in size from approximately 3,200 nucleotides (nt) (*e.g.* Hepadnaviruses) to approximately 1.2 million base pairs (Mbp, Mimivirus):

Unlike the genomes of *all* cells, which are composed of DNA, virus genomes may contain their genetic information encoded in *either DNA or RNA* (Figure 68).

Whatever the particular composition of a virus genome, all must conform to one condition. Since viruses are *obligate intracellular parasites* only able to replicate inside the appropriate host cells, the genome must contain information encoded in a form which can be recognised and decoded by the particular type of cell parasitized. Thus, the genetic code employed by the virus must match or at least be recognised by the host organism. Similarly, the control signals which direct the expression of virus genes must be appropriate to the host.

Many of the DNA viruses of eukaryotes closely resemble their host cells in terms of the biology of their genomes:

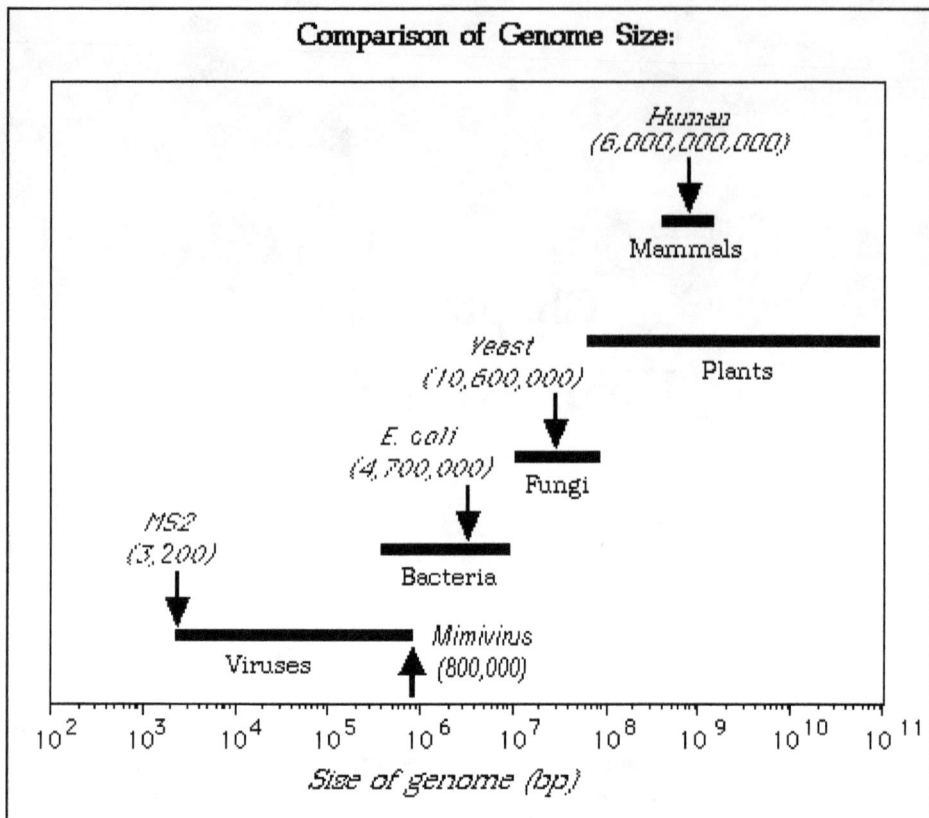

Figure 68. Comparision of Viral Genome Sizes

☆ Some DNA virus genomes are complexed with cellular histones to form a chromatin-like structure inside the virus particle.

☆ Vaccinia virus mRNAs were found to be polyadenylated at their 3' ends by Kates in 1970–the first observation of this phenomenon.

☆ Split genes containing non-coding introns, protein coding exons and spliced mRNAs were first discovered in adenoviruses by Sharp in 1977.

Molecular Genetics

As already described, the new techniques of molecular biology have had a major influence in concentrating much attention on the virus genome. Initially, the questions to be asked about any virus genome will usually include the following:

☆ Composition–DNA or RNA, single-stranded or double-stranded, linear or circular.

☆ Size and number of segments.

☆ Terminal structures.

☆ Nucleotide sequence.

☆ Coding capacity–open reading frames.

☆ Regulatory signals–transcription enhancers, promoters and terminators.

Direct analysis by *electron microscopy*, if calibrated with known standards, can be used to estimate the size of nucleic acid molecules.

The most important single technique has been *gel electrophoresis*. It is most common to use agarose gels to separate large nucleic acid molecules (several megabases or kilobases) and polyacrylamide gel electrophoresis (PAGE) to separate smaller pieces (a few hundred bp down to a few nucleotides).

Nucleotide sequencing is dependent on the ability to separate molecules which differ from each other by only one nucleotide in size

The relative simplicity of virus genomes (compared with even the simplest cell) offers a major advantage–the ability to 'rescue' infectious virus from purified or cloned nucleic acids. Infection of cells caused by nucleic acid alone is referred to as *transfection*:

Virus genomes which consist of (+)sense RNA (*i.e.* the same polarity as mRNA) are infectious when the purified vRNA is applied to cells in the absence of any virus proteins. This is because (+)sense vRNA is essentially mRNA and the first event in a normally-infected cell is to translate the vRNA to make the virus proteins responsible for genome replication. In this case, direct introduction of RNA into cells merely circumvents the earliest stages of the replicative cycle.

In most cases, virus genomes which are composed of double-stranded DNA are also infectious. The events which occur here are a little more complex, since the virus genome must first be transcribed by host polymerases to produce mRNA. Using these techniques, virus can be rescued from cloned genomes, including those which have been manipulated in vitro.

RNA Virus Genomes

Positive Strand RNA Viruses

The ultimate size of single-stranded RNA genomes is limited by the fragility of RNA and the tendency of long strands to break. In addition, RNA genomes tend to have higher mutation rates than those composed of DNA because they are copied less accurately, which also tends to drive RNA viruses towards smaller genomes.

Single-stranded RNA genomes vary in size from those of Coronaviruses at approximately 30kb long to those of bacteriophages such as MS2 and Qβ at about 3.5kb. Such genomes from different virus families share a number of common features:

☆ Purified (+)sense vRNA is directly *infectious* when applied to susceptible host cells in the absence of any virus proteins (although it is about one million times less infectious than virus particles).

☆ There is an *untranslated region (UTR)* at the 5' end of the genome which does not encode any proteins and a shorter UTR at the 3' end. These regions are functionally important in virus replication and are thus conserved in spite of the pressure to reduce genome size (Figure 69)

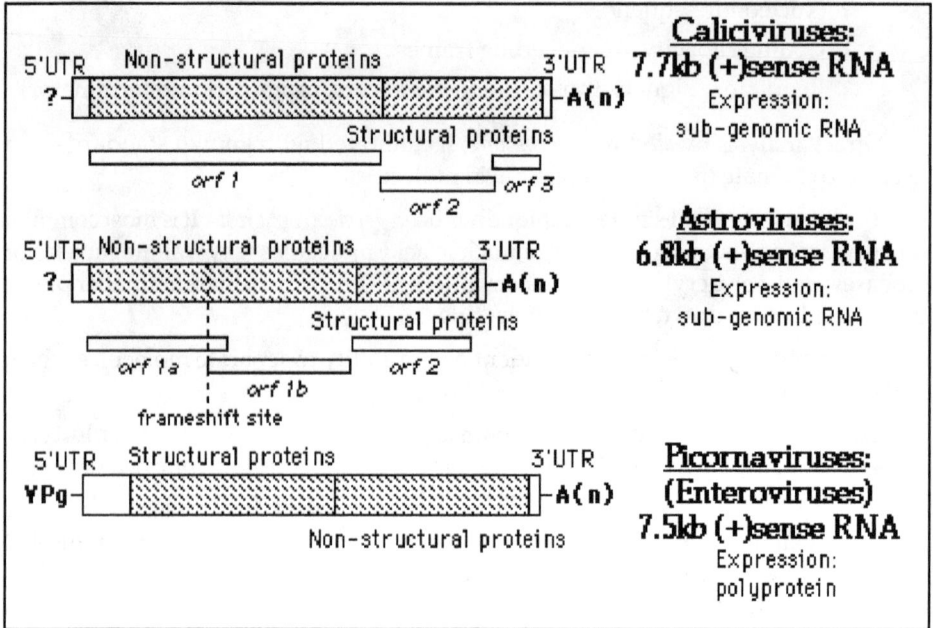

Figure 69. Positive Strand RNA Viruses

☆ Both ends of (+)stranded eukaryotic virus genomes are often *modified*, the 5' end by a small, covalently attached protein or a methylated nucleotide 'cap' structure and the 3' end by polyadenylation. These signals allow vRNA to be recognised by host cells and to function as mRNA.

Negative Strand RNA Viruses

Viruses with negative-sense RNA genomes are a little more diverse than positive-stranded viruses. Possibly because of the difficulties of expression, they tend to have larger genomes encoding more genetic information. Because of this, *segmentation* is a common though not universal feature of such viruses. (Figure 70).

Negative-sense RNA genomes are *not infectious* as purified RNA. This is because such virus particles all contain a *virus-specific polymerase*. The first event when the virus genome enters the cell is that the (-)sense genome is copied by the polymerase, forming either (+)sense transcripts which are used directly as mRNA, or a double-stranded molecule known either as the replicative intermediate (RI) or replicative form (RF), which serves as a template for further rounds of mRNA synthesis. Therefore, since purified negative-sense genomes cannot be directly translated and are not replicated in the absence of the virus polymerase, these genomes are inherently non-infectious.

Ambisense Genome Organization

Some RNA viruses are not strictly 'negative-sense' but ambisense, since they are part (-)sense and part (+)sense (Figure 71).

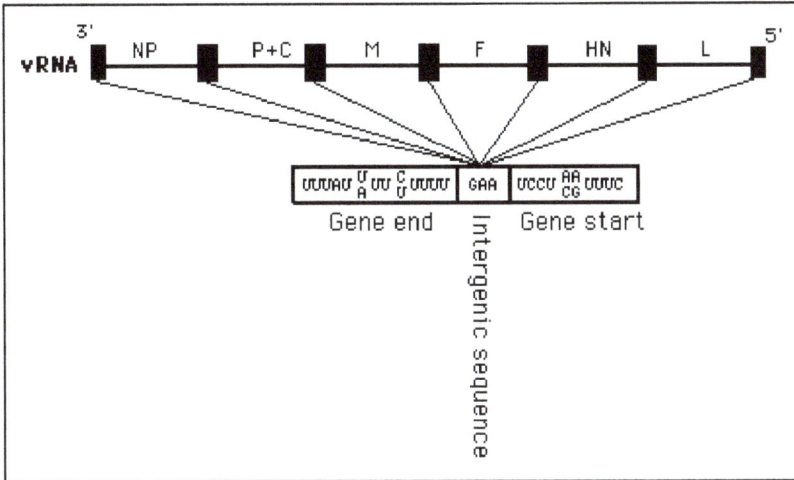

Figure 70. Negative-Strand RNA Viruses

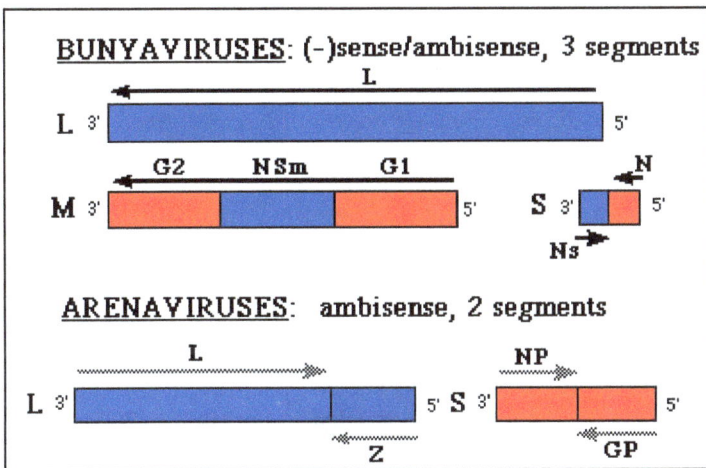

Figure 71. Ambisense Genome

DNA Virus Genomes

'Small' DNA Genomes

Bacteriophages have been extensively studied as examples of DNA virus genomes. Although they vary considerably in size, in general terms they tend to be relatively small.

The structure of the *bacteriophage M13* genome has been studied in great detail and modified extensively for use as a vector for DNA sequencing. The genome of this virus is:

☆ Circular

☆ Single-stranded DNA

☆ Approximately 7,200 nucleotides long.

Unlike other virion structures, the filamentous M13 capsid can be lengthened by the addition of further protein subunits. The genome size of this virus can also be increased by the addition of extra sequences in the non-essential intergenic region without the penalty of becoming incapable of being packaged into the capsid. This is very unusual. In other viruses, the packaging constraints are much more rigid, *e.g.* in *phage* λ, only DNA of between approximately 95 per cent–110 per cent (approximately 46kbp–54kbp) of the normal genome size (49kbp) can be packaged into the virus particle.

Not all bacteriophages have such simple genomes as M13, *e.g.* the genome of lambda is approximately 49kbp and that of phage T4 about 160kbp double-stranded DNA. These latter two bacteriophages also illustrate another common feature of linear virus genomes–the importance of the sequences present at the ends of the genome:

In the case of lambda, the substrate which is packaged into the phage heads during assembly consists of long *concatemers* of phage DNA which are produced during the later stages of vegetative replication. The DNA is 'reeled in' by the phage head and when a complete genome has been incorporated, cleaved at a specific sequence by a phage-encoded endonuclease. This enzymes leaves a 12bp 5' overhang on the end of each of the cleaved strands, known as the *cos* site. Hydrogen bond formation between these 'sticky ends' can result in the formation of a circular molecule. In a newly infected cell, the gaps on either side of the cos site are closed by DNA ligase and it is this circular DNA which is undergoes vegetative replication or integration into the bacterial chromosome.

Bacteriophage T4 molecular feature is certain linear virus genomes, *terminal redundancy*. Replication of the T4 genome also produces long concatemers of DNA. These are cleaved by a specific endonuclease, but unlike the lamda genome, the lengths of DNA incorporated into the particle are somewhat longer than a complete genome length. Therefore, some genes are repeated at each end of the genome, and the DNA packaged into the phage particles contains reiterated information.

As further examples of small DNA genomes, consider those of two families of animal viruses, the *parvoviruses* and *polyomaviruses* (Figure 72).

These are very small genomes, and even the replication-competent parvoviruses *contain only two genes*, rep, which encodes proteins involved in transcription and cap, which encodes the coat proteins. The ends of the genome have palindromic sequences of about 115nt, which form 'hairpins'. These structures are essential for the initiation of genome replication, once again emphasising the importance of the sequences at the ends of the genome.

The genomes of *polyomaviruses* consist of double-stranded, circular DNA molecules, approximately 5kbp in size (Figure 73).

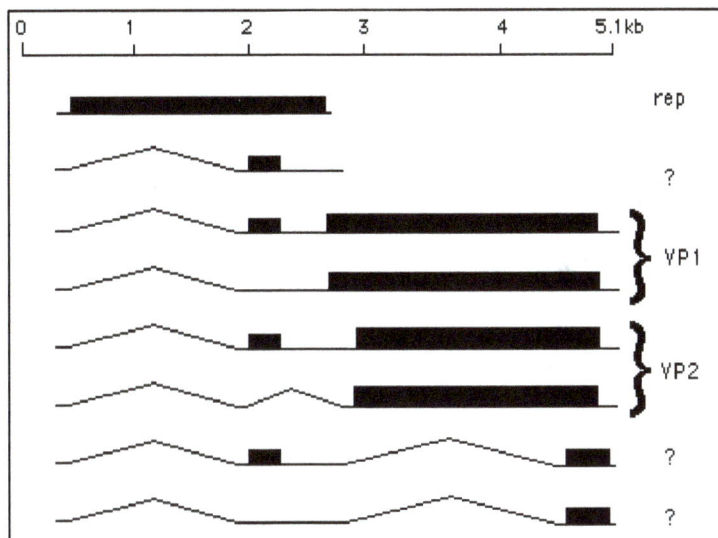

Figure 72. Parvovirus Genomes are
☆ Linear; ☆ Non-segmented; ☆ (+)sense; ☆ Single-stranded DNA; ☆ About 5kb long

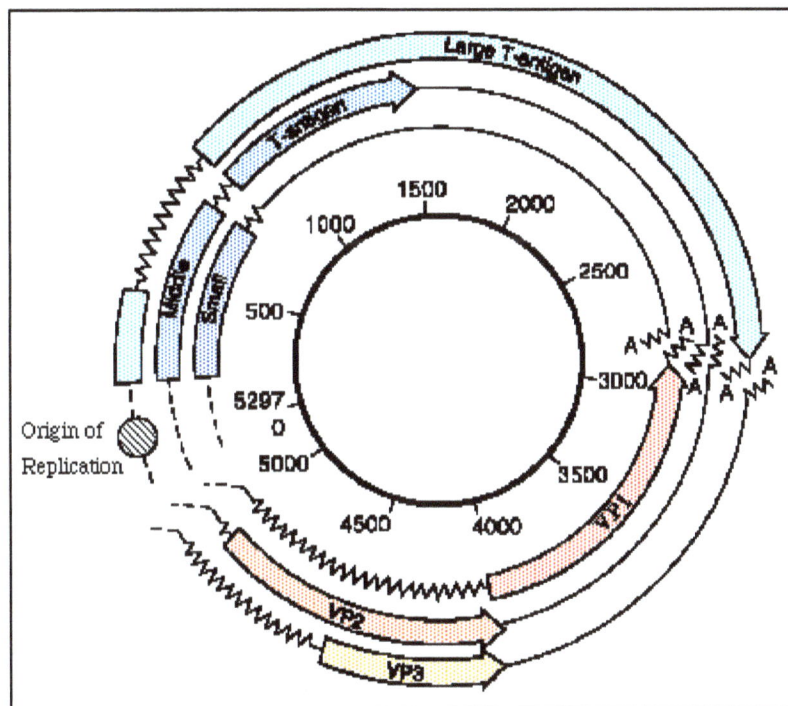

Figure 73. Genome of *Polyoma Virus*

The entire nucleotide sequence of all the viruses in the family is known and the architecture of the polyomavirus genome (*i.e.* number and arrangement of genes and function of the regulatory signals and systems) has been studied in great detail at a molecular level. Within the particles, the virus DNA is associated with four cellular histones. The genomic organization of these viruses has evolved to pack maximal information (6 genes) into minimal space (5kbp). This has been achieved by the use of both strands of the genome DNA and overlapping genes.

'Large' DNA Genomes

There are a number of virus groups which have double-stranded DNA genomes of considerable size and complexity. In many respects, these viruses are genetically very similar to the host cells which they infect. Two examples of such viruses are the *adenovirus* and *herpesvirus* families:

Herpesvirus Genomes

The herpesviruses are a large family containing more than 100 different members, at least one for most animal species which have been examined to date, including seven human herpesviruses.

Herpesviruses have very large genomes composed of *up to 230kbp* linear, double-stranded DNA. The different members of the family are widely separated in terms of genomic sequence and proteins, but all are similar in terms of structure and genome organization (Figure 74).

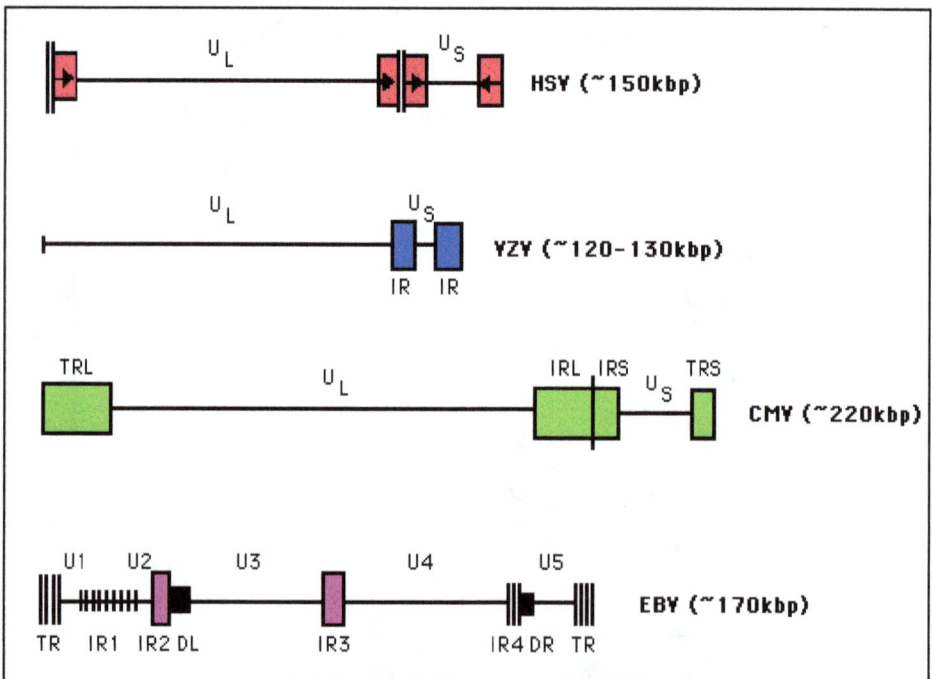

Figure 74. *Herpes Virus* Genome

Some herpesvirus genomes consist of two covalently joined sections, a unique long (UL) and a unique short (US) region, each bounded by inverted repeats. The repeats allow structural rearrangements of the unique regions and therefore, these genomes exist as a mixture of four isomers, all of which are functionally equivalent:

Herpesvirus genomes also contain multiple repeated sequences and depending on the number of these, the genome size of various isolates of a particular virus can vary by up to 10kbp.

Adenovirus Genomes

The genomes of adenoviruses consist of linear, double-stranded DNA of *30-38kbp*. These viruses contain 30-40 genes. The terminal sequences of each DNA strand are inverted repeats of 100-140bp and therefore, the denatured single strands can form 'panhandle' structures. These structures are important in DNA replication (Figure 75).

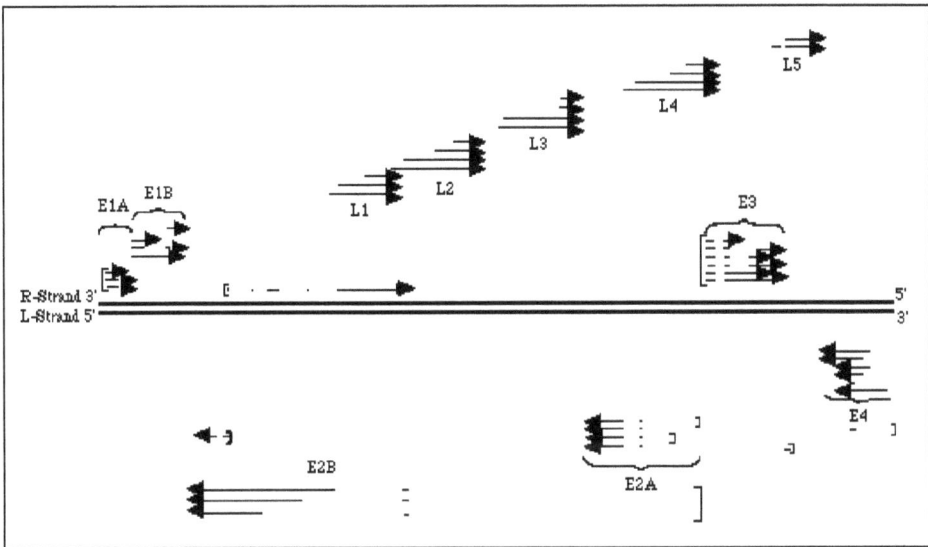

Figure 75. *Adenovirus* Genome

Although adenovirus genomes are considerably smaller than those of herpesviruses, the expression of the genetic information is rather more complex. Clusters of genes are expressed from a limited number of shared promoters. Multiply-spliced mRNAs and alternative splicing patterns are used to express a variety of polypeptides from each promoter. In contrast, herpesvirus genes each tend to be expressed from their own promoter–resulting in a much larger genome.

Segmented and Multipartite Virus Genomes (Figure 76)

☆ Segmented virus genomes are those which are divided into two or more physically separate molecules of nucleic acid, all of which are then packaged into a single virus particle.

☆ Multipartite genomes are those which are segmented and where each genome segment is packaged into a separate virus particle. These discrete

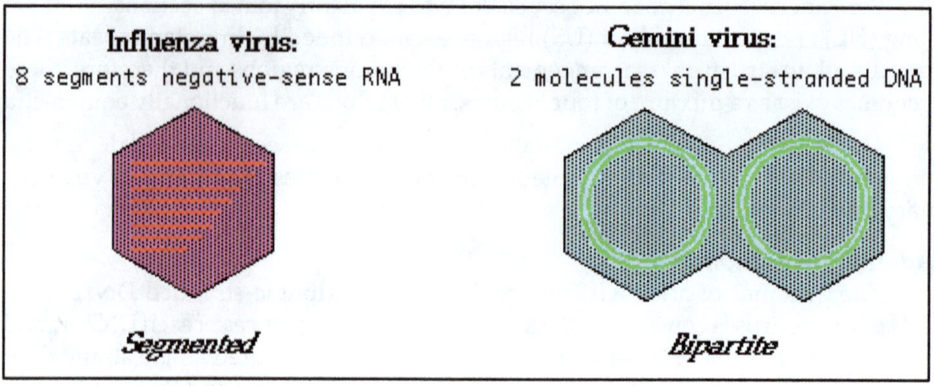

Figure 76. Segmented and Multipartite Virus Genomes

particles are structurally similar and may contain the same component proteins, but often differ in size depending on the length of the genome segment packaged.

There are many examples of segmented virus genomes, including many human, animal and plant pathogens such as orthomyxoviruses, reoviruses and bunyaviruses. There are rather fewer examples of multipartite viruses, all of which infect plants. These include:

☆ *Bipartite* viruses (which have two genome segments/virus particles)

☆ *Tripartite* viruses (three genome segments/virus particles)

Separating the genome segments into different particles removes the requirement for accurate sorting, but introduces a new problem in that all of the discrete virus particles must be taken up by a single host cell to establish a productive infection. This is perhaps the reason why multipartite viruses are only found in plants. Many of the sources of infection by plant viruses, such as inoculation by sap-sucking insects or after physical damage to tissues, results in a large input of infectious virus particles, providing the opportunity for infection of an initial cell by more than one particle.

Chapter 17
Vaccination

Introduction

1. B cells antigen receptors predominantly sees 3-Dimensional conformations.
2. T cells sees processed antigen in association with MHC molecules again as a 3-D conformation.
3. In an Antigen Presenting Cell, such as a macrophage, non-infectious proteins were endocytosed and degraded in lysosomes to peptides (endosomal pathway), some of which bound specifically to MHC II molecules. In contrast, if an agent such as a virus infected the macrophage, some newly synthesized viral antigens are likewise degraded in the cytoplasm to peptides (the cytoplasmic pathway), but these are associated with class I MHC molecules.

The Different Components of the Immune Response to Infection

Antibody has 3 important functions:

1. It is the only means to prevent an infection by neutralization of viral infectivity. The generation of protective Ab has usually been the only response measured by vaccine developers. Generally neutralizing Ab reacts with just a few epitopes on one or two surface antigens of the infecting agent. It is ineffective if the protective epitopes are subject to pronounced antigenic drift.
2. Infected cells which express viral antigen on their cell surface may be lysed by 2 antibody-dependent mechanisms–(a) complement pathway, or (b) antibody-dependent cell cytotoxicity.
3. Antibody may facilitate the removal of debris (a scavenging mechanism)

The main function of effector T cells, in particular Tc cells, is to clear an infection. Although antibody can contribute to recovery from infection, T cells are the main mechanism for achieving this. The most important feature of CMI, in direct contrast to Ab, is that it responds to many peptides from an agent. Some come from proteins that are not subject to antigenic variation. This broad T cell response is an important mechanism for overcoming antigenic variation of an agent and the genetic variability of the host population. Although such an agent may bypass the protection afforded by a preformed Ab after vaccination, the cell-mediated immune response allows most people to recover from an acute infection. Infectious agents causing chronic persistent infections have found a way of escaping a cell-mediated immune response. The mechanisms include:

1. Generation of cells that escape a cell-mediated immune response.
2. Down regulation of MHC production in infected cells so that they are not recognized and destroyed by T cells.
3. Infection of cells in immunoprivileged sites such as the brain.

The typical events that occurs after an acute virus infection *e.g.* murine influenza is as follows:

1. Virus replicates in the lungs, reaches maximum titres in 4–5 days,and decreases thereafter. About 12 days virus can no longer be recovered.
2. Effector Tc cell activity reaches a maximum activity after 6–8 days, becomes undetectable after 14 days. Tc memory cells reaches their highest level 2–6 weeks after infection and remain constant or decrease only slowly.
3. The number of Ab producing cells,producing first IgM then IgG and IgA peak at about 6 weeks and then steadily decline. Maximum B cell memory is found 10–15 weeks after infection but decreases thereafter but some are present at 18 months.

Both Ab-secreting and memory B cells are present after infection. In contrast Tc activity is generated only whilst infectious virus is present. Memory T cells persist but require further exposure to infectious virus for reactivation.

Requirements of a Vaccine

To be effective a vaccine should be capable of eliciting the following:

1. Activation of Antigen-Presenting Cells to initiate antigen processing and producing interleukins.
2. Activation of both T and B cells to give a high yield of memory cells.
3. Generation of Th and Tc cells to several epitopes, to overcome the variation in the immune response in the population due to MHC polymorphism.
4. Persistence of antigen, probably on dendritic follicular cells in lymphoid tissue, where B memory cells are recruited to form antibody-secreting cells that will continue to produce antibody.

Live vaccines fulfill these criteria par excellence. Neutralizing Abs are very important. Subunit vaccines induce poor immune responses and several doses with adjuvant are required to get an adequate response. The two main functions of the adjuvant antigen are to keep the antigen at or near the injection site and to activate antigen-presenting cells to achieve effective antigen processing and interleukin production. Vaccines composed simply of one T-cell epitope and one B-cell epitope is unlikely to be effective. This is exemplified by attempts to develop a malaria vaccine.

General Principles

The most successful immunization programs have been those directed against viral diseases such as smallpox, poliomyelitis and measles. Before embarking on any vaccination program, the need for a vaccine in a community must be evaluated. An epidemiological assessment of the incidence and severity of an infection will determine whether it is worth preventing. No vaccine is completely safe and potential benefits from immunization should be weighed against the risk of side effects. Two ingredients are needed for a successful immunization program.

1. A safe and effective vaccine
2. An appropriate strategy with adequate vaccine coverage

The strategy required depend on whether the aim of the program is eradication, elimination or containment. Eradication is the complete extinction of the organism in question. In elimination, the disease disappears but the organism remains. Containment is the control of the disease to the point at which it no longer constitutes a public health problem.

When eradication or elimination is the aim, mass immunization in early life of both sexes is usually necessary. In practice, it is very difficult to achieve 100 per cent coverage and the success of the program depends on the ability of the vaccine to interrupt transmission of the wild virus, thereby protecting the unvaccinated. When containment is the aim, selective immunization of those most at risk is normally sufficient. The usual indications for selective immunization are travel, occupational risk, outbreak control and for individuals at special risk of severe illness. Herd immunity plays little part, as the virus continues to circulate widely among the unvaccinated. In theory, selective immunization is less expensive than mass immunization. In practice, it is not always easy to identify and vaccinate those most at risk and mass immunization may be an easier option.

Vaccination Policy

1. Varies between different countries
2. Maternal antibodies present up to 6 months after birth, which may interfere with the induction of an effective immune response against the vaccine by the infant. This should be duly taken into account when formulating a vaccination policy.

WHO Expanded Program for Immunization (EPI)

1. Aimed at developing countries and initiated in 1974.

2. Aims to control and not eradicate 6 common disease through national health programs.

3. The six diseases are TB, DPT, polio and measles.

Different Types of Vaccine

Whole virus vaccines, either live or killed, constitute the vast majority of vaccines in use at present. However, recent advances in molecular biology had provided alternative methods for producing vaccines. Listed below are the possibilities:

1. Live whole virus vaccines
2. Killed whole virus vaccines
3. Subunit vaccines; purified or recombinant viral antigen
4. Recombinant virus vaccines
5. Anti-idiotype antibodies
6. DNA vaccines

1. Live Vaccines

Live virus vaccines are prepared from attenuated strains that are almost or completely devoid of pathogenicity but are capable of inducing a protective immune response. They multiply in the human host and provide continuous antigenic stimulation over a period of time, Primary vaccine failures are uncommon and are usually the result of inadequate storage or administration. Another possibility is interference by related viruses as is suspected in the case of oral polio vaccine in developing countries. Several methods have been used to attenuate viruses for vaccine production.

Use of a Related Virus from Another Animal

The earliest example was the use of cowpox to prevent smallpox. The origin of the vaccinia viruses used for production is uncertain.

Administration of Pathogenic or Partially Attenuated Virus by an Unnatural Route

The virulence of the virus is often reduced when administered by an unnatural route. This principle is used in the immunization of military recruits against adult respiratory distress syndrome using enterically coated live adenovirus type 4, 7 and (21).

Passage of the Virus in an "Unnatural Host" or Host Cell

The major vaccines used in man and animals have all been derived this way. After repeated passages, the virus is administered to the natural host. The initial passages are made in healthy animals or in primary cell cultures. There are several examples of this approach: the 17D strain of yellow fever was developed by passage in mice and then in chick embryos. Polioviruses were passaged in monkey kidney cells and measles in chick embryo fibroblasts. Human diploid cells are now widely used such as the WI-38 and MRC-5. The molecular basis for host range mutation is now beginning to be understood.

Development of Temperature Sensitive Mutants

This method may be used in conjunction with the above method.

2. Inactivated Whole Virus Vaccines

These were the easiest preparations to use. The preparation was simply inactivated. The outer virion coat should be left intact but the replicative function should be destroyed. To be effective, non-replicating virus vaccines must contain much more antigen than live vaccines that are able to replicate in the host. Preparation of killed vaccines may take the route of heat or chemicals. The chemicals used include formaldehyde or beta- propiolactone. The traditional agent for inactivation of the virus is formalin. Excessive treatment can destroy immunogenicity whereas insufficient treatment can leave infectious virus capable of causing disease. Soon after the introduction of inactivated polio vaccine, there was an outbreak of paralytic poliomyelitis in the USA use to the distribution of inadequately inactivated polio vaccine. This incident led to a review of the formalin inactivation procedure and other inactivating agents are now available, such as Beta-propiolactone. Another problem was that SV40 was occasionally found as a contaminant and there were fears of the potential oncogenic nature of the virus.

Potential Safety Problems

Live Vaccines

1. Underattenuation
2. Mutation leading to reversion to virulence
3. Preparation instability
4. Contaminating viruses in cultured cells
5. Heat lability
6. Should not be given to immunocompromized or pregnant patients

Killed Vaccines

1. Incomplete inactivation
2. Increased risk of allergic reactions due to large amounts of antigen involved

Present problems with vaccine development include:

1. Failure to grow large amounts of organisms in laboratory
2. Crude antigen preparations often give poor protection. *e.g.* Key antigen not identified, ignorance of the nature of the protective or the protective immune response.
3. Live vaccines of certain viruses can (1). induce reactivation, (2) be oncogenic in nature

3. Subunit Vaccines

Originally, non-replicating vaccines were derived from crude preparations of virus from animal tissues. As the technology for growing viruses to high titres in cell cultures advanced, it became practicable to purify virus and viral antigens. It is now possible to identify the peptide sites encompassing the major antigenic sites of viral antigens, from which highly purified subunit vaccines can be produced. Increasing purification may lead to loss of immunogenicity, and this may necessitate coupling to an immunogenic carrier protein or adjuvant, such as an aluminum salt. Examples of purified subunit vaccines include the HA vaccines for influenza A and B, and HBsAg derived from the plasma of carriers.

4. Recombinant Viral Proteins

Virus proteins have been expressed in bacteria, yeast, mammalian cells, and viruses. *E. Coli* cells were first to be used for this purpose but the expressed proteins were not glycosylated, which was a major drawback since many of the immunogenic proteins of viruses such as the envelope glycoproteins, were glycosylated. Nevertheless, in many instances, it was demonstrated that the non-glycosylated protein backbone was just as immunogenic. Recombinant hepatitis B vaccine is the only recombinant vaccine licensed at present.

An alternative application of recombinant DNA technology is the production of hybrid virus vaccines. The best known example is vaccinia; the DNA sequence coding for the foreign gene is inserted into the plasmid vector along with a vaccinia virus promoter and vaccinia thymidine kinase sequences. The resultant recombination vector is then introduced into cells infected with vaccinia virus to generate a virus that expresses the foreign gene. The recombinant virus vaccine can then multiply in infected cells and produce the antigens of a wide range of viruses. The genes of several viruses can be inserted, so the potential exists for producing polyvalent live vaccines. HBsAg, rabies, HSV and other viruses have been expressed in vaccinia.

Hybrid virus vaccines are stable and stimulate both cellular and humoral immunity. They are relatively cheap and simple to produce. Being live vaccines, smaller quantities are required for immunization. As yet, there are no accepted laboratory markers of attenuation or virulence of vaccinia virus for man. Alterations in the genome of vaccinia virus during the selection of recombinant may alter the virulence of the virus. The use of vaccinia also carries the risk of adverse reactions associated with the vaccine and the virus may spread to susceptible contacts. At present, efforts are being made to attenuate vaccinia virus further and the possibility of using other recombinant vectors is being explored, such as attenuated poliovirus and adenovirus.

5. Synthetic Peptides

The development of synthetic peptides that might be useful as vaccines depends on the identification of immunogenic sites. Several methods have been used. The best known example is foot and mouth disease, where protection was achieved by immunizing animals with a linear sequence of 20 aminoacids. Synthetic peptide vaccines would have many advantages. Their antigens are precisely defined and free from unnecessary components which may be associated with side effects. They are stable and relatively cheap to manufacture. Furthermore, less quality assurance is required. Changes due to natural variation of the virus can be readily accommodated, which would be a great advantage for unstable viruses such as influenza.

Synthetic peptides do not readily stimulate T cells. It was generally assumed that, because of their small size, peptides would behave like haptens and would therefore require coupling to a protein carrier which is recognized by T-cells. It is now known that synthetic peptides can be highly immunogenic in their free form provided they contain, in addition to the B cell epitope, T- cell epitopes recognized by T-helper cells. Such T-cell epitopes can be provided by carrier protein molecules, foreign antigens, or within the synthetic peptide molecule itself.

Synthetic peptides are not applicable to all viruses. This approach did not work in the case of polioviruses because the important antigenic sites were made up of 2 or more different viral capsid proteins so that it was in a concise 3-D conformation.

Advantages of Defined Viral Antigens or Peptides Include
1. Production and quality control simpler
2. No NA or other viral or external proteins, therefore less toxic.
3. Safer in cases where viruses are oncogenic or establish a persistent infection
4. Feasible even if virus cannot be cultivated

Disadvantages
1. May be less immunogenic than conventional inactivated whole-virus vaccines
2. Requires adjuvant
3. Requires primary course of injections followed by boosters
4. Fails to elicit CMI.

6. Anti-idiotype Antibodies

The ability of anti-idiotype antibodies to mimic foreign antigens has led to their development as vaccines to induce immunity against viruses, bacteria and protozoa in experimental animals. Anti-idiotypes have many potential uses as viral vaccines, particularly when the antigen is difficult to grow or hazardous. They have been used to induce immunity against a wide range of viruses, including HBV, rabies, Newcastle disease virus and FeLV, reoviruses and polioviruses.

7. DNA Vaccines

Recently, encouraging results were reported for DNA vaccines whereby DNA coding for the foreign antigen is directly injected into the animal so that the foreign antigen is directly produced by the host cells. In theory these vaccines would be extremely safe and devoid of side effects since the foreign antigens would be directly produced by the host animal. In addition, DNA is relatively inexpensive and easier to produce than conventional vaccines and thus this technology may one day increase the availability of vaccines to developing countries. Moreover, the time for development is relatively short which may enable timely immunization against emerging infectious diseases. In addition, DNA vaccines can theoretically result in more long-term production of an antigenic protein when introduced into a relatively nondividing tissue, such as muscle.

Indeed some observers have already dubbed the new technology the "third revolution" in vaccine development—on par with Pasteur's ground-breaking work with whole organisms and the development of subunit vaccines. The first clinical trials using injections of DNA to stimulate an immune response against a foreign protein began for HIV in 1995. Four other clinical trials using DNA vaccines against influenza, herpes simplex virus, T-cell lymphoma, and an additional trial for HIV were started in 1996.

The technique that is being tested in humans involves the direct injection of plasmids–loops of DNA that contain genes for proteins produced by the organism being targeted for immunity. Once injected into the host's muscle tissue, the DNA is taken up by host cells, which then start expressing the foreign protein. The protein serves as an antigen that stimulate an immune responses and protective immunological memory.

Enthusiasm for DNA vaccination in humans is tempered by the fact that delivery of the DNA to cells is still not optimal, particularly in larger animals. Another concern is the possibility, which exists with all gene therapy, that the vaccine's DNA will be integrated into host chromosomes and will turn on oncogenes or turn off tumor suppressor genes. Another potential downside is that extended immunostimulation by the foreign antigen could in theory provoke chronic inflammation or autoantibody production

Presentation of Immunogenic Proteins and Peptides

Proteins separated from virus particles are generally much less immunogenic than the intact particles. This difference in activity is usually attributed to the change in configuration of a protein when it is released from the structural requirements of the virus particle. Many attempts have been made to enhance the immunogenic activity of separated proteins.

Adjuvants

Certain substances, when administered simultaneously with a specific antigen, will enhance the immune response to that antigen. Such compounds are routinely included in inactivated or purified antigen vaccines.

Adjuvants in Common Use

1. Aluminium salts
    ☆ First safe and effective compound to be used in human vaccines.
    ☆ It promotes a good antibody response, but poor cell mediated immunity.
2. Liposomes and Immunostimulating complexes (ISCOMS)
3. Complete Freunds adjuvant is an emulsion of Mycobacteria, oil and water
    ☆ Too toxic for man
    ☆ Induces a good cell mediated immune response.
4. Incomplete Freund's adjuvant as above, but without Mycobacteria.
5. Muramyl di-peptide
    ☆ Derived from Mycobacterial cell wall.

6. Cytokines
   ☆ IL-2, IL-12 and Interferon-gamma.

Possible Modes of Action

☆ By trapping antigen in the tissues, thus allowing maximal exposure to dendritic cells and specific T and B lymphocytes.

☆ By activating antigen-presenting cells to secrete cytokines that enhance the recruitment of antigen-specific T and B cells to the site of inoculation.

DNA Vaccines

DNA vaccines are at present experimental, but hold promise for future therapy since they will evoke both humoral and cell-mediated immunity, without the dangers associated with live virus vaccines.

The gene for an antigenic determinant of a pathogenic organism is inserted into a plasmid. This genetically engineered plasmid comprises the DNA vaccine which is then injected into the host. Within the host cells, the foreign gene can be expressed (transcribed and translated) from the plasmid DNA, and if sufficient amounts of the foreign protein are produced, they will elicit an immune response.

Vaccines in General Use

Measles

Live attenuated virus grown in chick embryo fibroblasts, first introduced in the 1960's. Its extensive use has led to the virtual eradication of measles in the first world. In developed countries, the vaccine is administered to all children in the second year of life (at about 15 months). However, in developing countries, where measles is still widespread, children tend to become infected early (in the first year), which frequently results in severe disease. It is therefore important to administer the vaccine as early as possible (between six months and a year). If the vaccine is administered too early, however, there is a poor take rate due to the interference by maternal antibody. For this reason, when vaccine is administered before the age of one year, a booster dose is recommended at 15 months.

Mumps

Live attenuated virus developed in the 1960's. In first world countries it is administered together with measles and rubella at 15 months in the MMR vaccine. This is not a legal requirement in South Africa.

Rubella

Live attenuated virus. Rubella causes a mild febrile illness in children, but if infection occurs during pregnancy, the foetus may develop severe congenital abnormalities. Two vaccination policies have been adopted in the first world. In the USA, the vaccine is administered to all children in their second year of life (in an attempt to eradicate infection), while in Britain, until recently, only post pubertal girls were vaccinated. It was feared that if the prevalence of rubella in the community fell, then infection in the unimmunized might occur later–thus increasing the likelihood of infection occurring in the child-bearing years. This programme has since been abandoned in Britain and immunization of all children is the current practice.

Polio

Two highly effective vaccines containing all 3 strains of poliovirus are in general use:

   ☆ The *killed virus vaccine* (Salk, 1954) is used mainly in Sweden, Finland, Holland and Iceland.

   ☆ The *live attenuated oral polio vaccine* (Sabin, 1957) has been adopted in most parts of the world; its chief advantages being: low cost, the fact that it induces mucosal immunity and the possibility that, in poorly immunized communities, vaccine strains might replace circulating wild strains and improve herd immunity. Against this is the risk of reversion to virulence (especially of types 2 and 3) and the fact that the vaccine is sensitive to storage under adverse conditions.

   ☆ The inactivated Salk vaccine is recommended for children who are *immunosuppressed.*

Hepatitis B

Two vaccines are in current use: a serum derived vaccine and a recombinant vaccine. Both contain purified preparations of the hepatitis B surface protein.

The serum derived vaccine is prepared from hepatitis B surface protein, purified from the serum of hepatitis B carriers. This protein is synthesised in vast excess by infected hepatocytes and secreted into the blood of infected individuals. A vaccine trial performed on homosexual men in the USA has shown that, following three intra-muscular doses at 0, 1 and 6 months, the vaccine is at least 95 per cent protective.

A second vaccine, produced by recombinant DNA technology, has since become available. Previously, vaccine administration was restricted to individuals who were at high risk of exposure to hepatitis B, namely: infants of hepatitis B carrier mothers, health care workers, homosexual men and intravenous drug abusers. However, hepatitis B has been targetted for eradication, and since 1995 the vaccine has been included in the universal childhood immunization schedule. Three doses are given; at 6, 10, and 14 weeks of age. As with any killed viral vaccines, a booster will be required at some interval (not yet determined, but about 5 years) to provide protection in later life from hepatitis B infection as a venereal disease.

Hepatitis A

A vaccine for hepatitis A has been developed from formalin-inactivated, cell culture-derived virus. Two doses, administered one month apart, appear to induce high levels of neutralising antibodies. The vaccine is recommended for travellers to third world countries, and indeed all adults who are not immune to hepatitis A.

Yellow Fever

The 17D strain is a live attenuated vaccine developed in 1937. It is a highly effective vaccine which is administered to residents in the tropics and travellers to endemic areas. A single dose induces protective immunity to travellers and booster doses, every 10 years, are recommended for residents in endemic areas.

Rabies

No safe attenuated strain of rabies virus has yet been developed for humans. Vaccines in current use include:

☆ The neurotissue vaccine–here the virus is grown in the spinal cords of rabbits, and then inactivated with *beta*-propiolactone. There is a high incidence of neurological complications following administration of this vaccine due to a hypersensitivity reaction to the myelin in the preparation and largely it has been replaced by

☆ A human diploid cell culture-derived vaccine (also *inactivated*) which is much safer.

There are two situations where vaccine is given:

(a) *Post-exposure prophylaxis,* following the bite of a rabid animal: A course of 5-6 intramuscular injections, starting on the day of exposure. Hyperimmune rabies globulin may also administered on the day of exposure.

(b) *Pre-exposure prophylaxis* is used for protection of those whose occupation puts them at risk of infection with rabies; for example, vets, abbatoir and laboratory workers. This schedule is 2 doses one month apart, and a booster dose one year later. (Further boosters every 2-3 years should be given if risk of exposure continues).

Influenza

Repeated infections with influenza virus are common due to rapid antigenic variation of the viral envelope glycoproteins. Antibodies to the viral neuraminidase and haemagglutinin proteins protect the host from infection. However, because of the rapid antigenic variation, *new vaccines*, containing antigens derived from influenza strains currently circulating in the community, are produced *every year*.

Surveillance of influenza strains now allows the inclusion of appropriate antigens for each season. The vaccines consist of partially purified envelope proteins of inactivated current influenza A and B strains.

Individuals who are at risk of developing severe, life threatening disease if infected with influenza should receive vaccine. People at risk include the elderly, immunocompromised individuals, and patients with cardiac disease. In these patients, protection from disease is only partial, but the severity of infection is reduced.

Varicella-Zoster Virus

A live attenuated strain of varicella zoster virus has been developed. It is not licensed in South Africa for general use, but is used in some oncology units to protect immuno-compromised children who have not been exposed to wild-type varicella zoster virus. Such patients may develop severe, life threatening infections if infected with the wild type virus.

Chapter 18

Diagnostic Methods and Serology in Virology

Introduction

In general, diagnostic tests can be grouped into 3 categories.: (1) direct detection, (2) indirect examination (virus isolation), and (3) serology. In direct examination, the clinical specimen is examined directly for the presence of virus particles, virus antigen or viral nucleic acids. In indirect examination, the specimen into cell culture, eggs or animals in an attempt to grow the virus: this is called virus isolation. Serology actually constitute by far the bulk of the work of any virology laboratory. A serological diagnosis can be made by the detection of rising titres of antibody between acute and convalescent stages of infection, or the detection of IgM. In general, the majority of common viral infections can be diagnosed by serology. The specimen used for direction detection and virus isolation is very important. A positive result from the site of disease would be of much greater diagnostic significance than those from other sites. For example, in the case of herpes simplex encephalitis, a positive result from the CSF or the brain would be much greater significance than a positive result from an oral ulcer, since reactivation of oral herpes is common during times of stress.

1. Direct Examination of Specimen

1. Electron Microscopy morphology/immune electron microscopy
2. Light microscopy histological appearance–*e.g.* inclusion bodies
3. Antigen detection immunofluorescence, ELISA etc.
4. Molecular techniques for the direct detection of viral genomes

2. Indirect Examination

1. Cell Culture–cytopathic effect, haemadsorption, confirmation by neutralization, interference, immunofluorescence etc.
2. Eggs pocks on CAM–haemagglutination, inclusion bodies
3. Animals disease or death confirmation by neutralization

3. Serology

Detection of rising titres of antibody between acute and convalescent stages of infection, or the detection of IgM in primary infection.

4. Molecular Methods

Molecular methods detect viruses which cannot be cultured in *vitro*. Detection of viral Nucleic acid is the main base for molecular sequence.

1. Direct Examination

Direct examination methods are often also called rapid diagnostic methods because they can usually give a result either within the same or the next day. This is extremely useful in cases when the clinical management of the patient depends greatly on the rapid availability of laboratory results *e.g.* diagnosis of RSV infection in neonates, or severe CMV infections in immunocompromised patients. However, it is important to realize that not all direct examination methods are rapid, and conversely, virus isolation and serological methods may sometimes give a rapid result. With the advent of effective antiviral chemotherapy, rapid diagnostic methods are expected to play an increasingly important role in the diagnosis of viral infections.

1.1. Antigen Detection

Examples of antigen detection include immunofluorescence testing of nasopharyngeal aspirates for respiratory viruses *e.g.*. RSV, flu A, flu B, and adenoviruses, detection of rotavirus antigen in faeces, the pp65 CMV antigenaemia test, the detection of HSV and VZV in skin scrappings, and the detection of HBsAg in serum. (However, the latter is usually considered as a serological test). The main advantage of these assays is that they are rapid to perform with the result being available within a few hours. However, the technique is often tedious and time consuming, the result difficult to read and interpret, and the sensitivity and specificity poor. The quality of the specimen obtained is of utmost importance in order for the test to work properly.

1.2. Electron Microscopy (EM)

Virus particles are detected and identified on the basis of morphology. A magnification of around 50,000 is normally used. EM is now mainly used for the diagnosis of viral gastroenteritis by detecting viruses in faeces *e.g.* rotavirus, adenovirus, astrovirus, calicivirus and Norwalk-like viruses. Occasionally it may be used for the detection of viruses in vesicles and other skin lesions, such as herpesviruses and papillomaviruses. The sensitivity and specificity of EM may be enhanced by immune electron microscopy, whereby virus specific antibody is used to agglutinate virus particles together and thus making them easier to recognize, or to capture virus particles onto the EM grid. The main problem with EM is the expense involved in purchasing and maintaining the facility. In addition, the sensitivity of EM is often poor, with at least 10^5 to 10^6 virus particles per ml in the sample required for visualisation. Therefore the observer must be highly skilled. With the availability of reliable antigen detection and molecular methods for the detection of viruses associated with viral gastroenteritis, EM is becoming less and less widely used (Figure 77).

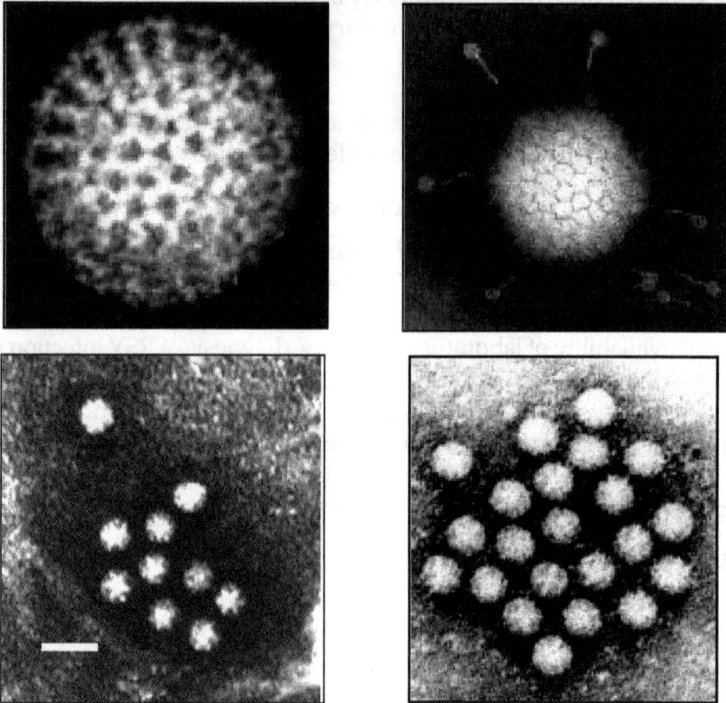

Figure 77. Electronmicrographs of Viruses Commonly Found in Stool Specimens from Patients Suffering from Gastroenteritis.
From left to right: rotavirus, adenovirus, astroviruses, Norwalk-like viruses.
(Courtesy of Linda M. Stannard, University of Cape Town,

1.3. Light Microscopy

Replicating virus often produce histological changes in infected cells. These changes may be characteristic or non-specific. Viral inclusion bodies are basically collections of replicating virus particles either in the nucleus or cytoplasm. Examples of inclusion bodies include the negri bodies and cytomegalic inclusion bodies found in rabies and CMV infections respectively. Although not sensitive or specific, histology nevertheless serves as a useful adjunct in the diagnosis of certain viral infections.

1.4. Viral Genome Detection

Methods based on the detection of viral genome are also commonly known as molecular methods. It is often said that molecular methods is the future direction of viral diagnosis. However in practice, although the use of these methods is indeed increasing, the role played by molecular methods in a routine diagnostic virus laboratory is still small compared to conventional methods. It is certain though that the role of molecular methods will increase rapidly in the near future. Classical molecular techniques such as dot-blot and Southern-blot depend on the use of specific DNA/RNA probes for hybridization. The specificity of the reaction depends on the conditions used for hybridization. These techniques may allow for the quantification of DNA/RNA present in the specimen. However, it is often found that the sensitivity of these techniques is not better than conventional viral diagnostic methods.

Newer molecular techniques such as the polymerase chain reaction (PCR), ligase chain reaction (LCR), nucleic acid based amplification (NASBA), and branched DNA (bDNA) depend on some form of amplification, either the target nucleic acid, or the signal itself. bDNA is essentially a conventional hybridization technique with increased sensitivity. However, it is not as sensitive as PCR and other amplification techniques. PCR is the only amplification technique which is in common use. PCR is an extremely sensitive technique: it is possible to achieve a sensitivity of down to 1 DNA molecule in a clinical specimen. However, PCR has many problems, the chief among which is contamination, since only a minute amount of contamination is needed to give a false positive result. In addition, because PCR is so sensitive compared to other techniques, a positive PCR result is often very difficult to interpret as it does not necessarily indicate the presence of disease. This problem is particular great in the case of latent viruses such as CMV, since latent CMV genomes may be amplified from the blood of healthy individuals. Despite all this, PCR is being increasingly used for viral diagnosis, especially as the cost of the assay come down and the availability of closed automated systems that could also perform quantification (Quantitative PCR) *e.g.* real-time PCR and Cobas Amplicor systems. Other amplification techniques such as LCR and NASBA are just as susceptible to contamination as PCR but that is ameliorated to a great extent by the use of propriatory closed systems. It is unlikely though that other amplification techniques will challenge the dominance of PCR since it is much easier to set up an house PCR assay than other assays.

2. Indirect Examination

Cultivation of Viruses

Viruses are obligate intracellular parasites. They multiply only inside the host cells. Animals, plants, human, bacteria, fungus, protozoa and algae are the natural hosts of viruses are host specific and grow only in selective hosts. Hence, virologists use only a suitable host system for cultivation of a virus. Viruses cannot be grown in artificial media in the laboratory.

Cultivation of Animal Viruses

Since animal viruses can multiply only in animal cells. It is hardly possible to grow them in chemically defined media; they can only be grown in relevant cells or tissues or animals. Isolation and culture of viruses from infected materials seem to be an indirect method in the clinical diagnois of infectious viral diseases.

Viruses infect only selective tissues and cause diseases them. So a proper sample is needed to isolate the virus for clinical diagnosis and culture. Samples can be collected from blood, CSF, stools, urine, body tissue, skin lesions, nasal discharges and amniotic fluid.

Animal viruses can be propagated in the laboratory using three methods. They are:

1. Animal inoculation
2. Embryonated egg
3. Cell/tissue cultures

1. Animal Inoculation

The cultivation of animal viruses in laboratory animals is called animal inoculation. It was the earliest method for the cultivation of animal viruses practised as early as 1900. Mice, rabbit, guinea pig, suckling mice, hamsters, monkeys, etc. are useful to grow animal viruses. Human volunteers may also be used for virus inoculation only when no other method is found suitable to grow the virus and the chosen virus is relatively harmless. Among these animals, mice is most generally employed in the cultivation of animal viruses. Theiler first cultivated animal viruses in mice in 1903.

Animal inoculation involves the following steps:

☆ Young mice free from viral infections are selected for animal inoculation.

☆ Virus to be grown is inoculated in the mice through *nasal instillation* or *intracerebral inoculation* or *intraperitoneal inoculation* or *subcutaneous inoculation*. The inoculation method is depending on the type of virus sample and its target site of infection. A sterile syringe may be used to inoculate the virus in the mice.

☆ After inoculation, the mice are reared in the laboratory with enough hygiene.

☆ Presence of virus in the inoculated mice is determined by examining the animal for visual symptoms of disease caused by the virus.

Animal viruses are cultivated in animals to understand the *viral pathogenesis, immune response, efficiency of vaccines and study the oncogeny.*

2. Embryonated Eggs

The cultivation of animal viruses in embryonated eggs in the laboratory is called embryonated egg method. Woodruff and Good Pasteur (1931) first used embryonated chicken egg for viral cultivation. This is a simpler technique than aminal inoculation, inexpensive and easily available. Usually eggs will not interfere with virus multiplication due to the absence of immune response. Suitable cells for the growth of viruses are available in embryo and its membrane, which may facilitate the growth of viruses.

A 8 to 11 days old chick embryo is used for the cultivation of virus. The egg is incubated for 2 to 9 days after inoculation. Duration of incubation depends on the type of virus and the route of inoculation.

Embryonated egg method involves the following steps:

☆ Hen's egg kept in incubator for 8-11 days is taken for virus cultivation

☆ Egg shell is disinfected with iodine.

☆ A hole is made in the egg shell with the help of a drill.

☆ Virus suspension or a piece of virus infected tissue is injected into the egg by using a sterile syringe. Viruses grow only in appropriate regions of the egg so that they are inoculated into a proper site. Therefore, the virus to be cultivated may be inoculated into *chorio allantoic membrane* or *amniotic fluid* or *allantoic cavity* or *yolk sac* depending upon the virus (Ex. Amniotic and Allantoic inoculation–*Influenza virus, Mumps virus*, Chrio Allantoic Membrane inoculation–*Herpes simplex virus* and *Pox virus*, Yolk sac inoculation–*Herpes simplex virus*) (Figure 78).

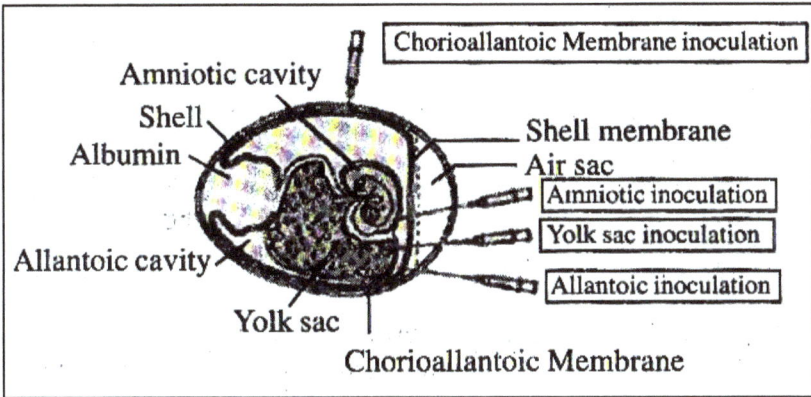

Figure 78. Virus Inoculation in an Embryonated Egg

☆ After inoculation, the hole in the egg shell is sealed with gelatin.

☆ The egg is kept in an incubator at hatching temperature.

☆ Death of embryo or formation of pocks or lesion on egg membrane is the indication for virus growth in the egg. *Pox virus* produces *Pock*.

☆ After confirmation of virus growth, the virus is identified by performing serological tests such as haemagglutination test. For instance, *influenza virus* can be identified by haemagglutination test.

Embryonated egg method is very useful to get viruses the preparation of vaccines for influenza, yellow fever and rabies.

3. Cell/Tisse Cultures

The cultivation of a tissue or cells in a chemically defined medium in vitro is called cell/tissue culture method. Now-a-days cell cultures are employed in cultivation or isolation of many animal viruses.

Cell Culture

The basic component of cell culture medium is a *Balanced Salt Solution* (BSS). It provides essential inorganic ions, correct osmolarity and correct pH. Two common *BSS are Hanks and Eagles Balanced Salt Solutions.* Some additional nutrients are also required for the proliferation of cells. They are 13 essential amino acids, eight vitamins, antibiotics to prevent bacterial contamination, glucose, phenol red and 5 per cent Foetal Calf Serum.

The cell and tissue culture method for virus cultivation includes three essential steps:

1. Culture of animal cells/tissues
2. Inoculation of virus
3. Detection of virus growth

1. Culture of Animal Cells/Tissues

Three tissue culture methods are employed in the virus cultivation. They are:

(*a*) Explant culture
(*b*) Organ culture
(*c*) Cell culture

(a) Explant Culture

A small portion of tissue excised from animal's body is called explant. The explants taken from laboratory animals such as mice, rabbit, guinea pigs, hamster and man can be grown in petri dishes. An explant is aseptically transferred into a sterile petri dish by using a fine tipped forceps and then a coverslip is placed over that explant. Enough volume of medium is poured into the peiri dish, which is then incubated at 37°C until cell growth appears.

(b) Organ Culture

The culture of a portion of a tissue taken from an organ is called organ culture. It produces a particular cell type in the culture. Organ cultures are particularly very useful to grow viruses which are specific to certain organs. For example, tracheal ring organ culture is used to isolate and grow *Corona virus* that infects respiratory system. The general method of organ culture is similar to the explant culture.

(c) Cell Culture

Animal tissue cultures established from individual cells are called cell cultures. Growth of cells dissociated from the parent tissue by spontaneous migration or mechanical processes or enzymatic process is called cell culture. Cell cultures are the sole system for virus isolation. To prepare cell cultures, tissue fragments are first dissociated with the help of trypsin. The cells are placed in a flat bottomed plastic container together with Eagles medium. After sometime, cells will attach to the bottom of the container and start dividing, giving rise to primary cultures. It is maintained by changing the medium 2 or 3 times per week. When the cells become crowded, the cells are detached from the vessel's wall by trypsin and a portions are used to initiate secondary cultures. Cells tend to adhere to the glass or plastic container and are reproduced to form a monolayer.

Animal cell cultures can be classified into the following types are:

(*a*) Primary cell culture

(*b*) Secondary cell culture

(*c*) Continuous cell lines

(a) Primary Cell Culture

The cell culture established directly from cells taken from animal's tissue is called primary culture. It is capable of only limited growth and hence it can be subcultured once or twice. Kidney cells of monkey and man, chick embryo cells, endothelial cells, hepatocytes, smooth muscle cells, alveolar cells, macrophages and amniotic cells are usually grown in primary cell cultures.

A small volume of cell suspension is aseptically transferred to a culture flask or petri dish containing nutrient medium, with the help of a pipette. The culture vessel is ticubated at 37°C for a few days to get a primary culture.

Primary cell cultures are widely used for the isolation of animal viruses and cultivation of viruses for vaccine production.

(b) Secondary Cell Culture

The cell cultures established from primary cell culture are called secondary cultures or subcultures. As secondary cell cultures can be maintained and subcultured for 20-50 times, they are called semi-continuous cells. Egs. *Human embryonic kidney cells* and *Skin fibroblast cells.*

Monolayer produced as a result of primary cell culture is detached from the bottom of the culture flask by adding trypsin or EDTA. It is then cut into small fragments. 2 or 3 fragments are inoculated into a roller drum containing nutrient medium and the roller drum is incubated at 37°C for a few days. The fragments of monoculture grow into large monolayers. These are called *secondary cultures.* Secondary cell cultures are used for the isolation of a wide group of animal viruses and growing *fastidious viruses.* Some secondary cultures are used for vaccine production.

(c) Continuous Cell Lines

Animal cells capable of indefinite growth are called continuous cell lines or cell lines. They are established from the secondary cell cultures, but they differ from the latter in morphology, higher growth rate, physiology and indefinite growth. The cell lines are generally derived from tumour cells of human tissues. Viruses grown in continuous cell lines have not been employed for vaccine production. *E.g.* HeLa, Vero, Hep2, LLC-MK2, BGM. These are immortalized cells *i.e.* tumour cell lines and may be passaged indefinitely.

Identification of Growing Virus

The presence of growing virus is usually detected by:

Cytopathic Effect (CPE)–may be specific or non-specific *e.g.* HSV and CMV produces a specific CPE, whereas enteroviruses do not (Figure 79a and b)

Haemadsorption–cells acquire the ability to stick to mammalian red blood cells. Haemadsorption is mainly used for the detection of influenza and parainfluenza viruses.

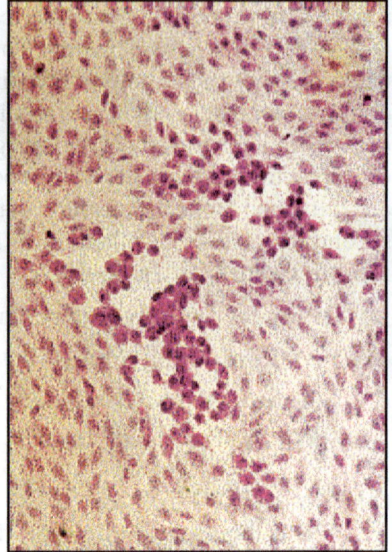

Figure 79a. Cytopathic Effect of HSV, Enterovirus

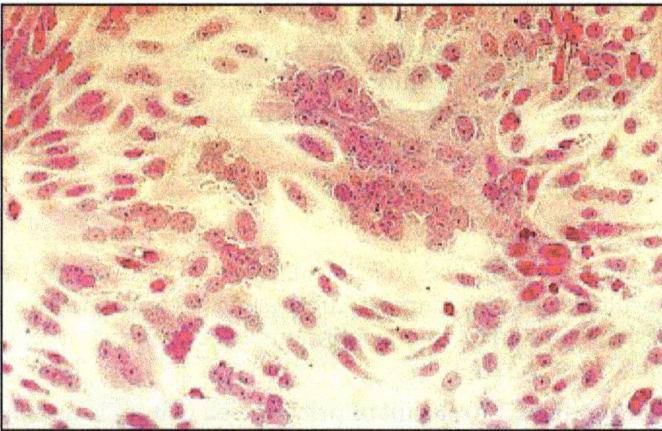

Figure 79b. Cytopathic Effect RSV in Cell Culture

Confirmation of the identity of the virus may be carried out using neutralization, haemadsorption- inhibition, immunofluorescence, or molecular tests.

Problems with Cell Culture

The main problem with cell culture is the long period (up to 4 weeks) required for a result to be available. Also, the sensitivity is often poor and depends on many factors, such as the condition of the specimen, and the condition of the cell sheet. Cell cultures are also very susceptible to bacterial contamination and toxic substances in the specimen. Lastly, many viruses will not grow in cell culture at all *e.g. Hepatitis B* and *C, Diarrhoeal viruses, parvovirus etc.*

Rapid Culture Techniques

Rapid culture techniques are available whereby viral antigens are detected 2 to 4 days after inoculation. Examples of rapid culture techniques include shell vial cultures and the CMV DEAFF test. In the CMV DEAFF test, the cell sheet is grown on individual cover slips in a plastic bottle. After inoculation, the bottle then is spun at a low speed for one hour (to speed up the adsorption of the virus) and then incubated for 2 to 4 days. The cover slip is then taken out and examined for the presence of CMV early antigens by immuno-fluorescence (Figure 80).

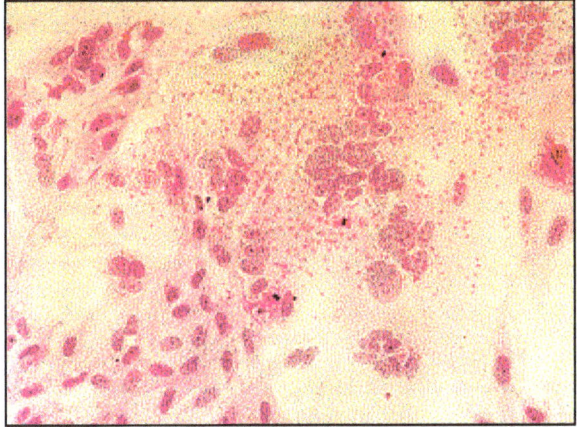

Figure 80. Haemadsorption of Red Blood Cells onto the Surface of a Cell Sheet Infected by Mumps Virus. Also note the presence of syncytia which is indistinguishable from that of RSV (Courtesy of Linda Stannard, University of Cape Town).

The role of cell culture (both conventional and rapid techniques) in the diagnosis of viral infections is being increasingly challenged by rapid diagnostic methods *i.e.* antigen detection and molecular methods. Therefore, the role of cell culture is expected to decline in future and is likely to be restricted to large central laboratories.

3. Serology

Serology forms the mainstay of viral diagnosis. This is what happens in a primary humoral immune response to antigen. Following exposure, the first antibody to appear is IgM, which is followed by a much higher titre of IgG. In cases of reinfection, the level of specific IgM either remain the same or rises slightly. But IgG shoots up rapidly and far more earlier than in a primary infection. Many different types of serological tests are available. With some assays such as EIA and RIA, one can look specifically for IgM or IgG, whereas with other assays such as CFT and HAI, one can only detect total antibody, which comprises mainly IgG. Some of these tests are much more sensitive than others: EIAs and radioimmunoassays are the most sensitive tests available, whereas CFT and HAI tests are not so sensitive. Newer techniques such as EIAs offer better sensitivity, specificity and reproducibility than classical techniques such as CFT and HAI. The sensitivity and specificity of the assays depend greatly on the antigen used. Assays that use recombinant protein or synthetic peptide antigens tend to be more specific than those using whole or disrupted virus particles.

Criteria for Diagnosing Primary Infection

A Significant Rise in Titre of IgG/Total Antibody Between Acute and Convalescent Sera

However, a significant rise is very difficult to define and depends greatly on the assay used. In the case of CFT and HAI, it is normally taken as a four-fold or greater increase in titre. The main problem is that diagnosis is usually retrospective because by the time the convalescent serum is taken, the patient had probably recovered.

Presence of IgM

EIA, RIA, and IF may be are used for the detection of IgM. This offers a rapid means of diagnosis. However, there are many problems with IgM assays, such as interference by rheumatoid factor, re-infection by the virus, and unexplained persistence of IgM years after the primary infection.

Seroconversion

This is defined as changing from a previously antibody negative state to a positive state *e.g.* seroconversion against HIV following a needle-stick injury, or against rubella following contact with a known case.

Serological Examination

Detection of rising titers of antibody between acute and convalescent stages of infection or the detection of IgM in primary infection is called Serology. Blood serum is used in the examination of virus infections by:

(*a*) Neutralization test

(*b*) Complement fixation test

(*c*) Haemagglutination inhibition test

(*d*) Haemagglutination test

(*e*) Immunofluorescence technique

(*f*) Single radial analysis

(*g*) Radioimmunoassay

(*h*) ELISA.

(*a*) Neutralization Test

Neutralization of a virus is defined as the loss of infectivity through reaction of the virus with specific antibody. Virus and the serum are mixed under appropriate condition and then inoculated into cell culture, eggs or animals. The presence of un-neutralized virus may be detected by reations such as cytopathic effect, haemadsorption, interference and immunofluorescence tests.

(*b*) Complement Fixation Test

Complement fixation test (CFT) was first introduced by Wasserman in 1909 for the diagnosis of *Syphilis*. Now this technique is also used for the diagnosis of viral infections. CFT is convenient and rapid to perform, which can performed with simple equipment and reagents. Several varieties of test antigens are readily available for CFT to detect viral infections.

It is a simple test that has two antigens–antibody reaclions, one of which is the indicator system.

☆ Two test tubes are taken.

☆ Antigen is added to both tubes.

☆ Test serum is added to one tube.

☆ If antibody is present, Ag-Ab complex is formed.

☆ When complement is added, if Ag-Ab complex is present, it fixes complement and consume it.

☆ Indicator cells and antierythrocytic antibodies are added.

☆ If the complement is present, the indicator cell will he lysed (negative result).

☆ If the complement is consumed, no lysis of cells (positive result). The positive result indicates the presence of the particular virus in the specimen.

First reaction takes place between a known virus and a specific antibody in the presence of complement. Complement is fixed by antigen-antibody complex.

Second antigen-antibody reaction consists of reacting sheep RBC with haemolysin. When this indicator system is added to the reactants, the sensitized RBCs will only lyse in the presence of free complement.

(c) Haemagglutination Inhibition Test (HAI)

A wide variety of different viruses possess the ability to agglutinate the erythrocytes of mammalian or avian species. The actual animal species whose erythrocytes could be agglutinated depends on the actual virus. Examples of viruses which could haemagglutinate include influenza, parainfluenza, adenoviruses, rubella, alphaviruses, bunyaviruses, flaviviruses and some strains of picornaviruses. Antibodies against the viral protein responsible for haemagglutination can prevent haemagglutination; this is the basis behind the haemagglutination-inhibition test (HAI). The specificity of the HAI test varies with different viruses. With some viruses such as influenza A, the haemagglutination antigen is the same as the antigen responsible for virus adsorption and thus virus neutralization, and therefore the HAI test is highly specific for the different strains of the virus. With other viruses, the HAI test is less specific *e.g. flaviviruses*, where HAI antibodies against one flavivirus may cross-react with other related flaviviruses. HAI tests are more sensitive than complement-fixation tests but are less sensitive than EIAs and RIAs.

The HAI test is simple to perform and requires inexpensive equipment and reagents. Serial dilutions of patient's sera are allowed to react with a fixed dose of viral haemagglutinin, followed by the addition of agglutinable erythrocytes. In the presence of antibody, the ability of the virus to agglutinate the erythrocytes is inhibited. The HAI test may be complicated by the presence of non-specific inhibitors of viral haemagglutination. and naturally occurring agglutinins of the erthrocytes. Therefore, the sera should be treated before use or false positive or negative results may arise. HAI tests are widely used for the diagnosis of rubella and influenza virus infections. The following is a brief description of the HAI test for rubella.

For rubella HAI testing, one day old chick or goose erythrocytes are used. Bovine albumin veronal buffer (BAVB) is used as the diluent. The HAI test should be carried out using 4 haemagglutination units of rubella antigen. The actual concentration of antigen required should be determined before each HAI test by carrying out a rubella antigen titration from 1:2 to 1:1024. One HA unit is defined as the highest dilution of antigen that gives complete haemagglutination of cells.

In the actual HAI test, the patients' sera are diluted in BAVB from 1:8 to 1:1024. Either V-shaped or U-shaped 96 -well microtitre plate may be used. Non-specific inhibitors of viral haemagglutination may be removed by the treatment of sera before testing by kaolin, RDE, potassium periodate (KIO) or by heat inactivation. Non-specific agglutinins for erythrocytes may be removed by the addition of erythrocytes to the sera prior to testing to allow the erythrocytes to absorb the non-specific agglutinins. This procedure may be carried out for each serum before testing or may be carried out for sera which had shown agglutination in the serum control wells (serum and erythrocytes only) in a previous HAI test. 4HA of rubella antigen is then added to each well containing diluted test sera except for the serum control wells. A back titration of rubella antigen should be incorporated into the test from 4 HA units to 0.25 HA units. The plate is then allowed to stand at room temperature for 60 minutes after which either 0.5 per cent goose cells or 0.4 per cent chick cells are added to each well and incubated at 4°C for 60 minutes. The plate is then read.

The erythrocytes only control should show a button at the bottom of the well. The serum controls for each serum should show the absence of agglutination. The haemagglutinin back titration should show agglutination at 4, 2 and 1 HA units. A fourfold or greater rise in HAI antibody between acute and convalescent phase sera is indicative of a recent rubella infection.

The advantages of HAI tests are that they are relatively easy and inexpensive to perform. The disadvantages are that HAI tests are not as sensitive as EIAs or RIAs, the actual reading of results is subjective and the reagents should be fresh or else abnormal agglutination patterns may arise which makes the reading and interpretation of the test very difficult. As a result the HAI test for rubella had been replaced by more sensitive and reliable EIA and RIA tests for rubella IgG in many virus diagnostic laboratory.

(d) Haemagglutination Test

All strains of Newcastle disease virus will agglutinate chicken red blood cells. This is the result of the haemagglutinin part of the haemagglutinin/neuraminidase viral protein binding to receptors on the membrane of red blood cells. The linking together of the red blood cells by the viral particles results in clumping. This clumping is known as haemagglutination.

Haemagglutination is visible macroscopically and is the basis of haemagglutination tests to detect the presence of viral particles. The test does not discriminate between viral particles that are infectious and particles that are degraded and no longer able to infect cells. Both can cause the agglutination of red blood cells.

Note that some other viruses and some bacteria will also agglutinate chicken red blood cells. To demonstrate that the haemagglutinating agent is Newcastle disease virus, it is necessary to use a specific Newcastle disease virus antiserum to inhibit the haemagglutinating activity.

Substances that agglutinate red blood cells are referred to as haemagglutinins.

Red Blood Cell Control in the Haemagglutination Test
Every time a haemagglutination test is carried out, it is necessary to test the settling pattern of the suspension of red blood cells. This involves mixing diluent with red blood cells and allowing the cells to settle.

1. Dispense diluent.
2. Add red blood cells and mix by gently shaking.
3. Allow the red blood cells to settle and observe the pattern.
4. Observe if the cells have a normal settling pattern and there is no auto-agglutination. This will be a distinct button of cells in the micro test and an even suspension with no signs of clumping in the rapid test.

Control Allantoic Fluid Samples
Negative and positive control samples are tested in both the rapid and micro haemagglutination tests to ensure the validity of the test.

Negative control allantoic fluid is harvested from 14-day old embryonated eggs that have not been inoculated with Newcastle disease virus. It should always test negative for the presence of haemagglutinins. There should not be any sign of haemagglutination.

Positive control allantoic fluid is known to contain a high infectivity titre of Newcastle disease virus. It should always test positive for the presence of haemagglutinins. Haemagglutination should be visible.

Rapid Haemagglutination Test
This test can determine the presence of a haemagglutinating agent in one minute. If testing many samples at the same time, it is necessary to test the negative and positive control samples only once.

Materials
   ☆ Clean glass microscope slide or a clean white ceramic tile.
   ☆ 10 per cent suspension of washed chicken red blood cells.
   ☆ Micropipette and tips, glass Pasteur pipette or a wire loop.
   ☆ PBS.
   ☆ Negative and positive control allantoic fluid samples.
   ☆ Sample to be tested for the presence of Newcastle disease virus, for example allantoic fluid.

Method
1. Place 4 separate drops of 10 per cent chicken red blood cells onto a glass slide or a white tile.

2. To each drop of blood, add one drop of the control and test samples as follows. Use separate tips, pipettes or a flamed loop to dispense each sample.
    ☆ Drop 1 PBS
    ☆ Drop 2 Negative control allantoic fluid (no haemagglutinin)
    ☆ Drop 3 Positive control allantoic fluid (contains haemagglutinin)
    ☆ Drop 4 Unknown sample to be tested
3. Mix by rotating the slide or tile for one minute.
4. Observe and record results. Compare results of the test samples with the control samples.

Results
☆ Agglutinated red blood cells in suspension have a clumped appearance distinct from non-agglutinated red blood cells.
☆ The red blood cells mixed with the positive control allantoic fluid will clump within one minute.
☆ The red blood cells mixed with the PBS and negative control allantoic fluid remain as an even suspension and do not clump.
☆ Judge the results of the test sample by comparison with the positive and negative controls.
☆ The PBS and negative allantoic fluid controls are used to detect clumping of the red blood cells in the absence of virus. This is unlikely to occur. If it does occur, the test is invalid.

Micro Haemagglutination Test in a V-bottom Microwell Plate
This method is convenient when testing allantoic fluid from a large number of embryonated eggs for the presence or absence of haemagglutinin. A 1 percent solution of red blood cells is used. The cells settle faster in V-bottom plates and there is a better contrast between positive and negative results than observed in U-bottom plates. The method for preparing eggs for and harvesting of allantoic fluid.

Materials
☆ Inoculated eggs, chilled for at least 2 hours, preferably overnight
☆ Negative and positive control samples
☆ V-bottom microwell plate and lid
☆ Micropipette and tips to measure 50 μL
☆ 1 percent suspension of red blood cells
☆ 70 percent alcohol solution
☆ Cotton wool
☆ Forceps and/or small scissors
☆ Absolute alcohol
☆ Discard tray
☆ Microwell plate recording sheet.

Method

1. Fill in the details of samples being tested on a recording sheet. Samples and controls will be distributed into the wells as indicated on this sheet.
2. Use a micropipette to remove 50 mL of allantoic fluid from each egg and dispense into a well of the microwell plate. Use a separate tip for each sample.
3. Include negative and positive control allantoic fluid samples on one of the plates.
4. Dispense 50 mL of PBS into two wells. These wells will be the red blood cells controls for auto-agglutination.
5. Add 25 mL of 1 percent red blood cells to each well.
6. Gently tap sides of the plate to mix. Place a cover on the plate.
7. Allow the plate to stand for 45 minutes at room temperature.
8. Observe and record the results.

Results

☆ The settling patterns of single and agglutinated red blood cells are different. Single cells roll down the sides of the V-bottom well and settle as a sharp button. Agglutinated cells do not roll down the sides of the well to form a button. Instead they settle as a diffuse film.

☆ Negative HA result = a sharp button

☆ Positive HA result = a diffuse film

☆ Red blood cell control = a sharp button

☆ Mark the HA results on microwell recording sheet.

Summary of Haemagglutination tests

Tests	Result	Interpretation
Rapid HA, Micro HA	Positive	Presence of viral particles that may or may not be infectious.
Rapid HA, Micro HA	Negative	Absence of viral particles or presence of viral particles in levels too low to detect.

(e) Immunofluorescence Technique

The purpose of immunofluorescence is to detect the location and relative abundance of any protein for which you have an antibody. Once you have antibodies to your favorite protein, you can use them to indicate where the protein is located. In this example, we will use antibodies for the *calcium ATPase*, or pump, that is located in the *endoplasmic reticulum* (ER) of very cell. The antibody used here only recognized the chicken calcium ATPase but immunofluorescence can be used on any protein.

The key to this entire process is the ability to visualize the antibody when looking through a microscope. Since antibodies are smaller than calcium ATPases, you cannot see the antibody directly. Therefore, you have to use a fluorescent dye that is covalently attached to the antibody. When a light illuminates the fluorescent dye, it absorbs the light an emits a different color light which is visible to the investigator and can be photographed (Figure 81).

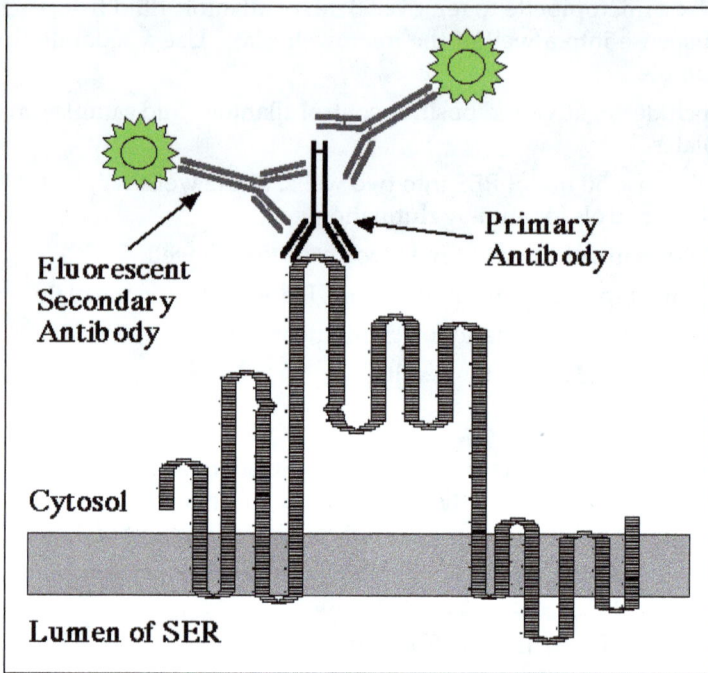

**Figure 81. In Most Immunofluorescence Experiments,
Two Antibodies are Employed.**

The first one, called the *primary antibody*, is typically generated in a mouse and binds to your favorite protein, which in this case is the chicken calcium ATPase (shown as a series undulating striped line that zigzags through the ER membrane 10 times). The *secondary antibody* was purchased from a company that sells antibodies that bind to mouse antibodies and have a fluorescent dye covalently attached to it. As illustrated here, the secondary antibodies can bind to multiple sites on the primary antibody and thus produce a brighter signal since more dyes are brought to a single location.

The first step is to choose your cells of interest. In this case, we will look at a chicken fibroblast, or skin cell. It was grown in tissue culture and so it appears as an isolated cell with no visible neighbors.

The cell was fixed with formaldehyde to retain the shape and location of all cellular proteins. The cell was treated with a mild detergent to disolve small holes in the membranes so the antibodies could have access to the cytoplasm. Because the calcium ATPase is located in the ER, the antibodies must have access to the cytoplasm or they could not bind to the target protein (Figure 82)

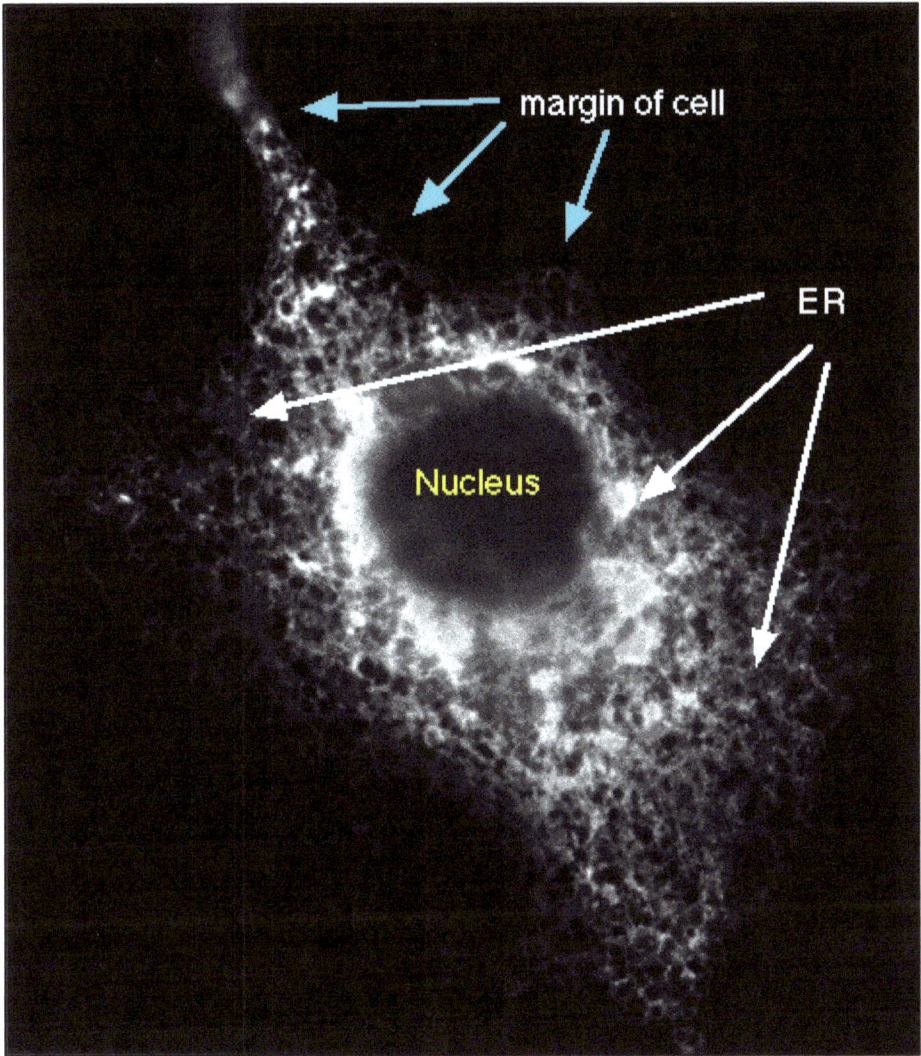

Figure 82. This Immunofluorescence Micrograph Shows the ER being Labeled with a Monoclonal Antibody Against the the Chicken Calcium ATPase.

This chicken cell was fixed, permeabilized, and processed for immunofluorescence. White indicates the location of the fluorescent antibody and thus the calcium ATPase to which the antibody was bound. Immunofluorescence photomicrograph by A. Malcolm Campbell.

Using immunofluorescence, investigators can see when, where and how much of their favorite protein is expressed in any cell or tissue. On this web page, you can see other examples where the same antibody was used to labled chicken calcium ATPase in other chicken tissues.

Immunofluorescence Staining Protocol

1. Preparation of Slides

A. Cell Lines

☆ Grow cultured cells on sterile glass cover slips or slides overnight at 37 ° C

☆ Wash briefly with PBS

☆ Fix as desired. Possible procedures include:

- 10 minutes with 10 per cent formalin in PBS (keep wet)
- 5 minutes with ice cold methanol, allow to air dry
- 5 minutes with ice cold acetone, allow to air dry

☆ Wash in PBS

B. Frozen Sections

Snap frozen fresh tissues in liquid nitrogen or isopentane pre-cooled in liquid nitrogen, embedded in OCT compound in cryomolds. Store frozen blocks at–80 °C.

Cut 4-8 um thick cryostat sections and mount on superfrost plus slides or gelatin coated slides. Store slides at–80 °C until needed.

Before staining, warm slides at room temperature for 30 minutes and fix in ice cold acetone for 5 minutes. Air dry for 30 minutes.

Wash in PBS

C. Paraffin Sections

☆ Deparaffinize sections in xylene, 2 × 5 min.

☆ Hydrate with 100 per cent ethanol, 2 × 3 min.

☆ Hydrate with 95 per cent ethanol, 1 min.

☆ Rinse in distilled water.

☆ Follow procedure for pretreatment as required.

2. Pretreatments of Tissue Sections

Antigenic determinants masked by formalin-fixation and paraffin-embedding often may be exposed by epitope umasking, enzymatic digestion or saponin, etc. *Do not use this pretreatment with frozen sections or cultured cells that are not paraffin-embedded.*

3. Procedure

1. Rinse sections in PBS-Tween 20 for 2x2min

2. *Serum Blocking*: incubate sections with normal serum block–species same as secondary antibody, for 30 minutes to block non-specific binding of immunoglobulin. Note: since this protocol uses avidin-biotin detection system, avidin/biotin block may be needed based on tissue type. If you do, the avidin/biotin block should be done after normal serum block.

3. *Primary Antibody*: incubate sections with primary antibody at appropriate dilution in primary antibody dilution buffer for 1 hour at room temperature or overnight at 4 °C.

4. Rinse in PBS-Tween 20.

5. *Secondary Antibody:* incubate sections with biotinylated secondary antibody at appropriate dilution in PBS for 30 minutes at room temperature.

6. Rinse in PBS-Tween 20 for 3x2min.

7. *Detection:* incubate sections in FITC-Avidin D in PBS for 30 minutes at room temperature. Protecting slides from light starting from this step to the end by covering slides with aluminum foil or black box.

8. Rinse in PBS-Tween 20 for 3x2min.

9. *Counterstain* with PI or DAPI if desire.

10. Rinse in PBS-Tween 20.

11. Dehydrate through 95 per cent ethanol for 2 min, 100 per cent ethanol for 2x3min.

12. Coverslip with anti-fade mounting medium.

(*f*) Single Radial Analysis

It is routinely used for the detection of Rubella and Mumps. Test sera are placed in wells on a plate containing Rubella antigen- coated RBC and complement. The presence of Rubella specific IgB is detected by the lysis of rubella antigen–coated RBC. The zone of lysis around the well depends on the level of specific antibody present.

(*g*) Radioimmunoassay

Radioimmunoassay (RIA) is a scientific method used to test antigens (for example, hormone levels in the blood) without the need to use a bioassay. It was developed by Rosalyn Yalow and Solomon Aaron Berson in the 1950s. In 1977, Rosalyn Sussman Yalow received the Nobel Prize in Medicine for the development of the RIA for insulin: the precise measurement of minute amounts of such a hormone was considered a breakthrough in endocrinology.

Although the RIA technique is extremely sensitive and extremely specific, it requires a sophisticated apparatus and is costly. It also requires special precautions, since radioactive substances are used. Therefore, today it has been largely supplanted by the ELISA method, where the antigen-antibody reaction is measured using colorometric signals instead of a radioactive signal.

To perform a radioimmunoassay, a known quantity of an antigen is made radioactive, frequently by labeling it with gamma-radioactive isotopes of iodine attached to tyrosine. This radiolabeled antigen is then mixed with a known amount of antibody for that antigen, and as a result, the two chemically bind to one another. Then, a sample of serum from a patient containing an unknown quantity of that same antigen is added. This causes the unlabeled (or "cold") antigen from the serum to compete with the radiolabeled antigen for antibody binding sites.

As the concentration of "cold" antigen is increased, more of it binds to the antibody, displacing the radiolabeled variant, and reducing the ratio of antibody-bound radiolabeled antigen to free radiolabeled antigen. The bound antigens are then separated from the unbound ones, and the radioactivity of the free antigen remaining in the supernatant is measured. A binding curve can then be plotted, and the exact amount of antigen in the patient's serum can be determined.

With this technique, separating bound from unbound antigen is crucial. Initially, the method of separation employed was the use of a second "anti-antibody" directed against the first for precipitation and centrifugation. The use of charcoal suspension for precipitation was extended but replaced later by Drs. Werner and Acebedo at Columbia University for RIA of T3 and T4 (Figure 83).

The technique of radioimmunoassay has revolutionized research and clinical practice in many areas, *e.g.*,

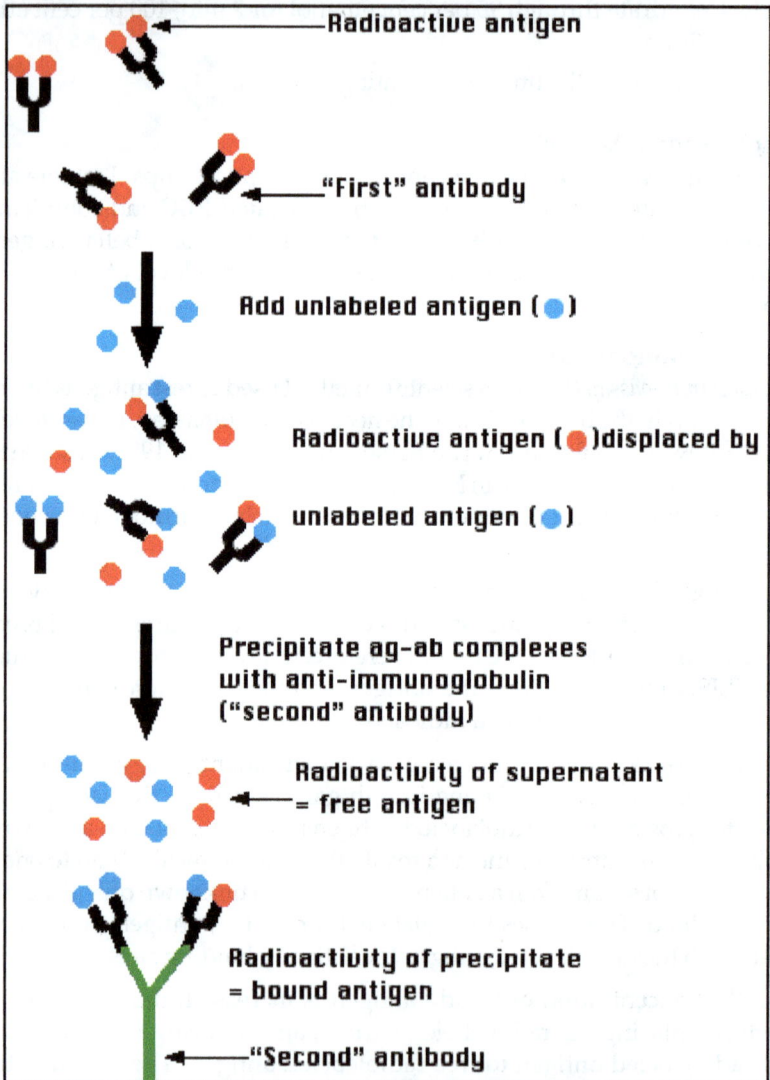

Figure 83. Principle of Radioimmunoassay

☆ Blood banking

☆ Diagnosis of allergies

☆ Endocrinology

The technique was introduced in 1960 by Berson and Yalow as an assay for the concentration of insulin in plasma. It represented the first time that hormone levels *in the blood* could be detected by an in vitro assay.

The Technique

☆ A mixture is prepared of:

• Radioactive antigen

Because of the ease with which iodine atoms can be introduced into tyrosine residues in a protein, the radioactive isotopes ^{125}I or ^{131}I are often used.

• Antibodies against that antigen.

☆ Known amounts of unlabeled ("cold") antigen are added to samples of the mixture. These compete for the binding sites of the antibodies.

☆ At increasing concentrations of unlabeled antigen, an increasing amount of radioactive antigen is displaced from the antibody molecules.

☆ The antibody-bound antigen is separated from the free antigen in the supernatant fluid, and

☆ The radioactivity of each is measured.

☆ From these data, a standard binding curve, like this one shown in red, can be drawn.

☆ The samples to be assayed (the unknowns) are run in parallel.

☆ After determining the ratio of bound to free antigen in each unknown, the antigen concentrations can be read directly from the standard curve (as shown above).

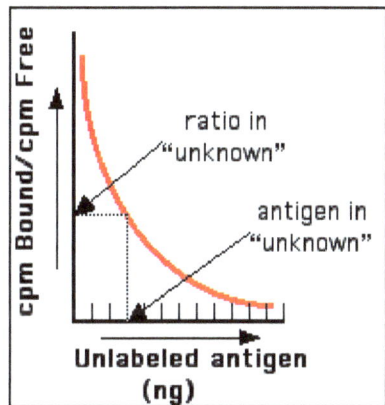

Separating Bound from Free Antigen

There are several ways of doing this.

☆ Precipitate the antigen-antibody complexes by adding a "second" antibody directed against the first. For example, if a rabbit IgG is used to bind the antigen, the complex can be precipitated by adding an antirabbit-IgG antiserum (*e.g.*, raised by immunizing a goat with rabbit IgG). This is the method shown in the diagram above.

☆ The antigen-specific antibodies can be coupled to the inner walls of a test tube After incubation,

- The contents ("free") are removed;
- The tube is washed ("bound"), and
- The radioactive of both is measured.

☆ The antigen-specific antibodies can be coupled to particles, like Sephadex. Centrifugation of the reaction mixture separates

- The bound counts (in the pellet) from
- The free counts in the supernatant fluid.

Radioimmunoassay is widely-used because of its great sensitivity. Using antibodies of high affinity ($K_0 = 10^8–10^{11} M^{71}$), it is possible to detect a few picograms (10^{712} g) of antigen in the tube.

The enzyme-linked immunosorbent assay (ELISA) has many of the advantages (*e.g.*, sensitivity, ease of handling multiple samples) without the disadvantages of dealing with radioactivity.

Despite these drawbacks, RIA has become a major tool in the clinical laboratory where it is used to assay

☆ Plasma levels of:

- Most of our hormones;
- Digitoxin or digoxin in patients receiving these drugs;
- Certain abused drugs

☆ For the presence of hepatitis B surface antigen (HBsAg) in donated blood;

☆ Anti-DNA antibodies in systemic lupus erythematosus (SLE).

Uses for RIA

RIA has many uses, including narcotics (drug) detection, blood bank screening for the hepatitis (a highly contagious condition) virus, early cancer detection, measurement of growth hormone levels, tracking of the leukemia virus, diagnosis and treatment of peptic ulcers, and research with brain chemicals called neurotransmitters.

The presence of antibodies to New castle disease virus in chickens is detected by serological testing. The results of these tests are used for three purposes.

1. To assess the efficacy of New castle disease vaccine in laboratory and field trials.

2. To assess the level of New castle disease virus antibodies in the field.

3. Serum known to contain antibodies to Newcastle disease virus is used to confirm the presence of Newcastle disease virus in a test sample of allantoic fluid. Such a sample would be obtained during the isolation of virulent Newcastle disease virus.

(*h*) ELISA

Enzyme-Linked ImmunoSorbent Assay, also called *ELISA, Enzyme ImmunoAssay* or *EIA,* is a biochemical technique used mainly in immunology to detect the presence of an antibody or an antigen in a sample. The ELISA has been used as a diagnostic tool in medicine and plant pathology, as well as a quality control check in various industries. In simple terms, in ELISA an unknown amount of antigen is affixed to a surface, and then a specific antibody is washed over the surface so that it can bind to the antigen. This antibody is linked to an enzyme, and in the final step a substance is added that the enzyme can convert to some detectable signal. Thus in the case of fluorescence ELISA, when light of the appropriate wavelength is shone upon the sample, any antigen/antibody complexes will fluoresce so that the amount of antigen in the sample can be inferred through the magnitude of the fluorescence.

Performing an ELISA involves at least one antibody with specificity for a particular antigen. The sample with an unknown amount of antigen is immobilized on a solid support (usually a polystyrene microtiter plate) either non-specifically (via adsorption to the surface) or specifically (via capture by another antibody specific to the same antigen, in a "sandwich" ELISA). After the antigen is immobilized the detection antibody is added, forming a complex with the antigen. The detection antibody can be covalently linked to an enzyme, or can itself be detected by a secondary antibody which is linked to an enzyme through bioconjugation. Between each step the plate is typically washed with a mild detergent solution to remove any proteins or antibodies that are not specifically bound. After the final wash step the plate is developed by adding an enzymatic substrate to produce a visible signal, which indicates the quantity of antigen in the sample. Older ELISAs utilize chromogenic substrates, though newer assays employ fluorogenic substrates enabling much higher sensitivity.

Applications

Because the ELISA can be performed to evaluate either the presence of antigen or the presence of antibody in a sample, it is a useful tool both for determining serum antibody concentrations (such as with the HIV test or West Nile Virus) and also for detecting the presence of antigen. It has also found applications in the food industry in detecting potential food allergens such as milk, peanuts, walnuts, almonds, and eggs. ELISA can also be used in toxicology as a rapid presumptive screen for certain classes of drugs.

The ELISA test, or the enzyme immunoassay (EIA), was the first screening test commonly employed for HIV. It has a high sensitivity. In an ELISA test, a person's serum is diluted 400-fold and applied to a plate to which HIV antigens have been attached. If antibodies to HIV are present in the serum, they may bind to these HIV antigens. The plate is then washed to remove all other components of the serum. A specially prepared "secondary antibody" — an antibody that binds to other antibodies — is then applied to the plate, followed by another wash. This secondary antibody is chemically linked in advance to an enzyme. Thus the plate will contain enzyme in proportion to the amount of secondary antibody bound to the plate. A substrate for the enzyme is applied, and catalysis by the enzyme leads to a change in color or fluorescence. ELISA results are reported as a number; the most controversial aspect of this test is determining the "cut-off" point between a positive and negative result.

One method of determining a cut-off point is by comparison with a known standard. For example, if an ELISA test will be used in workplace drug screening, a cut-off concentration (*e.g.*, 50 ng/mL of drug) will be established and a sample will be prepared that contains that concentration of analyte. Unknowns that generate a signal that is stronger than the known sample are called "positive"; those that generate weaker signal are called "negative."

Types
"Indirect" ELISA
The steps of the general, "indirect," ELISA for determining serum antibody concentrations are:

1. Apply a sample of known antigen of known concentration to a surface, often the well of a microtiter plate. The antigen is fixed to the surface to render it immobile. Simple adsorption of the protein to the plastic surface is usually sufficient. These samples of known antigen concentrations will constitute a standard curve used to calculate antigen concentrations of unknown samples. Note that the antigen itself may be an antibody.

2. A concentrated solution of non-interacting protein, such as bovine serum albumin (BSA) or casein, is added to all plate wells. This step is known as blocking, because the serum proteins block non-specific adsorption of other proteins to the plate.

3. The plate wells or other surface are then coated with serum samples of unknown antigen concentration, diluted into the same buffer used for the antigen standards. Since antigen immobilization in this step is due to non-specific adsorption, it is important for the total protein concentration to be similar to that of the antigen standards.

4. The plate is washed, and a detection antibody specific to the antigen of interest is applied to all plate wells. This antibody will only bind to immobilized antigen on the well surface, not to other serum proteins or the blocking proteins.

5. Secondary antibodies, which will bind to any remaining detection antibodies, are added to the wells. These secondary antibodies are conjugated to the substrate-specific enzyme. This step may be skipped if the detection antibody is conjugated to an enzyme.

6. Wash the plate, so that excess unbound enzyme-antibody conjugates are removed.

7. Apply a substrate which is converted by the enzyme to elicit a chromogenic or fluorogenic or electrochemical signal.

8. View/quantify the result using a spectrophotometer, spectrofluorometer, or other optical/electrochemical device.

The enzyme acts as an amplifier; even if only few enzyme-linked antibodies remain bound, the enzyme molecules will produce many signal molecules. A major disadvantage of the indirect ELISA is that the method of antigen immobilization is non-specific; any proteins in the sample will stick to the microtiter plate well, so small concentrations of analyte in serum must compete with other serum proteins when binding to the well surface. The sandwich ELISA provides a solution to this problem.

ELISA may be run in a qualitative or quantitative format. Qualitative results provide a simple positive or negative result for a sample. The cutoff between positive and negative is determined by the analyst and may be statistical. Two or three times the standard deviation is often used to distinguish positive and negative samples. In quantitative ELISA, the optical density or fluorescent units of the sample is interpolated into a standard curve, which is typically a serial dilution of the target.

Sandwich ELISA

A sandwich ELISA. (1) Plate is coated with a capture antibody; (2) sample is added, and any antigen present binds to capture antibody; (3) detecting antibody is added, and binds to antigen; (4) enzyme-linked secondary antibody is added, and binds to detecting antibody; (5) substrate is added, and is converted by enzyme to detectable form (Figure 84).

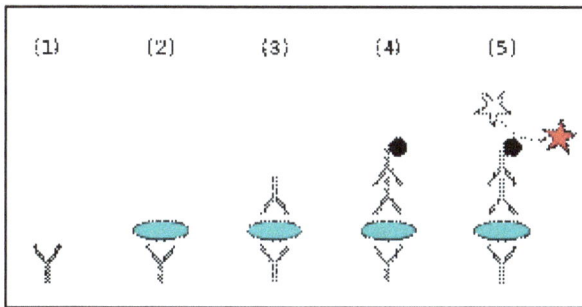

Figure 84. Sandwich ELISA

A less-common variant of this technique, called "sandwich" ELISA, is used to detect sample antigen. The steps are as follows:

1. Prepare a surface to which a known quantity of capture antibody is bound.
2. Block any non specific binding sites on the surface.
3. Apply the antigen-containing sample to the plate.
4. Wash the plate, so that unbound antigen is removed.
5. Apply primary antibodies that bind specifically to the antigen.
6. Apply enzyme-linked secondary antibodies which are specific to the primary antibodies.
7. Wash the plate, so that the unbound antibody-enzyme conjugates are removed.
8. Apply a chemical which is converted by the enzyme into a color or fluorescent or electrochemical signal.

9. Measure the absorbance or fluorescence or electrochemical signal (*e.g.*, current) of the plate wells to determine the presence and quantity of antigen.

The image to the right includes the use of a secondary antibody conjugated to an enzyme, though technically this is not necessary if the primary antibody is conjugated to an enzyme. However, use of a secondary-antibody conjugate avoids the expensive process of creating enzyme-linked antibodies for every antigen one might want to detect. By using an enzyme-linked antibody that binds the Fc region of other antibodies, this same enzyme-linked antibody can be used in a variety of situations. The major advantage of a sandwich ELISA is the ability to use crude or impure samples and still selectively bind any antigen that may be present. Without the first layer of "capture" antibody, any proteins in the sample (including serum proteins) may competitively adsorb to the plate surface, lowering the quantity of antigen immobilized.

Competitive ELISA

A third use of ELISA is through competitive binding. The steps for this ELISA are somewhat different than the first two examples:

1. Unlabeled antibody is incubated in the presence of its antigen.
2. These bound antibody/antigen complexes are then added to an antigen coated well.
3. The plate is washed, so that unbound antibody is removed. (The more antigen in the sample, the less antibody will be able to bind to the antigen in the well, hence "competition.")
4. The secondary antibody, specific to the primary antibody is added. This second antibody is coupled to the enzyme.
5. A substrate is added, and remaining enzymes elicit a chromogenic or fluorescent signal.

For competitive ELISA, the higher the original antigen concentration, the weaker the eventual signal.

(Note that some competitive ELISA kits include enzyme-linked antigen rather than enzyme-linked antibody. The labeled antigen competes for primary antibody binding sites with your sample antigen (unlabeled). The more antigen in the sample, the less labeled antigen is retained in the well and the weaker the signal).

ELISA Reverse Method and Device (ELISA-R m&d)

A newer technique uses a solid phase made up of an immunosorbent polystyrene rod with 4-12 protruding ogives. The entire device is immersed in a test tube containing the collected sample and the following steps (washing, incubation in conjugate and incubation in chromogenous) are carried out by dipping the ogives in microwells of standard microplates pre-filled with reagents.

Advantages

1. The ogives can each be sensitized to a different reagent, allowing the simultaneous detection of different antibodies and different antigens for multi-target assays;

2. The sample volume can be increased to improve the test sensitivity in clinical (saliva, urine), food (bulk milk, pooled eggs) and environmental (water) samples;

3. One ogive is left unsensitized to measure the non-specific reactions of the sample;

4. The use of laboratory supplies for dispensing sample aliquots, washing solution and reagents in microwells is not required, facilitating ready-to-use lab-kits and on-site kits.

Virus Assay

The analysis of a sample to determine the presence virus particles and relative proportion of those particles the sample is called virus assay. Since viruses are too small particles that can only visualized through an electron microscope, we cannot sure whether the sample contains virus particles or not. To demonstrate the presence of virus in a sample, it is necessary to perform certain confirmative tests. They are called virus assay methods.

Virus assay methods are of two types. They are:

☆ Direct virus assay

☆ Virus infectivity assay.

Direct Virus Assay

The direct virus assay includes methods to visualize viruses or viral antigens directly in specimens or sample. There are two methods of direct virus assay:

1. Electron microscopic enumeration
2. Haemagglutination

1. Electron Microscopic Enumeration

The sample or specimen is stained with a negative stain and examined under an electron microscope. For practical purpose, the virus suspension is mixed with a known volume of latex particles and then the ratio between the virus and latex particles are counted. Latex mixing makes the virus enumeration easy.

2. Haemagglutination

This method is useful to measure haemagglutination viruses in the samples. Here, the sample is treated with chicken erythrocytes and serially diluted with water. Then each dilution is incubated for establishing haemagglutination. Later it is visualized under a microscope to see cell clumps. The dilution that gives one cell clump is called *titre*. If 32 dilution gives one cell clump,the titre is 32HA/ units/ml. This is a simple technique that is very convenient for virus assay. The virus is mixed with RBCs. The viruses cross-link the RBC and cause agglutination. (Figure 85)

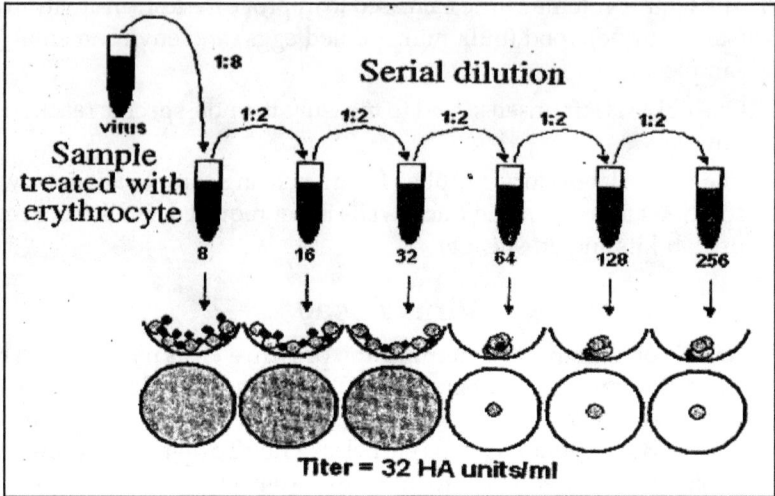

Figure 85. Haemagglutination Assay

Virus Infectivity Assay

In this method, presence of virus is confirmed by looking at the specimens for infection. Number of viral particles required to cause infection in a particular quantity of cells is called *Multiplicity of Infection* (MOI).

MOI = Number of Infectious Units (IU)/ Number of cells

The following methods are used for quantification of viruses

(a) Local Lesion Assay

Infected patch on the tissue or body is called local lesion. It may be a vesicle or blebs or pustule or tubercle. Animal and plant viruses produce lesions at the site of infection. Number of plaques and lesions are directly proportional to the number of viral particles. This is used for diagnosis and quantification of viruses.

(b) Plaque Assay

Plaque is a clear bacteria-free area caused by the lysis of bacterial cells. Bacteriophages, plant and animal form plaques on the infected sites.

★ It was introduced by Dulbecco (1952).

★ A viral suspension is added to a monolayer of tured cells in a bottle or petridish.

★ After adsorption, the medium is removed and with a solid agar gel.

★ Growth of each infectious viral particle gives rise a localized focus (pl.foci) of infected cells. Such foci are known as plaques. Each plaque indicates an infected virus.

★ The number of plaques is counted from each inoculation and the result is expressed in terms of number of plaque forming units/ml. *e.g. Phages, herpes viruses.*

For clear counting of plaques, it is necessary to differentiate plaques from the cell layer. The cell layer is stained with neutral red or crystal violet. Living cells absorb the stain. Plaques appear clear against a red or purple background healthy cells (Figure 86).

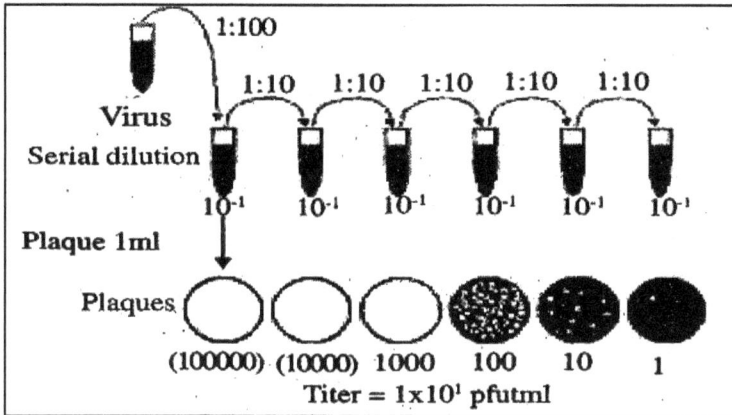

Figure 86. Plaque Assay Method

(c) Pock assay

Small depressions on the surface of sites of infection are called pock. They are formed after pox pustules. Viruses that can form pocks on Chorio Allantoic Membrane (CAM) can be assayed by counting the number of pocks formed on CAM. *e.g. Vaccinia virus.*

(d) Fluorescent–Focus Assay

It is a modification of plaque assay. It is used to assay viruses that do not kill cells. After adsorption and propagation, cells are treated with methanol or acetone and incubated with an antibody raised against the virus. A second antibody coupled with an indicator such as fluorescein is added to it. This recognizes the first antibody. The cells are then examined under UV microscope. Infected cells fluoresces against dark background. Viruses call be expressed as fluorescent–focus forming units/millimeter.

(e) Transformation Assay

It is used for determining filters of some retroviruses that do not form plaques on cells. Cells when grown on a monolayer exhibit contact inhibit 10 and become heaped upon one another. Transformed cells form small piles or foci that can be distinguish from the rest of monolayer. Infectivity is expressed in focus — forming units/ml.

(f) Endpoint Dilution Assay

The maximum dilution of a virus that cannot produce an infection or disease is called endpoint dilution. It is used for determining virulence of a virus in animals. Serial dilutions of virus stock are inoculated into test units. Test units can be cell cultures, embryonated eggs or animals. The number of test units that have become infected is then enumerated for each virus dilution. At a high dilution, none of the cells or animals is infected because no infectious particles are delivered in the cell. At a low dilution, every cell in the test unit is infected. The end point of the dilution assay is the dilution of virus that affects 50 per cent of the test units. It is expressed as per cent infectious dose (ID_{50}) per ml. When this assay is used to assess the virulence of a virus, the result of the assay can be expressed in terms of 0 per cent lethal dose (LD_{50}) per ml or 50 per cent infective dose (ID_{50}). LD_{50} is the amount of virus required to kill (mortality) 50 per cent cells of the host. ID_{50} is the amount of virus required to cause infection in 50 per cent of the host. Concentration of viruses can be enumerated as PFU/ml. The dilution at which no infection was demonstrated known as *Dilution End Point* (DEP).

Molecular Methods

Molecular methods detect viruses which cannot be cultured *in vitro*. Detection of viral nucleic acid is the main base for molecular diagnosis.

Analysis of nucleic acid involves the separation of nucleic acid, annealing of nucleic acid by base pairing with known sequence.

☆ Specificity and sensitivity are achieved in diagnosis by genome detection.

☆ Nucleic acid probes are useful for the genomic diagnosis.

Identification of viruses by detecting their genome in the sample is called viral genome detection or molecular methods. Molecular methods like PCR and nucleic acid based amplification are used for the detection of viral genome.

☆ The classical techniques include *Dot blot, Southern blot* and *Northern blot* • New molecular techniques are Polymerase Chain Reaction, Ligase Chain Reaction, Nucleic acid based Amplification and Branched DNA.

1. Dot Blot Technique

Dot blot technique is the identification of recombinant animals by blotting the DNA or RNA samples not subjected to gel electrophoresis.

It involves the following steps:

☆ Sample DNAs or RNAs of transgenic animal cells or tissues are, separated.

☆ It is transferred to a nitrocellulose filter in the form of a dot (Figure 87).

☆ Samples of several animals are blotted in the same filter.

☆ The dot-blotted DNAs are denatured.

☆ The DNA samples are fixed to the nitrocellulose fluter by baking it at 80°C.

☆ The filter is subject to pretreatment to prevent nonspecific binding of the probe.

☆ The pretreated filter is placed in a solution of radioactive single stranded DNA probe.

☆ The probe hybridizes with the complementary DNA (RNA) on the filter.

☆ After hybridization, the filter membrane is washed to remove the unbound probes.

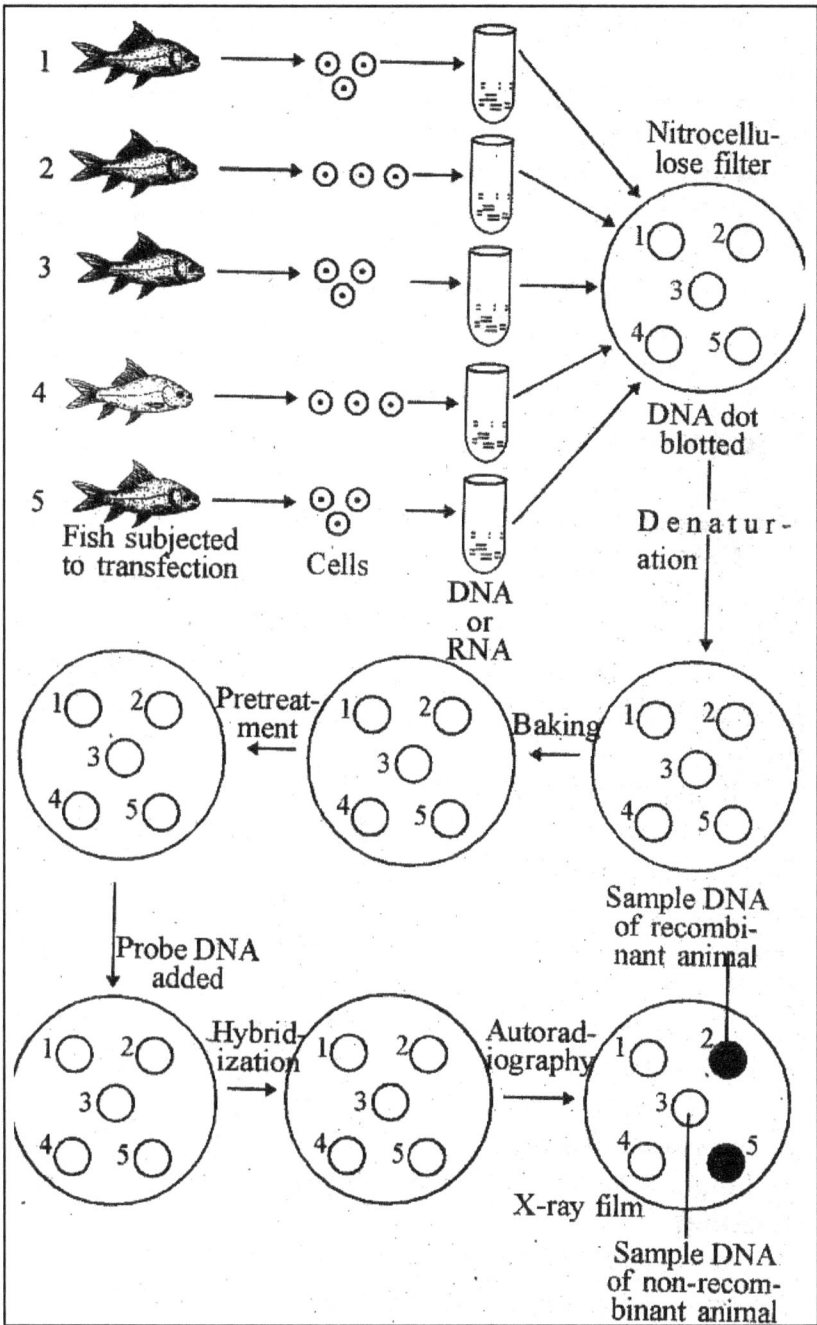

Figure 87. Dot Blotting Technique

☆ The filter membrane is then placed in close contact with an X-ray film.

☆ The X-ray film is developed.

☆ The dots in the X-ray film denote the animals containing recombinant DNA.

☆ In our experiment, fishes numbered 2 and 5 are recombinant animals.

2. Southern Blotting

Southern blotting is the separation of desired DNA fragments of a particular size. This technique was devised Southern in 1975. The name 'Southern' is the name of inventor. Blotting is the transfer of DNA from gel to cellulose filter resembling blotting.

Southern blotting is also called Southern hybridization because DNA-DNA base pairing is the basis of this technique.

It involves the following steps:

1. The DNA is cut with a restriction enzyme to produce DNA fragments of different sizes.

2. It is subjected to agarose gel electrophoresis. This results in the separation of DNA fragments on the gel based on their size.

3. It is denatured by alkali treatment to produce single stranded DNA.

4. The DNA strands are transferred from agarose gel a nitrocellulose filter with the help of a Southern technique which involves the following steps:

   Blotting

   ● A stack of filter paper is placed in a plastic tray.

   ● A buffer is poured about 2/3 of the height of filter paper stack.

   ● The agarose gel, containing denatured DNA, is placed on the filter paper stack.

   ● A nitrocellulose filter membrane is placed over the agarose gel.

   ● Dry filter papers are placed on the nitrocellulose membrane.

   ● At the top a glass plate is placed.

   ● A paper weight is placed over the glass plate.

   ● The experimental set up is kept as such for 4 to 8 hours.

   ● The buffer moves from the tray through the agarose gel and the nitrocellulose filter.

   ● The buffer carries with it the DNA from the agarose gel.

   ● The DNA is trapped in the nitrocellulose membrane.

   ● The positions of bands on the membrane remain same as in the agarose gel (Figure 88).

5. The DNA strands are firmly fixed to the nitrocellulose filter by baking at 80°C.

6. Then the filter is placed in a solution containing radiolabelled RNA or single stranded DNA probe of known sequence.

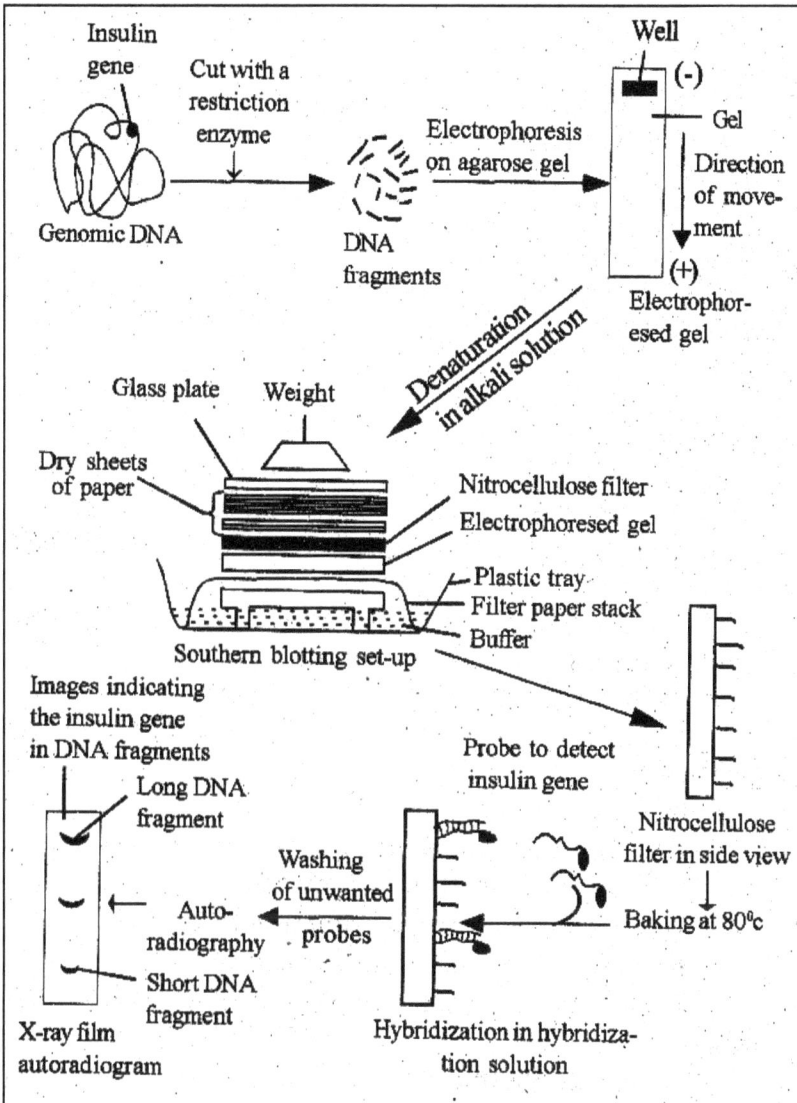

Figure 88. Southern Blotting

7. The radiolabelled nucleic acid base pairs with the complementary DNA fragments of nitrocellulose filter.

8. The hybridized regions are recorded on an X-ray film.

9. The X-ray film reveals bands indicating positions in the agarose gel of the DNA fragments that are complementary to the probe.

10. Thus the sequences of DNA are recognized following the sequences of nucleic acid probe.

Uses of Southern Blotting

It is used to identify the rDNA, map the restriction sites of a gene, DNA finger printing, preparation of RFLP maps and also its helps to isolate a desired DNA for the construction of rDNA and identify a transformed cell.

3. Northern Blotting

Northern blotting is a method of separation of mRNA from a sample. It was devised by Alwine and his colleagues in 1979. In this technique, mRNA is transferred from gel to a filter paper, similar to blotting. The term 'Northern' is a misnomer. It is just the reverse of Southern. It has no historical or scientific significance. It is a slight variant of Southern blotting. It involves the following steps:

☆ mRNA is isolated by gel electrophoresis.

☆ The electrophoresed agarose gel is immersed in a depurination buffer for 10 minutes and then washed with water.

☆ The mRNAs present in the agarose gel are transferred to an *aminobenzyloxymethyl filter paper* by the blotting method.

Blotting

- A stack of filter paper is placed in a plastic tray.
- A buffer is poured about 2/3 of the height of filter paper stack.
- The agarose gel, containing denatured DNA, is placed on the filter paper stack.
- A nitrocellulose filter membrane placed over the agarose gel.
- Dry filter papers are placed on the nitrocellulose membrane.
- At the top, a glass plate is placed.
- A paper weight is placed over the glass plate.
- The experimental set up is kept as such for 4 to 8 hours.
- The buffer moves from the tray through the agarose gel and the nitrocellulose filter.
- The buffer carries with it the DNA from the gel.
- The DNA is trapped in the nitrocellulose membrane.
- The positions of bands on the membrane remain same as in the agarose gel.
- The mRNA blotted filter paper is baked at 80°C.
- The blotted filter is treated with a pre-hybridization solution and then placed in a heat resistant bag.
- A particular DNA probe and hybridization solution are filled in the bag and the bag is sealed.
- The bag is kept at 42°C for 4-8 hours to hybridization.
- The filter is then washed with a wash solution to remove unbound probes.
- An autoradiogram is taken from the filter to know the exact position of the filter having hybrid mRNA-DNA (Figure 89).

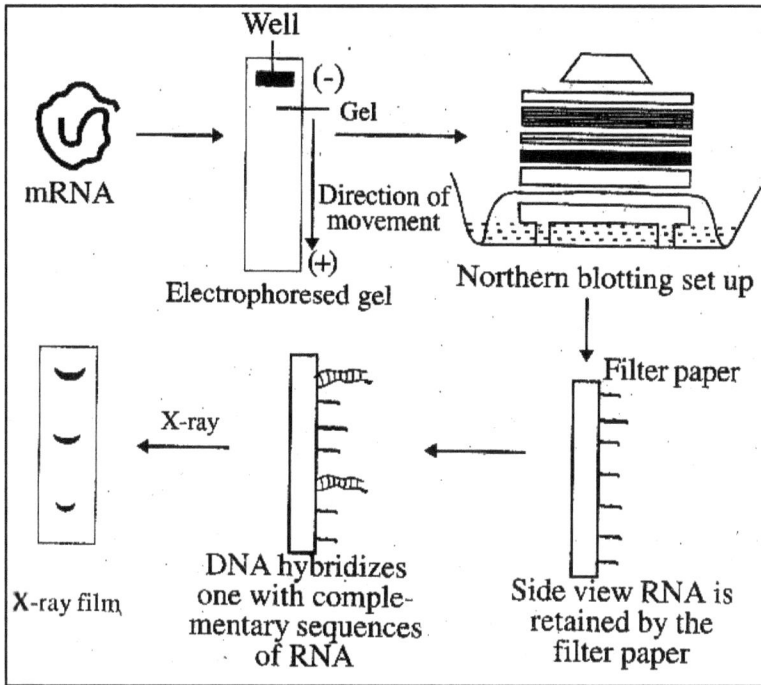

Figure 89. Northern Blotting

Uses

To Identification and separation of RNA and Detection of transcription of DNA.

Molecular Techniques

Polymerase Chain Reaction (PCR)

The *polymerase chain reaction* (*PCR*) is a technique widely used in molecular biology. It derives its name from one of its key components, a DNA polymerase used to amplify a piece of DNA by *in vitro* enzymatic replication. As PCR progresses, the DNA generated is used as a template for replication. This sets in motion a chain reaction in which the DNA template is exponentially amplified. With PCR it is possible to amplify a single or few copies of a piece of DNA across several orders of magnitude, generating millions or more copies of the DNA piece. PCR can be extensively modified to perform a wide array of genetic manipulations.

Almost all PCR applications employ a heat-stable DNA polymerase, such as Taq polymerase, an enzyme originally isolated from the bacterium *Thermus aquaticus*. This DNA polymerase enzymatically assembles a new DNA strand from DNA building blocks, the nucleotides, by using single-stranded DNA as a template and DNA oligonucleotides (also called DNA primers), which are required for initiation of DNA synthesis. The vast majority of PCR methods use thermal cycling, *i.e.*, alternately heating and cooling the PCR sample to a defined series of temperature steps. These thermal cycling steps are necessary to physically separate the strands (at high temperatures) in a DNA double helix (DNA melting) used as the template during DNA synthesis (at lower temperatures) by the DNA polymerase to selectively amplify the target DNA. The selectivity of PCR results from the use of primers that are complementary to the DNA region targeted for amplification under specific thermal cycling conditions.

Developed in 1984 by Kary Mullis, PCR is now a common and often indispensable technique used in medical and biological research labs for a variety of applications. These include DNA cloning for sequencing, DNA-based phylogeny, or functional analysis of genes; the diagnosis of hereditary diseases; the identification of genetic fingerprints (used in forensic sciences and paternity testing); and the detection and diagnosis of infectious diseases. In 1993 Mullis was awarded the Nobel Prize in Chemistry for his work on PCR.

PCR Principles and Procedure

PCR is used to amplify specific regions of a DNA strand (the DNA target). This can be a single gene, a part of a gene, or a non-coding sequence. Most PCR methods typically amplify DNA fragments of up to 10 kilo base pairs (kb), although some techniques allow for amplification of fragments up to 40 kb in size.

A basic PCR set up requires several components and reagents. These components include:

☆ *DNA template* that contains the DNA region (target) to be amplified.

☆ Two *primers*, which are complementary to the DNA regions at the 5' (five prime) or 3' (three prime) ends of the DNA region.

☆ *Taq polymerase* or another DNA polymerase with a temperature optimum at around 70°C.

☆ *Deoxynucleoside triphosphates* (dNTPs; also very commonly and erroneously called deoxynucleotide triphosphates), the building blocks from which the DNA polymerases synthesizes a new DNA strand.

☆ *Buffer solution*, providing a suitable chemical environment for optimum activity and stability of the DNA polymerase.

☆ *Divalent cations*, magnesium or manganese ions; generally Mg^{2+} is used, but Mn^{2+} can be utilized for PCR-mediated DNA mutagenesis, as higher Mn^{2+} concentration increases the error rate during DNA synthesis.

☆ *Monovalent cation* potassium ions. (Figures 90a and b).

The PCR is commonly carried out in a reaction volume of 10-200 µl in small reaction tubes (0.2-0.5 ml volumes) in a thermal cycler. The thermal cycler heats and cools the reaction tubes to achieve the temperatures required at each step of the reaction. Many modern thermal cyclers make use of the Peltier effect which permits both heating and cooling of the block holding the PCR tubes simply by reversing the electric current. Thin-walled reaction tubes permit favorable thermal conductivity to allow for rapid thermal equilibration. Most thermal cyclers have heated lids to prevent

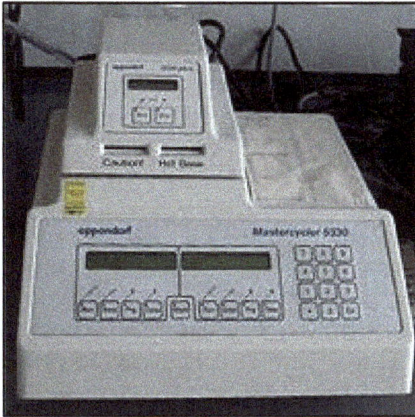

Figure 90a: An Old Thermal Cycler for PCR

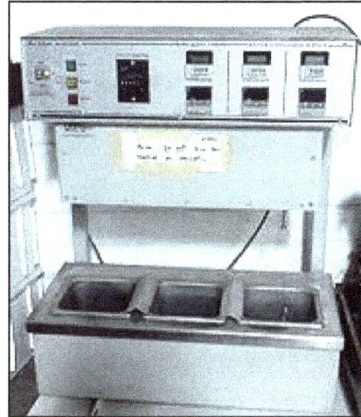

Figure 90b: An Older Model Three-Temperature Thermal Cycler for PCR

condensation at the top of the reaction tube. Older thermocyclers lacking a heated lid require a layer of oil on top of the reaction mixture or a ball of wax inside the tube.

Procedure

1. Denaturing at 94–96°C
2. Annealing at ~65°C
3. Elongation at 72°C.

The PCR usually consists of a series of 20 to 40 repeated temperature changes called cycles; each cycle typically consists of 2-3 discrete temperature steps. Most commonly PCR is carried out with cycles that have three temperature steps. The cycling is often preceded by a single temperature step (called *hold*) at a high temperature (>90°C), and followed by one hold at the end for final product extension or brief storage. The temperatures used and the length of time they are applied in each cycle depend on a variety of parameters. These include the enzyme used for DNA synthesis, the concentration of divalent ions and dNTPs in the reaction, and the melting temperature (Tm) of the primers.

1. Initialization Step

This step consists of heating the reaction to a temperature of 94-96°C (or 98°C if extremely thermostable polymerases are used), which is held for 1-9 minutes. It is only required for DNA polymerases that require heat activation by hot-start PCR.

2. Denaturation Step

This step is the first regular cycling event and consists of heating the reaction to 94-98°C for 20-30 seconds. It causes melting of DNA template and primers by disrupting the hydrogen bonds between complementary bases of the DNA strands, yielding single strands of DNA.

3. Annealing Step

The reaction temperature is lowered to 50-65°C for 20-40 seconds allowing annealing of the primers to the single-stranded DNA template. Typically the annealing temperature is about 3-5 degrees Celsius below the Tm of the primers used. Stable DNA-DNA hydrogen bonds are only formed when the primer sequence very closely matches the template sequence. The polymerase binds to the primer-template hybrid and begins DNA synthesis.

4. Extension/Elongation Step

The temperature at this step depends on the DNA polymerase used; Taq polymerase has its optimum activity temperature at 75-80°C, and commonly a temperature of 72°C is used with this enzyme. At this step the DNA polymerase synthesizes a new DNA strand complementary to the DNA template strand by adding dNTPs that are complementary to the template in 5' to 3' direction, condensing the 5'-phosphate group of the dNTPs with the 3'-hydroxyl group at the end of the nascent (extending) DNA strand. The extension time depends both on the DNA polymerase used and on the length of the DNA fragment to be amplified. As a rule-of-thumb, at its optimum temperature, the DNA polymerase will polymerize a thousand bases per minute. Under optimum conditions, *i.e.*, if there are no limitations due to limiting substrates or reagents, at each extension step, the amount of DNA target is doubled, leading to exponential (geometric) amplification of the specific DNA fragment.

5. Final Elongation

This single step is occasionally performed at a temperature of 70-74°C for 5-15 minutes after the last PCR cycle to ensure that any remaining single-stranded DNA is fully extended.

6. Final Hold

This step at 4-15°C for an indefinite time may be employed for short-term storage of the reaction.

To check whether the PCR generated the anticipated DNA fragment (also sometimes referred to as the amplimer or amplicon), agarose gel electrophoresis is employed for size separation of the PCR products. The size(s) of PCR products is determined by comparison with a DNA ladder (a molecular weight marker), which contains DNA fragments of known size, run on the gel alongside the PCR products (Figure 91).

PCR stages

The PCR process can be divided into three stages:

1. Exponential Amplification

At every cycle, the amount of product is doubled (assuming 100 per cent reaction efficiency). The reaction is very specific and precise.

2. Levelling Off Stage

The reaction slows as the DNA polymerase loses activity and as consumption of reagents such as dNTPs and primers causes them to become limiting.

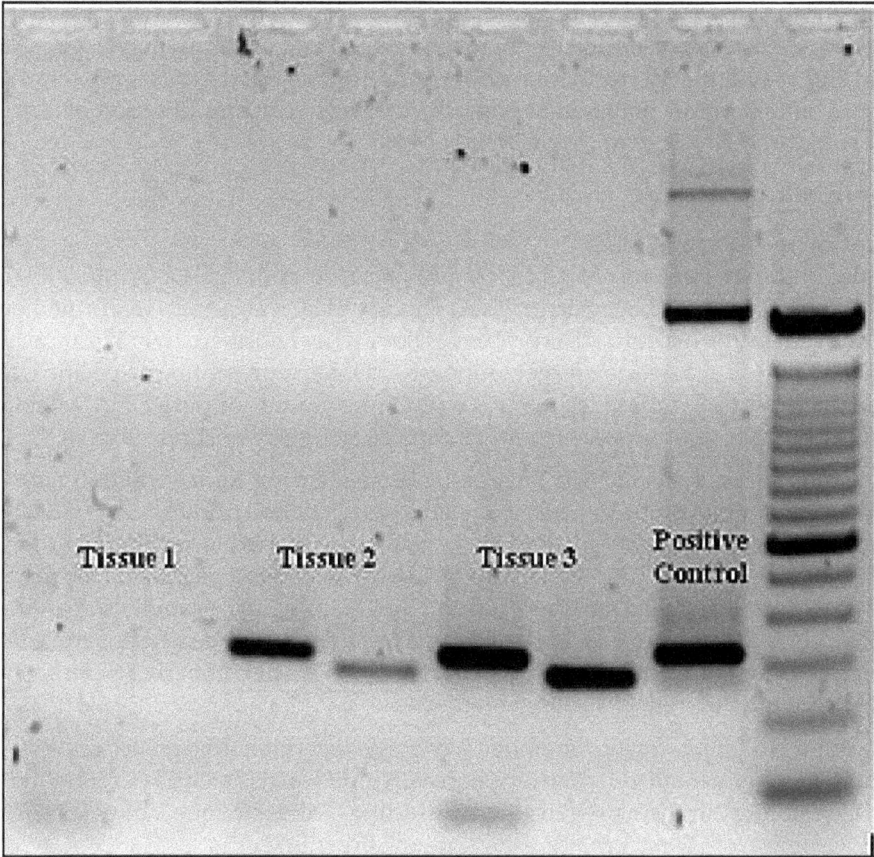

Figure 91. Ethidium Bromide-Stained PCR Products After Gel Electrophoresis.
Two sets of primers were used to amplify a target sequence from three different tissue samples. No amplification is present in sample number 1; DNA bands in sample number 2 and number 3 indicate successful amplification of the target sequence. The gel also shows a positive control, and a DNA ladder containing DNA fragments of defined length for sizing the bands in the experimental PCRs.

3. Plateau
No more product accumulates due to exhaustion of reagents and enzyme.

PCR Optimization
In practice, PCR can fail for various reasons, in part due to its sensitivity to contamination causing amplification of spurious DNA products. Because of this, a number of techniques and procedures have been developed for optimizing PCR conditions. Contamination with extraneous DNA is addressed with lab protocols and procedures that separate pre-PCR mixtures from potential DNA contaminants. This usually involves spatial separation of PCR-setup areas from areas for analysis

or purification of PCR products, and thoroughly cleaning the work surface between reaction setups. Primer-design techniques are important in improving PCR product yield and in avoiding the formation of spurious products, and the usage of alternate buffer components or polymerase enzymes can help with amplification of long or otherwise problematic regions of DNA.

Application of PCR

Isolation of Genomic DNA

PCR allows isolation of DNA fragments from genomic DNA by selective amplification of a specific region of DNA. This use of PCR augments many methods, such as generating hybridization probes for Southern or northern hybridization and DNA cloning, which require larger amounts of DNA, representing a specific DNA region. PCR supplies these techniques with high amounts of pure DNA, enabling analysis of DNA samples even from very small amounts of starting material.

Other applications of PCR include DNA sequencing to determine unknown PCR-amplified sequences in which one of the amplification primers may be used in Sanger sequencing, isolation of a DNA sequence to expedite recombinant DNA technologies involving the insertion of a DNA sequence into a plasmid or the genetic material of another organism. Bacterial colonies (*E. coli*) can be rapidly screened by PCR for correct DNA vector constructs PCR may also be used for genetic fingerprinting; a forensic technique used to identify a person or organism by comparing experimental DNAs through different PCR-based methods.

Some PCR 'fingerprints' methods have high discriminative power and can be used to identify genetic relationships between individuals, such as parent-child or between siblings, and are used in paternity testing. This technique may also be used to determine evolutionary relationships among organisms.

Amplification and Quantitation of DNA

Because PCR amplifies the regions of DNA that it targets, PCR can be used to analyze extremely small amounts of sample. This is often critical for forensic analysis, when only a trace amount of DNA is available as evidence. PCR may also be used in the analysis of ancient DNA that is tens of thousands of years old. These PCR-based techniques have been successfully used on animals, such as a forty-thousand-year-old mammoth, and also on human DNA, in applications ranging from the analysis of Egyptian mummies to the identification of a Russian Tsar.

Quantitative PCR methods allow the estimation of the amount of a given sequence present in a sample–a technique often applied to quantitatively determine levels of gene expression. Real-time PCR is an established tool for DNA quantification that measures the accumulation of DNA product after each round of PCR amplification.

PCR in Diagnosis of Diseases

PCR allows early diagnosis of malignant diseases such as leukemia and lymphomas, which is currently the highest developed in cancer research and is already being used routinely. PCR assays can be performed directly on genomic

DNA samples to detect translocation-specific malignant cells at a sensitivity which is at least 10,000 fold higher than other methods.

PCR also permits identification of non-cultivatable or slow-growing microorganisms such as mycobacteria, anaerobic bacteria, or viruses from tissue culture assays and animal models. The basis for PCR diagnostic applications in microbiology is the detection of infectious agents and the discrimination of non-pathogenic from pathogenic strains by virtue of specific genes.

Viral DNA can likewise be detected by PCR. The primers used need to be specific to the targeted sequences in the DNA of a virus, and the PCR can be used for diagnostic analyses or DNA sequencing of the viral genome. The high sensitivity of PCR permits virus detection soon after infection and even before the onset of disease. Such early detection may give physicians a significant lead in treatment. The amount of virus ("viral load") in a patient can also be quantified by PCR-based DNA quantitation techniques.

Variations on the Basic PCR Technique

1. Allele-specific PCR

This diagnostic or cloning technique is used to identify or utilize single-nucleotide polymorphisms (SNPs) (single base differences in DNA). It requires prior knowledge of a DNA sequence, including differences between alleles, and uses primers whose 3' ends encompass the SNP. PCR amplification under stringent conditions is much less efficient in the presence of a mismatch between template and primer, so successful amplification with an SNP-specific primer signals presence of the specific SNP in a sequence.

2. Assembly PCR or Polymerase Cycling Assembly (PCA)

Assembly PCR is the artificial synthesis of long DNA sequences by performing PCR on a pool of long oligonucleotides with short overlapping segments. The oligonucleotides alternate between sense and antisense directions, and the overlapping segments determine the order of the PCR fragments thereby selectively producing the final long DNA product.

3. Asymmetric PCR

Asymmetric PCR is used to preferentially amplify one strand of the original DNA more than the other. It finds use in some types of sequencing and hybridization probing where having only one of the two complementary stands is required. PCR is carried out as usual, but with a great excess of the primers for the chosen strand. Due to the slow (arithmetic) amplification later in the reaction after the limiting primer has been used up, extra cycles of PCR are required. A recent modification on this process, known as Linear-After-The-Exponential-PCR (LATE-PCR), uses a limiting primer with a higher melting temperature (Melting temperature | Tm) than the excess primer to maintain reaction efficiency as the limiting primer concentration decreases mid-reaction.

4. Helicase-dependent Amplification

This technique is similar to traditional PCR, but uses a constant temperature rather than cycling through denaturation and annealing/extension cycles. DNA Helicase, an enzyme that unwinds DNA, is used in place of thermal denaturation.

5. Hot-start PCR

This is a technique that reduces non-specific amplification during the initial set up stages of the PCR. The technique may be performed manually by heating the reaction components to the melting temperature (*e.g.*, 95°C) before adding the polymerase. Specialized enzyme systems have been developed that inhibit the polymerase's activity at ambient temperature, either by the binding of an antibody or by the presence of covalently bound inhibitors that only dissociate after a high-temperature activation step. Hot-start/cold-finish PCR is achieved with new hybrid polymerases that are inactive at ambient temperature and are instantly activated at elongation temperature.

6. Intersequence-specific PCR (ISSR)

A PCR method for DNA fingerprinting that amplifies regions between some simple sequence repeats to produce a unique fingerprint of amplified fragment lengths.

7. Inverse PCR

A method used to allow PCR when only one internal sequence is known. This is especially useful in identifying flanking sequences to various genomic inserts. This involves a series of DNA digestions and self ligation, resulting in known sequences at either end of the unknown sequence.

8. Ligation-mediated PCR

This method uses small DNA linkers ligated to the DNA of interest and multiple primers annealing to the DNA linkers; it has been used for DNA sequencing, genome walking, and DNA footprinting.

9. Methylation-specific PCR (MSP)

The MSP method was developed by Stephen Baylin and Jim Herman at the Johns Hopkins School of Medicine, and is used to detect methylation of CpG islands in genomic DNA. DNA is first treated with sodium bisulfite, which converts unmethylated cytosine bases to uracil, which is recognized by PCR primers as thymine. Two PCRs are then carried out on the modified DNA, using primer sets identical except at any CpG islands within the primer sequences. At these points, one primer set recognizes DNA with cytosines to amplify methylated DNA, and one set recognizes DNA with uracil or thymine to amplify unmethylated DNA. MSP using qPCR can also be performed to obtain quantitative rather than qualitative information about methylation.

10. Miniprimer PCR

Miniprimer PCR uses a novel thermostable polymerase (S-Tbr) that can extend from short primers ("smalligos") as short as 9 or 10 nucleotides, instead of the approximately 20 nucleotides required by Taq. This method permits PCR targeting smaller primer binding regions, and is particularly useful to amplify unknown, but conserved, DNA sequences, such as the 16S (or eukaryotic 18S) rRNA gene. 16S rRNA miniprimer PCR was used to characterize a microbial mat community growing in an extreme environment, a hypersaline pond in Puerto Rico. In that study, deeply divergent sequences were discovered with high frequency and included representatives that defined two new division-level taxa, suggesting that miniprimer PCR may reveal new dimensions of microbial diversity. By enlarging the "sequence space" that may be queried by PCR primers, this technique may enable novel PCR strategies that are not possible within the limits of primer design imposed by Taq and other commonly used enzymes.

11. Multiplex Ligation-dependent Probe Amplification (MLPA)

Permits multiple targets to be amplified with only a single primer pair, thus avoiding the resolution limitations of multiplex PCR.

12. Multiplex-PCR

The use of multiple, unique primer sets within a single PCR mixture to produce amplicons of varying sizes specific to different DNA sequences. By targeting multiple genes at once, additional information may be gained from a single test run that otherwise would require several times the reagents and more time to perform. Annealing temperatures for each of the primer sets must be optimized to work correctly within a single reaction, and amplicon sizes, *i.e.*, their base pair length, should be different enough to form distinct bands when visualized by gel electrophoresis.

13. Nested PCR

Increases the specificity of DNA amplification, by reducing background due to non-specific amplification of DNA. Two sets of primers are being used in two successive PCRs. In the first reaction, one pair of primers is used to generate DNA products, which besides the intended target, may still consist of non-specifically amplified DNA fragments. The product(s) are then used in a second PCR with a set of primers whose binding sites are completely or partially different from and located 3′ of each of the primers used in the first reaction. Nested PCR is often more successful in specifically amplifying long DNA fragments than conventional PCR, but it requires more detailed knowledge of the target sequences.

14. Overlap-extension PCR

It is a genetic engineering technique allowing the construction of a DNA sequence with an alteration inserted beyond the limit of the longest practical primer length.

15. Quantitative PCR (Q-PCR)

It is used to measure the quantity of a PCR product (preferably real-time). It is the method of choice to quantitatively measure starting amounts of DNA, cDNA or RNA. Q-PCR is commonly used to determine whether a DNA sequence is present in a sample and the number of its copies in the sample. The method with currently the highest level of accuracy is *Quantitative real-time PCR*. It is often confusingly known as RT-PCR (Real Time PCR) or RQ-PCR. QRT-PCR or RTQ-PCR are more appropriate contractions. RT-PCR commonly refers to reverse transcription PCR, which is often used in conjunction with Q-PCR. QRT-PCR methods use fluorescent dyes, such as Sybr Green, or fluorophore-containing DNA probes, such as TaqMan, to measure the amount of amplified product in real time.

16. RT-PCR (Reverse Transcription PCR)

It is a method used to amplify, isolate or identify a known sequence from a cellular or tissue RNA. The PCR is preceded by a reaction using reverse transcriptase to convert RNA to cDNA. RT-PCR is widely used in expression profiling, to determine the expression of a gene or to identify the sequence of an RNA transcript, including transcription start and termination sites and, if the genomic DNA sequence of a gene is known, to map the location of exons and introns in the gene. The 5' end of a gene (corresponding to the transcription start site) is typically identified by an RT-PCR method, named RACE-PCR, short for *Rapid Amplification of cDNA Ends*.

17. Solid Phase PCR

Encompasses multiple meanings, including Polony Amplification (where PCR colonies are derived in a gel matrix, for example), 'Bridge PCR' (the only primers present are covalently linked to solid support surface), conventional Solid Phase PCR (where Asymmetric PCR is applied in the presence of solid support bearing primer with sequence matching one of the aqueous primers) and Enhanced Solid Phase PCR (where conventional Solid Phase PCR can be improved by employing high Tm solid support primer with application of a thermal 'step' to favour solid support priming).

18. TAIL-PCR: Thermal Asymmetric Interlaced

PCR is used to isolate unknown sequence flanking a known sequence. Within the known sequence TAIL-PCR uses a nested pair of primers with differing annealing temperatures; a degenerate primer is used to amplify in the other direction from the unknown sequence.

19. Touchdown PCR

A variant of PCR that aims to reduce nonspecific background by gradually lowering the annealing temperature as PCR cycling progresses. The annealing temperature at the initial cycles is usually a few degrees (3-5°C) above the T_m of the primers used, while at the later cycles, it is a few degrees (3-5°C) below the primer T_m. The higher temperatures give greater specificity for primer binding, and the lower temperatures permit more efficient amplification from the specific products formed during the initial cycles.

20. PAN-AC

This method uses isothermal conditions for amplification, and may be used in living cells.

21. Universal Fast Walking

This method allows genome walking and genetic fingerprinting using a more specific 'two-sided' PCR than conventional 'one-sided' approaches (using only one gene-specific primer and one general primer—which can lead to artefactual 'noise') by virtue of a mechanism involving lariat structure formation. Streamlined derivatives of UFW are LaNe RAGE (lariat-dependent nested PCR for rapid amplification of genomic DNA ends), 5'RACE LaNe and 3'RACE LaNe.

Glossary

ABDOMEN: The area of the body between the thorax and pelvis. The abdomen contains the liver, the spleen and most of the digestive organs.

ACETAMINOPHEN (TYLENOL): An analgesic and antipyretic

ACQUIRED IMMUNE DEFICIENCY SYNDROME (AIDS): Severe manifestation of infection with the human immunodeficiency virus (HIV). The Centers for Disease Control and Prevention lists numerous opportunistic infections and neoplasms which, in the presence of HIV infection, constitute an AIDS diagnosis. In addition, a CD4 count below $200/mm^3$ in the presence of HIV infection constitutes an AIDS diagnosis.

ACID FAST BACILLUS: Bacteria capable of living in acidic environments, such as inside macrophages. Tuberculosis and MAI/MAC are acid fast bacilli.

ACTIVE CONTROL TREATMENT: A control treatment that involves use of a pharmacologically or medically active substance. Active controls are used when there is already an available treatment, so the use of a placebo or no treatment would be unethical.

ACUTE: Short term, intense symptomatology or pathology, as distinct from chronic. Many diseases have an acute phase (like HIV seroconversion disease) and a chronic phase. This distinction is sometimes used in treatments, as in highdose ganciclovir for acute (or induction) treatment of CMV retinitis, followed by a lower dose for chronic (or maintenance) treatment.

ACYCLOVIR (ZOVIRAX): An antiviral drug used in the treatment of herpes simplex virus 1 (HSV-1, fever blisters, cold sores), herpes simplex virus 2 (HSV-2, genital herpes) and herpes zoster (shingles), and sometimes for acute varicella-zoster virus (Chickenpox).

ADENOPATHY: Enlargement of glands, especially the lymph nodes.

ADHERENCE: Degree to which patient care exactly follows study protocol.

ADMINISTRATION (ROUTE OF ADMINISTRATION): How a drug or therapy is introduced into the body (*e.g.*, intravenously, or orally).

ADVERSE EXPERIENCE (AE): A toxic reaction to a treatment under study. Typically, all Grade 4, life-threatening toxicities and deaths are adverse experiences, irrespective of whether or not they are believed to be due to study drug, as are any toxicities that require permanent discontinuation of study drug.

ADVERSE REACTIONS: Any undesirable effect of a medication. All drugs may cause such reactions, so that periodic monitoring is necessary to detect any that do occur, even though their occurrence may be uncommon.

AER: Adverse experience report: A document in a prespecified format used to collect information on all reportable adverse experiences.

AEROSOLIZED: A form of administration in which a drug, such as pentamidine, is turned into a fine spray or mist by a nebulizer, and inhaled.

AFEBRILE: Without a fever.

AIDS CLINICAL TRIALS GROUP (ACTG): A clinical trials network of 59 medical centers, sponsored by the National Institute of Allergy and Infectious Disease, which conducts multicenter trials of treatments for AIDS/HIV and opportunistic infections.

AIDS DEMENTIA COMPLEX: A degenerative neurological condition, with a wide variety of clinical presentations, including loss of coordination, mood swings and loss of inhibitions, and finally wide spread cognitive deficit, AIDS dementia complex is generally thought to be caused by HIV itself

AlANINE AMINOTRANSFERASE (ALT): A liver enzyme, measured through a blood test, that indicates the health of the liver. Lower counts are better. Levels may go up because of hepatitis and other infections, or because of drug toxicities

ALKALINE PHOSPHATASE: A liver enzyme, measured through a blood test, that indicates the health of the liver. Lower counts are better. Levels may go up because of hepatitis and other infections, or because of drug toxicities

ALLOCATION RATIO: Treatment allocation ratio. For example, 1:1 is equal allocation, 3:1 is 75 per cent in one group, and 25 per cent in the other

ALLOCATION STRATA: Treatment allocation strata. Allocations are generally balanced within strata.

ALTERNATIVE HYPOTHESIS: 1. In a trial, the alternative hypothesis might be that a treatment of unknown efficacy has more benefit than the standard treatment. The alternative hypothesis is an alternative to the null hypothesis of no difference that specifies some true underlying difference of set of differences between two or more populations or groups with regard to some function, trait, characteristic, or effect. It may be stated in such a way so as to be concerned with a difference(s) in only one direction (one-sided alternative hypothesis) or in either direction (two-sided alternative hypothesis) relative to the null value. 2. Alternative treatment hypothesis.

AMIKACIN: An antibiotic used as a component in combination therapy to treat tuberculosis or *mycobacterium avium* complex.

AMINO ACID: Any one of 20 or more organic acids, some of which are the building blocks for proteins and are necessary for metabolism and growth.

AMPHOTERICIN B (FUNGIZONE): An antifungal drug that is used to treat fungal infections in persons with HIV, including candida, cryptococcus, histoplasmosis, and others.

AMYLASE: A pancreatic enzyme. High levels in the blood may indicate pancreatic damage.

ANALGESIC: A compound used to reduce or treat pain. Examples of analgesics include aspirin, morphine, and acetaminophen.

ANALOG (ANALOGUE): A chemical compound with a structure similar to that of another but differing from it in respect to a certain component; it may have a similar or opposite action metabolically.

ANALYSIS BY INTENTION TO TREAT: A method of data analysis in which the primary tabulations and companion summaries of outcome data are by assigned treatment, regardless of treatment adherence.

ANALYSIS BY TREATMENT ADMINISTERED: A method of data analysis in which the primary tabulations and companion summaries of outcome data are by treatment administered, not be treatment assigned.

ANAPHYLACTIC SHOCK: A life-threatening allergic reaction characterized by a swelling of body tissues (including the throat) and a sudden decline in blood pressure; can be (rarely) triggered by medication.

ANCILLARY TRIAL: An investigation, stimulated by the trial and intended to generate information of interest to the trial, that is designed and carried out by investigators from one or more of the centers in the trial and that utilizes resources of the trials (*e.g.*, money, study patients, staff time, etc.), but that is not a required part of the design or data collection procedures of the main trial.

ANDROGEN: A masculinizing hormone, *e.g.* testosterone

ANGIOGENESIS: The process of forming new blood vessels. Angiogenesis is essential for the growth of tumors, especially KS

ANEMIA: A condition in which there is a decreased volume of red cells in the blood. There are many causes for anemia, including drug toxicities and chronic infections. The most common way in which anemia is measured is by the titer of hemoglobin (Hgb) in peripheral venous blood.

ANERGIC: Refers to the state of being so immunologically suppressed that one is unable to produce cutaneous delayed type hypersensitivity reaction (DTH). Such patients will usually not test positive for TB on a PPD (mantoux) test.

ANOREXIA: Lack of or complete loss of appetite for food.

ANTAGONISM: The opposite of synergy. One factor (treatment) reduces or cancels the effect of another

ANTENATAL: Before the time of birth.

ANTERIOR: The front, or forward part of something

ANTIBIOTIC: A chemical substance that kills or inhibits the growth of bacteria; some antibiotics are used to treat infectious diseases.

ANTIBODY: A protein molecule in the blood serum or other body fluids that destroys or neutralizes bacteria, viruses, or other harmful toxins. Antibody production occurs in response to the presence of an antagonistic, usually foreign substance (antigen) in the body. They are members of a class of proteins known as immunoglobulins that are produced and secreted by B-lymphocytes in response to stimulation by an antigen. The antigen/antibody reaction forms the basis of humoral (non-cellular) immunity.

ANTIBODY-DEPENDENT CELL-MEDIATED CYTOTOXICITY (ADCC): An immune response in which antibodies bind to target cells, identifying them for attack by the immune system.

ANTIEMETIC: An agent that prevents nausea and vomiting.

ANTIGEN: Any substance that antagonizes or stimulates the immune system to produce antibodies, proteins that fight antigens. Antigens are often foreign substances such as bacteria or viruses that invade the body.

ANTIGEN PRESENTING CELL (APC): A white blood cell that devours foreign bodies, breaks them down, and carries characteristic antigen peptides to it's surface. The foreign antigen, complexed with MHC I or II is presented to CD4 or CD8 to initiate an immune response specific to that peptide.

ANTIGENEMIA: The presence of detectable amounts of an antigen in the blood.

ANTIGENICITY: The ability of an antigen to combine with antibodies and T-cell receptors to invoke a reaction from the immune system.

ANTISENSE: Complementary: an antisense compound is the mirror image of the genetic sequence that it is suppose to inactivate.

ANTIOXIDANT: A substance that inhibits oxidation or reactions promoted by oxygen or peroxides. Antioxidant nutrients protect human cells from damage caused by "free radicals": highly reactive oxygen compounds.

ANTIPYRETIC: A compound that reduces fever

ANTIRETROVIRAL: A substance, drug, or process that destroys a retrovirus, or suppresses it's replication. Often used to describe a drug active against HIV

ANTISEPTIC: Sterilized, or clean of any microorganisms.

ANTIVIRAL: A substance, drug, or process which destroys a virus or suppresses its replication. Can apply to anti-HIV activity, or other viruses, such as herpes or CMV

APHASIA: Complete or partial loss of the ability to speak, or understand speech.

APTHOUS ULCER: A sore of indeterminate origin in the mouth or esophagus. Some apthous ulcers have been linked to CMV or ddC use.

APOPTOSIS: "Cell suicide". Thought to be primarily a way that the body clears out immune cells that respond to the body's own proteins, apoptosis involves a complete physical destruction of a cell, driven by enzymes. Apoptosis also occurs when one receptor on a CD4 cell is triggered without the normal "co-activation" signal. Abnormal apoptosis may be elevated in persons with HIV.

ARC: A term never officially defined by the CDC which has been used to describe a variety of symptoms and signs found in some persons infected with HIV. These may include a decrease in CD4 cells, recurrent fevers, unexplained weight loss, swollen lymph nodes, and/or fungus infection of the mouth and throat. Most of the clinical findings which were formerly denoted as ARC are now in groups 3 or 4 of the CDC AIDS classification system, although the term ARC is still popularly used to describe these symptoms and diagnoses. ARC is also commonly described as symptomatic HIV infection.

ASEPTIC: Without the presence of disease causing microorganisms.

ASPARTATE AMINOTRANSFERASE: A liver enzyme, measured through a blood test, that indicates the health of the liver. Lower counts are better. Levels may go up because of hepatitis and other infections, or because of drug toxicities

ASPERGILLOSIS: A fungal infection resulting from *Aspergillus*, it is also known as aspergillomycosis.

ASPERGILLUS: An ubiquitous fungus, most commonly found in compost heaps. Aspergillus is a frequent cause of disease in transplant patients, and is increasingly seen in persons with HIV.

ASSOCIATION: Synonyms for association include correlation and relationship. An association between two conditions or states means that if one is present, the other is likely to be so as well. Association does not necessarily imply a causal relationship. In addition, association does not necessarily imply a statistically significant relationship.

ASTHENIA: Weakness, debilitation

ASYMPTOMATIC: Without signs or symptoms of disease.

ASYMPTOMATIC INFECTION: An infection or phase of infection, without symptoms.

ATAXIA: Problems with coordination or proper use of muscles

ATOVAQUONE (MEPRON): Antibiotic sometimes used in treatment of PCP. In trials for prophylaxis of PCP, and for treatment of toxoplasmosis

ATTENUATED VIRUS: A weakened virus with reduced ability to infect or produce disease. Some vaccines are based on attenuated viruses.

AUTOIMMUNE DISEASES: Diseases caused when an organism's own immune system attacks its own cells.

AUTOLOGOUS: Pertaining to the same organism or one of its parts; originating within an organism itself.

AZITHROMYCIN (ZITHROMAX): A macrolide antibiotic sometime used as a component in combination therapy for *mycobacterium avium* complex.

AZOLES: A family of antifungal drugs (Fluconazole, itraconazole, etc.) that are used to treat fungal infections in persons with HIV, including candida, cryptococcus, histoplasmosis, and others.

AZT: Also called zidovudine, Retrovir, or azidothymidine. A thymidine (genetic building block) analog that suppresses replication of HIV. It is the only drug FDA-approved for the initial treatment of HIV infection. Adverse side effects may include anemia, leukopenia, muscle fatigue, muscle wasting, nausea and headaches.

BACTERICIDAL: Capable of killing bacteria.

BACTERIOPHAGES: Viruses that infect bacteria.

BACTEREMIA: The presence of bacteria in the blood

BACTERIOLOGICAL SPECIMEN: Refers to any body fluid, secretion, or tissue sent to the laboratory where smears and cultures for bacteria will be performed. The specimen may consist of blood, sputum, urine, spinal fluid, material obtained at biopsy, etc.

BACTERIOSTATIC: Capable of inhibiting bacterial growth (but not necessarily capable of killing bacteria).

BACTRIM: Brand name of trimethoprim-sulphamethoxazole

BASELINE CHARACTERISTIC: A variable that is measured, observed, or assessed on a patient at or shortly before treatment assignment and the initiation of treatment.

BASE PAIR: A pair of nucleotides contained in a double stranded nucleic acid which are linked together by hydrogen bonds.

bDNA (BRANCHED DNA ASSAY): A DNA test for detecting and measuring HIV in the blood plasma of people with HIV. The bDNA test is faster and probably more accurate than plasma culture, the test currently used‹it is less sensitive than PCR, another new test. bDNA testing may eventually be useful to monitor the effectiveness of anti-HIV drugs and to gauge HIV disease progression. It is not yet FDA-approved, nor is it widely available.

BETA CAROTENE: A form of carotene, precursor to vitamin A; a red-orange pigment found in plants and plant-eating animals, and also found in dark green and dark yellow fruits and vegetables. Beta carotene may have beneficial effects on the immune system.

BIAS: Deviation of results from the truth or mechanisms leading to such deviation, *e.g.*, analysis bias, confounding factors, measurement bias, selection bias, withdrawal bias, and others.

BIAS, CONFOUNDING FACTOR: A confounding factor is an variable which is related to one or more of the variables defined in a study. A confounding factor may (1) mask an actual association or (2) falsely demonstrate an apparent association between the study variables where no real association between them exists. For example, alcohol intake may appear to be positively associated with laryngeal cancer but the actual association may be with the confounding factor of cigarette smoking, *i.e.*, people who drink alcohol may be at increased risk for laryngeal cancer because they also smoke cigarettes. If confounding factors are not measured and considered, bias may result.

BIAXIN: Brand name of clarithromycin.

BID: Common abbreviation for "twice a day"

BILATERAL: Having, or being distributed on, two sides.

BILIRUBIN: A bile pigment, bilirubin measurement indicates the health of the liver.

BINARY OUTCOME MEASURE: An outcome measure that can assume only one of two values, such as in a trial with death as the outcome measure.

BIOAVAILABILITY: The rate and extent to which a substance is absorbed and circulated in the body.

BIOLOGICAL RESPONSE MODIFIERS (BRMs): Substances, either natural or synthesized, that boost, direct, or restore normal immune defenses. BRMs include interferons, interleukins, thymic hormones, and monoclonal antibodies.

BIOPSY: A diagnostic technique that involves the surgical removal of a small piece of tissue for microscopic examination and sometimes culture(s).

BIOTECHNOLOGY: The use of living organisms or their products to make or modify a substance. These include recombinant DNA techniques (genetic engineering) and hybridoma technology.

BLIND: A condition imposed on an individual (or group of individuals) for the purpose of keeping that individual or group of individuals from knowing or learning of some fact or observation, such as treatment assignment. Also called a "mask"

BLOOD BRAIN BARRIER: A selective barrier between brain blood vessels and brain tissues whose effect is to restrict what may pass from the blood into the brain. Certain compounds readily cross the blood brain barrier. Others are completely blocked.

B-LYMPHOCYTES: B-lymphocytes are blood cells of the immune system derived from the bone marrow and spleen involved in the production of antibodies. B-lymphocytes float through all body fluids, are able to detect the presence of foreign invaders, and produce antibodies on their own and when primed by T-lymphocytes. B-lymphocytes can later differentiate into plasma and memory cells. B-cells mediate the "humoral" immune response.

BODY FLUIDS: Term used for a number of fluids manufactured within the body. Usually used when referring to semen, blood, urine, and saliva.

BONE MARROW: Soft tissue located in the cavities of the bones where blood cells are formed, including erythrocytes, leukocytes, and platelets.

BRONCHITIS: An inflammation of the bronchial tubes, generally accompanied by coughing, pain, or shortness of breath.

BRONCHOSCOPY:Procedure for examining the respiratory tract by means of a fiber-optic instrument (bronchoscope) which is inserted through the mouth or nose into the trachea. Diagnostic specimens such as bronchial washings and transbronchial biopsies of lung tissue can be obtained during bronchoscopy. (This is also known as bronchoscopy and lavage, or "BAL.") *BUDDING* A step in the replication of some viruses in which the virus leaves the host cell encapsulated with a portion of the cell's membrane without killing the host cell in the process.

BURSA: A tissue space lined by joint tissue; bursas are found in between tendon and bone, skin and bone and muscles.

CANDIDA ALBICANS: A yeast-like fungi, commonly found in the normal flora of the mouth, skin, intestinal tract, and vagina. Generally, candida is harmless, but can become clinically infectious in immune compromised people.

CANDIDA KRUSEII: Another candida species, similar to *Candida albicans*, but often less susceptible to the common drugs used to treat *C Albicans*.

CANDIDEMIA: *Candida albicans* in the blood.

CANDIDIASIS: An infection with a fungus of the *Candida* family, generally *C. albicans*. The most common sites for candidiasis are the mouth, the throat, and the vagina.

CASPID: The protein covering of some viruses–made up of capsomeres; may stimulate the body's immune response.

CARCINOGEN: Any cancer-producing substance or agent.

CARDIOMYOPATHY: A degenerative condition of the heart muscle, cardiomyopathy may be caused by HIV, or by some drugs.

CARRIER: Organism that carries a virus either in form of an infection or while it is in incubation.

CASE-CONTROL STUDY: A study that involves the identification of persons with the disease or condition of interest (cases) and a suitable group of persons without the disease or condition of interest (controls). Cases and controls are compared with respect to some existing or past attribute or exposure believed to be causally related to the disease or condition. Also referred to as a retrospective study because the research approach proceeds from effect to cause. The term applies even if cases and controls are accumulated in a prospective manner.

CASE REPORT FORM: A standardized data entry form used in a clinical trial. Generally, all information collected in trials appears on case report forms, or is referred to and explained by case report forms (as in the case of attached lab slips). Even in circumstances where there is other documentation in addition to

CRFs (like the lab slips), generally all key values that will be analyzed appear on the CRF.

CATHETER: A semi-permanently installed venous line used to inject fluids into the body, or to drain fluids out.

CAUSATION: There is causation only when one factor necessarily alters the possibility of a second. Statistical methods alone cannot establish a causal relationship between factors. Examples of criteria to test causation include: (1) strength of the association, (2) biologic credibility of the association, (3) consistency of the findings with other investigations, (4) temporal relationship of the association, (5) presence of a dose-response relationship. Randomization allows assessment of causation.

CAVITATING: Eroding, or creating holes. Often used to describe the characteristic destruction of lung tissue accompanying active tuberculosis.

CD4 (T4): A protein embedded in the cell surface of helper T-lymphocytes; also found to a lesser degree on the surface of monocyte/macrophage, langerhans cells, astrocytes, keratinocytes, and glial cells. One of the ways HIV invades cells is by first attaching to the CD4 molecule (CD4 receptor).

CD4 CELL: "Helper" T-cell, responsible for coordinating much of the immune response. CD4 cells are one of the main targets damaged by HIV.

CD4 COUNT: The number of T-helper lymphocytes per cubic millimeter of blood. The CD4 count is a good predictor of immune health. A CD4 count less than 200 qualifies as a diagnosis of AIDS.

CD8 (T8): A protein embedded in the cell surface of killer and suppresser T-lymphocytes.

CD8 COUNT: The number of killer/suppresser T-lymphocytes in a cubic millimeter of blood.

CDC: Centers for Disease Control (a part of the United States Public Health Service), Atlanta, Georgia.

CELL: A small, enclosed unit containing the DNA, proteins, and chemicals needed for life functions. The fundamental unit of life.

CELL LINES: Specific cell types artificially maintained in the laboratory (in-vitro) for scientific purposes.

CELL- MEDIATED IMMUNITY (CMI): A branch of the immune system responsible for the reaction to foreign material by specific defense cells (T-lymphocytes, killer cells, macrophage and other white blood cells) rather than antibodies.

CENSORING: A term used in survival or time-to-event analyses to denote an individual who has not experienced the event of interest as of a specific point in follow-up, *e.g.* time of interim analysis, end of study, or time at lost to follow-up. The process by which patient outcome data cannot be obtained beyond a specific point in time.

CENTRAL NERVOUS SYSTEM (CNS): Composed of the brain, spinal cord, and its coverings (meninges).

CEREBRAL: Relating to the brain.

CEREBROSPINAL FLUID (CSF): Fluid that bathes the brain and spinal cord.

CERVICAL DYSPLASIA: The development of abnormal tissue on the cervix, the lower part of the uterus; may progress to cancer of the uterus.

CERVIX: The cylindrical, lower part of the uterus leading to the vagina.

CHANCE: Random variation, *i.e.*, the happening of events without an apparent cause.

CHEMOPROPHYLAXIS: Prevention of disease by chemical means.

CHEMOTHERAPY: The treatment of disease by chemical agents; usually, but not always refers to cancer treatment.

CHROMATOGRAPHY: The separation of chemical substances and particles (originally plant pigments and other highly colored compounds).

CHROMOSOMES: A condensed DNA structure normally found in the nucleus of a cell..

CHRONIC: Referring to a process, such as a disease process, that occurs slowly and persists over a long period of time; opposite of acute.

CIPROFLOXACIN (CIPRO): A flouroquinolone antibiotic sometimes used in combination therapy for treatment of *mycobacterium avium* complex.

CIDOFOVIR (HPMPC): An experimental treatment for CMV

CLARITHROMYCIN (BIAXIN): A macrolide antibiotic sometime used as a component in combination therapy for *mycobacterium avium* complex.

CLINDAMYCIN: An antibiotic sometimes used in the treatment of PCP and toxoplasmosis. Clindamycin usage has been associated with severe diarrhoea caused by the bacteria *C. dificile.*

CLINIC COORDINATOR: The study nurse or other staff person who is primary administrator and contact person for a research effort.

CLINICAL: Pertaining to or founded on actual observation and treatment of patients, as distinguished from theoretical or basic science.

CLINICAL EVENT COMMITTEE: A group of physicians used to review endpoints in trials, and to evaluate their certainty.

CLINICAL SCIENCE REVIEW COMMITTEE: An internal committee at the division of AIDS in charge of prioritizing research, approving trial designs, coordinating intra- and extramural efforts, and clearing specific protocols for implementation.

CLOFAZIMINE: An anti-leprosy drug that is used as a component in combination therapy of *Mycobacterium avium* complex.

CLONE: A group of genetically identical cells or organisms descended from a common ancestor. To produce such genetically identical copies.

CLOSTRIDIUM DIFICILE (C. DIFICILE): A normal gastrointestinal tract bacteria, antibiotics can cause overgrowth of *C. dificile*, and accompanying perforating enterocolitis

CLOTRIMAZOLE: An topical antifungal drug that is used to treat fungal infections in persons with HIV, particularly candida.

COCCIDIOIDOMYCOSIS: A fungal disease which results from infection with *Coccidioides immitis*. Coccidioidomycosis or "valley fever" is common only in a limited geographic area.

CODON: A sequence of 3 bases that specifies a particular amino acid; a building block of the cell's genetic material, DNA and RNA.

COFACTOR: A substance, microorganism or environmental factor that activates or enhances the action of another entity such as a disease-causing agent. Cofactors may influence the progression of a disease or the likelihood of becoming ill. Possible cofactors that have been suggested in AIDS are the herpes viruses, parasites, mycoplasma (a form of life intermediate between bacteria and viruses) and non-HIV retrovirii.

COGNITIVE: Pertaining to thought, awareness, or the ability to rationally apprehend the world and abstract meaning.

COHORT: A group of individuals with some characteristic in common.

COLITIS: Inflammation of the colon

COLON: A division of the lower intestine, extending from the cecum to the rectum; also called the large intestine.

COLONIZATION: Residence of bacteria in, or on, part of the body and causing neither disease nor a response by the individual's immune system.

COLORECTAL: Relating to the colon and rectum, or to the entire large bowel (large intestine).

COLPOSCOPY: A type of examination of a living tissue surface, under magnification, to identify location and extent of lesions.

COMMUNITY ADVISORY BOARD: A lay panel. consisting of patients and affected others, who provide guidance and feedback to clinical trials sites with respect to accrual, retention, compliance, access, and ethical issues surrounding clinical trials

COMPASSIONATE USE: A method of providing unapproved drugs to very sick patients who have no other treatment options. Often, case-by-case approval must be obtained from the FDA for "compassionate use" of a drug. See also "Expanded Access."

COMPLEMENT: A group of proteins in normal blood serum and plasma that, in combination with antibodies, causes the destruction of antigens (particularly bacteria and foreign blood corpuscles).

COMPLEMENT CASCADE: A precise sequence of events, usually triggered by an antigen-antibody complex, in which each component of the complement system is activated in turn, inactivating and occasionally destroying pathogens.

COMPLETE BLOOD COUNT (CBC): A breakdown of the various cells in a sample of blood into white cells, red blood cells, platelets, etc.

COMPLIANCE: How closely a particular protocol is followed. May be influenced by the willingness and/or ability of patients to conform to treatment by taking medications as prescribed and keeping necessary clinic appointments. Often the resources available to the patient, and the resources and flexibility of the provider will have as great an implication for compliance as any specific patient behavior.

COMPUTED TOMOGRAPHY (CAT/CT SCAN): A kind of computer aided imaging that assembles multiple X-rays of "slices' of the body to produce a three dimensional picture.

CONFIDENCE INTERVAL: The range of values that includes, with a stated probability (*e.g.* 95 per cent), the actual population descriptor of interest.

CONJUNCTIVITIS: Inflammation of the protective membrane surrounding the eye.

CONNECTIVE TISSUE: Tissue that surrounds other more highly ordered tissues and organs; blood, cartilage and bone.

CONTAGIOUS: Any infectious disease capable of being transmitted by casual contact from one person to another.

CONTRAINDICATION ("TO INDICATE AGAINST"): A specific circumstance when the use of certain treatments could be harmful.

CONTROLLED CLINICAL TRIAL: A clinical trial involving one or more test treatments, at least one control treatment, and concurrent enrollment, treatment, and follow-up of all patients in the trial.

COOPERATIVE CLINICAL TRIAL: Term frequently used to denote a multicenter trial.

COORDINATING CENTER: A center in the structure of a study that is responsible for receiving, editing, processing, analyzing, and storing data generated in a study and that, in addition, has responsibility for coordination of activities required for execution of the study. See "Statistical Center", "Coordinating center for biometric research"

CORRELATION, PEARSON r: A statistical technique used to assess the magnitude and the direction of the relationship between two variables. Values of the Pearson r can range between -1 and +1. The continuum of -1 to 0 indicates the degree of strength of an inverse relationship. The continuum of 0 to +1 indicates the degree of strength of a positive relationship. A Pearson r of 0 indicates no linear relationship. For example, bicycle helmet use and bicycle accident head injury are negatively correlated; asbestos exposure and cumulative incidence of mesothelioma are positively correlated.

CORTEX: The exterior, or surrounding portion of an organ.

CORTICOSTEROIDS: Any of a number of steroid substances obtained from the cortex of the adrenal gland or manufactured synthetically. Corticosteroids are immunosuppressive and people with HIV should be cautious about taking them for longer than a few weeks. Steroids appear to be an effective adjunct to standard PCP treatments in people with moderate to severe PCP.

CREATININE: A protein found in muscles and blood and excreted by the kidneys in the urine. The level of creatinine in the blood and urine provides a measure of kidney function.

CREATININE KINASE (CREATININE PHOSPHO-KINASE): An enzyme found in the muscles. High levels in the blood indicate breakdown of muscle tissue. In AIDS, may be diagnostic of myopathy.

CROSSED TREATMENTS: Two or more study treatments that are used in sequence (*e.g.*, as in a crossover design) or in combination (*e.g.*, as in a factorial treatment structure).

CROSSOVER: Treatment crossover. A patient who does not comply to assigned treatment and begins to adhere to one of the other treatments. Patient may be a drop-in or drop-out, depending on the direction of the crossover.

CROSSOVER DESIGN; Crossover treatment design. Patients are given treatments in sequence, and crossover is determined by time, not clinical outcomes.

CRYOTHERAPY: The use of intense cold as a treatment, as in freezing off warts.

CRYPTOCOCCAL MENINGITIS: Inflammation of the membranes surrounding the brain and spinal cord by the fungus *c. neoformans.* Cryptococcal meningitis can have symptoms of headache, stiff neck, visual and other sensory distortions, and if untreated, coma and death.

CRYPTOCOCCOSIS: An infectious disease seen in HIV-infected patients due to the fungus *Cryptococcus neoformans.*

CRYPTOCOCCUS NEOFORMANS: A fungus, pathogenic in the immune suppressed, which is acquired via the respiratory tract. Cryptococcosis most frequently causes meningitis, with symptoms of headache and stiff neck.

CRYPTOSPORIDIOSIS: An opportunistic infection caused by a protozoan parasite (*Cryptosporidium parvum*). Cryptosporidiosis causes diarrhoea and abdomen pain.

CRYPTOSPORIDIUM PARVUM: A waterborne enterocyte protozoan, *c. parvum* is often deposited in water supply by wild or domestic animals. *C. parvum* has caused at least one large epidemic of diarrhoeal disease through contamination of a municipal water supply.

CT; Computed Tomography (CAT Scan). A kind of three-dimensional x-ray.

CULTURE: The process of growing bacteria or other microorganisms in the laboratory so that organisms can be identified.

CYTOKINE: A chemical messenger protein released by certain white blood cells, including macrophages, monocytes or lymphocytes, the cytokines include the interferons, the interleukins, Tumor necrosis factor, and many others. Cytokines produced by lymphatic cells are also called "Lymphokines"

CYTOMEGALOVIRUS (CMV): A herpes virus which is a common cause of opportunistic diseases in people with AIDS and other people with immune suppression. While CMV can infect most organs of the body, people with AIDS are most susceptible to CMV retinitis and colitis.

CYTOMETRY: The counting of cells, especially blood cells, using a cytometer (a standardized, ruled glass slide or small glass chamber of known volume).

CYTOPENIA: A lack of specific cellular components in the blood.

CYTOSKELETON: Protein filaments that extend through the cytoplasm of cells and enable them to move and change shape.

CYTOTOXIC: An agent or process which is toxic to cells that results in suppression of function or cell death.

CYTOTOXIC T LYMPHOCYTE (CTL): A lymphocyte that is able to kill foreign cells that have been marked for destruction by the cellular immune system.

DAIDS: The Division of AIDS at the National Institute of Allergy and Infectious Diseases, DAIDS coordinates all of NIAID's AIDS research, and supervises extramural networks, including ACTG, DATRI, AVEU, and SPIRAT.

DAPSONE: An antibiotic active against *Pneumocystis carinii*, leprosy, and other pathogens, and to a lesser degree against *toxoplasma gondii* and MAC. Often used as a second line PCP prophylaxis in those unable to tolerate trimethoprim-sulphamethoxazole.

DARAPRIM: Brand name of Pyrimethamine

DEGENERATION: Deterioration; change from a higher to lower form, especially as in change to less functional or healthy tissue.

DELAYED-TYPE HYPERSENSITIVITY (DTH): A cell-mediated immune response that produces a cellular infiltrate and edema (swelling), redness and induration (hardness) between 48 and 72 hours after exposure to an antigen. DTH response is the basis for PPD testing of tuberculosis exposure.

DEMARCATE: To separate, or to indicate the interface between two areas.

DEMENTIA: Chronic intellectual impairment (loss of mental capacity) with organic origins, that affects a person's ability to function in a social or occupational setting. See "AIDS Dementia complex"

DENDRITIC CELL: A type of antigen-presenting immune cell. Dendritic cells have elongated, tentacle like branches in which they trap foreign objects.

DESENSITIZATION: The reduction or abolition of allergic sensitivity or reactions to the specific antigen (allergen).

DESQUAMATING: Shedding or losing skin, particularly as in grade 3 and 4 rashes

DIAGNOSIS: The evaluation of a patient's medical history, clinical symptoms and laboratory tests which confirms or establishes the nature/origin of an illness.

DIARRHEA: Abnormally frequent and liquid stools.

DIFFUSION: To spread out evenly, as in a liquid.

DIRECT PATIENT CONTACT: Patient contacts that are initiated by the study clinic for the purpose of patient recruitment or data collection and that are directed at specified patients without any reliance on interviewing persons, agencies, institutions, or generalized advertising campaigns to make contacts.

DISCRETE VARIABLE: A variable is capable of assuming only certain values over a defined range. See also "Continuous Variable."

DI-HYDROFOLATE REDUCTASE (DHR): The bacterial (or protozoal) enzyme targeted by trimethoprim-sulphamethoxazole: DHR is necessary for pneumocystis or Toxoplasmosis to survive.

DNA (DEOXYRIBONUCLEIC ACID): A complex protein that is the carrier of genetic information. HIV can insert itself into a cell's DNA and use cellular mechanisms for replication.

DOSAGE: A specific quantity of a biologically active compound.

DOSE RANGING: The establishment of the optimal dosage of a new drug by repeated trials of varying dosages

DOUBLE BLIND: 1. A procedure in a clinical trial for issuing and administering treatment assignments by code number in order to keep study patients and all members of the clinic staff, especially those responsible for patient treatment and data collection, from knowing the assigned treatments. 2. Any condition in which two different groups of people are purposely denied access to a piece of information in order to keep that information from influencing some measurement, observation, or process.

DRUG TRIAL: A clinical trial in which the test treatments are drugs.

DTH: Delayed Type Hypersensitivity.

DYSMENORRHEA: Difficult and painful menstruation.

DYSPEPSIA: Digestive upset.

DYSPHAGIA: Difficulty swallowing.

DYSPLASIA: The abnormal development of tissue. In disease, the alteration of size, shape, and organization of adult cells.

DYSPNEA: Difficult or labored breathing.

ED50: Short for "Effective dose 50"; the dosage of a drug or poison that kills, or stops replication in 50 per cent of the organisms it is tested against. Often applies to does of drug against microorganisms.

EDEMA: Swelling.

ELECTRON MICROSCOPY: An imaging method, which uses a focused beam of electrons to enlarge the image of an object on a screen or photographic plate.

ELISA (ENZYME LINKED IMMUNOSORBENT ASSAY): A laboratory test to determine the presence of antibodies to HIV in the blood. See also "Western Blot."

ENCEPHALITIS: A general term denoting inflammation of the brain.

ENCEPHALOPATHY: Lesions in the brain, or general degeneration of brain matter

ENDEMIC: Pertaining to diseases associated with particular locales or population groups.

ENDOCRINE: Relating to the internal secretion of hormones into systemic circulation.

ENDOCRINE GLANDS: The organs in the body that produce hormones. They are ductless glands that empty hormonal secretions directly into the bloodstream.

ENDOCYTOSIS: The process in which cells take in fluids or other large molecules.

ENDOGENOUS: Relating to or produced by the body.

ENDOSCOPY: Viewing the inside of a body cavity with a device using flexible fiber optics.

ENTERIC: Relating to, or of the intestines or gastrointestinal tract.

ENZYME: A protein that triggers or accelerates chemical reactions, without itself being consumed in the reaction.

EOSINOPHIL: One of white blood cell called granulocytes that can digest microorganisms.

EPIDEMIOLOGY: The science concerned with the determination of the specific causes of a disease or the interrelation between various factors determining a disease, as well as disease trends in a specific population.

EPITHELIAL: The cell linings covering most of the internal and external surface of the body and its organs.

EPITOPE: A unique shape or marker carried on an antigen's surface which triggers a corresponding antibody response.

EPOGEN: Brand name of Erythropoietin

EPSTEIN-BARR VIRUS (EBV): A herpes virus that causes infectious mononucleosis and hairy leukoplakia. EBV also has been associated with Burkitt's lymphoma, a cancer of the lymph nodes and the nasopharynx, the part of the pharynx (throat) which lies above the soft palate.

EQUAL TREATMENT ALLOCATION: A scheme in which the assignment probability in the randomization process for any one treatment is the same as for every other treatment in the trial. A process that ensures that approximately equal numbers of patients receive each treatment

ERYTHEMA: Redness, usually of the skin.

ERYTHEMATOUS: Red or reddened.

ERYTHROCYTES: Red blood cells. The primary function of erythrocytes is to carry oxygen to cells.

ERYTHROPOIETIN (EPO, PROCRIT): A recombinant version of a natural hormone that induces growth of red blood cells, Erythropoietin is used in the treatment of anemia

ESTIMATED SAMPLE SIZE: The number of patients required for a study, as derived from a sample size calculation or in some other way.

ETHAMBUTOL (MYAMBUTOL): An antibiotic used as a component in combination therapy to treat tuberculosis or *mycobacterium avium* complex.

ETIOLOGY: The study or theory of factors which cause disease.

EXOGENOUS: Developed or originating outside the body.

EXPANDED ACCESS: A general term for methods of distributing experimental drugs to patients who are unable to participate in ongoing clinical efficacy trials and have no other treatment options. Specific types of expanded-access mechanisms include parallel track, Treatment IND, and compassionate use.

EXPLANATORY TRIAL: Trials that are designed to explain how a treatment works, in which patients are typically analyzed by treatment received and not as assigned.

FACTORIAL TREATMENT STRUCTURE: A treatment structure in which one study treatment is used in combination with at least one other study treatment in a trial, or where multiples of a defined dose of a specified treatment are used in the same trial. See "Partial" and "Full Factorial Treatment Structure."

FALSE-POSITIVE ERROR: The probability of rejecting the null hypothesis of no treatment difference when there is no treatment difference, that is, falsely claiming a treatment difference. Also called type I error.

FAMCICLOVIR (FAMVIR): A drug chemically related to acyclovir, used in the treatment of herpetic diseases

FANSIDAR: A combination of two antibiotics, Pyrimethamine and sulfadoxine, Fansidar was once used in the treatment of PCP. It proved highly toxic, however, with a high rate of allergic reactions and rash, including Stevens-Johnson syndrome. Rarely used anymore.

FDA: Food and Drug Administration (a regulatory agency of the United States government, located in Rockville, Maryland). FDA decides which drugs may be approved for sale in the United States.

FEASIBILITY STUDY: A preliminary study designed to determine the practicality of a larger study.

FEBRILE: With a fever.

FILGRASTIM (GRANULOCYTE COLONY STIMULATING FACTOR, G-CSF): A recombinant version of an endogenous cytokine that stimulates the production of white blood cells. Often used for treatment of neutropenia attendant upon chemotherapy, or treatment with AZT.

FILOVIRUS: The thread-like virus family which includes such viruses as Ebola and Marburg; very deadly.

FINAL DATA ANALYSIS: The term given to data analyses carried out at the end of the trial, normally in the termination stage, for describing results of the trial.

FIRST-LINE TREATMENT: The preferred therapy for a particular condition (*e.g.*, trimethoprim-sulfamethoxazole (TMP-SMX) is the first-line treatment for PCP).

FLOUROQUINOLONES: A family of antibiotics, including Ciprofloxacin, ofloxacin, and Sparfloxacin. Sometimes used as components of combination therapy for *mycobacterium avium* complex.

FLUCONAZOLE: An antifungal drug that is used to treat fungal infections in persons with HIV, including candida, cryptococcus, histoplasmosis, and others.

FLUCYTOSINE (5-FC): Antifungal drug used as an adjunct to amphotericin for the treatment of cryptococcal meningitis.

FOLATE: A salt or ester of folic acid, a crystalline vitamin of the B complex. Drugs such as trimethoprim sulfa and trimetrexate work by "starving" bacteria of folate, without starving the human body.

FOLLICULITIS: An infection or inflammation of the follicle, at the root of a hair. Folliculitis may be aseptic, as in eosinophilic folliculitis.

FOSCARNET (FOSCAVIR): An antiviral drug FDA-approved for the treatment of cytomegalovirus (CMV) retinitis and other diseases caused by CMV. It is also used to treat acyclovir-resistant herpes virus infections. Adverse side effects may include kidney toxicity, muscle twitching, nausea and skin ulcers.

FULMINATING: Serious acute, active infection

FUNDOSCOPIC EXAM: Visual inspection of the interior of the eye. Often used to diagnose CMV retinitis.

FUNGEMIA: The presence of fungus in the blood.

FUNGIZONE: Brand name for Amphotericin B

GALLIUM SCAN: A diagnostic procedure where mildly radioactive gallium particles are ingested, and then disease (often pneumocystis carinii pneumonia) is diagnosed by scanning the body for radioactive signature.

GANCICLOVIR (CYTOVENE): An antiviral drug FDA-approved to treat cytomegalovirus (CMV) retinitis and to prevent CMV disease in transplant patients at risk for CMV. It has also been used to treat CMV colitis, CMV esophagitis, AIDS-related meningoencephalitis and AIDS-related polyradiculopathy; generally administered intravenously. An oral form of the drug is under study for the treatment and prevention of CMV disease in people with HIV infection.

GASTRIC: Relating to the stomach.

GASTROENTERITIS: Inflammation of the stomach and/or intestines.

GASTROINTESTINAL: Relating to the stomach and intestines.

GENE: A unit of DNA that carries information for the bio-synthesis of a specific product.

GENOME: The DNA code that comprises the complete genetic composition of an organism.

GIARDIASIS: A common protozoal infection of the small intestine spread via contaminated food and water and direct person-to-person contact.

GLYCOPROTEIN: A protein molecule coated with sugars.

GONORRHEA: A sexually transmitted disease; inflammation of genital mucous membranes caused by the bacteria *gonococcus*.

gp41: A protein on the outer shell, or envelope, of HIV. gp41 is the portion of HIV pierces a helper T-cell's surface protein, CD4, allowing viral entry. The "120" refers to its molecular weight.

gp120: Another protein on the outer shell, or envelope, of HIV. gp120 is the portion of HIV that binds to a helper T-cell's surface protein, CD4. The "120" refers to its molecular weight.

gp160: The "precursor" protein to both gp41 and gp120, gp160 is cleaved by viral enzymes into the two surface proteins at a late stage of viral assembly.

GRAM: A metric unit of weight measure. There are approximately 454 grams to an US Pound.

GRANULOCYTES: A white blood cell type of the immune system filled with granules of toxic chemicals that enable them to digest microorganisms. Basophils, neutrophils, eosinophils, and PMNs are examples of granulocytes.

GRANULOCYTOPENIA: A lack or low level of granulocytes in the blood. Often used interchangeably with "neutropenia"

HAIRY LEUKOPLAKIA: A whitish, slightly raised lesion that appears on the side of the tongue. Thought to be related to Epstein-Barr virus infection, it was not observed before the HIV epidemic.

HALF LIFE: The amount of time required for half of a given substance to be eliminated from the body.

HAPHAZARD: A process occurring without any apparent order or pattern. Distinct from random, in that there is no mathematical basis for characterizing a haphazard process.

HARD ENDPOINT: Any outcome measure that is not subject to serious errors of interpretation or measurement. Usually death, infection or some other explicit clinical event.

HEALTH AND HUMAN SERVICES ADMINISTRATION (HHS): A large branch of government that encompasses many other branches. The NIH and the Public Health service (among many others) are parts of HHS.

HEALTH REIMBURSEMENT SERVICES AGENCY (HRSA): The branch of the federal government with primary responsibility for paying for health care, HRSA runs Medicaid and Medicare.

HELPER-SUPPRESSER RATIO: The ratio of helper (CD4+) T-cells to suppresser (CD8+) T-cells.

HEMATOCRIT: A laboratory measurement which determines the percentage of packed red blood cells in a given volume of blood.

HEMATOLOGIC: Pertaining to, or involving the blood, or it's constituent cells.

HEMATOMA: Bruise

HEMATOPOIETIC: Pertaining to the formation of blood cells.

HEMIPARESIS: Paralysis of only one side of the body.

HEMOLYSIS: Destruction of blood cells.

HEMORRHAGIC FEVER: A condition characterized by non-stop internal or external bleeding resulting from a viral infection which has caused blood vessel damage.

HEPARIN: A chemical that prevents blood from clotting.

HEPATIC: Pertaining to the liver.

HEPATITIS: An inflammation of the liver caused by any of several causes. Often accompanied by jaundice, enlarged liver, fever, fatigue and nausea, and abnormal liver function blood tests.

HEPATITIS B: A viral liver disease that can be acute or chronic and even life-threatening, particularly in people with poor immune resistance. Like HIV, the hepatitis B virus can be transmitted by sexual contact, contaminated needles or contaminated blood or blood products. Unlike HIV, it is also transmissible through close casual contact.

HEPATITIS C: A recently recognized viral disease that causes inflammation of the liver, and may cause severe, life-threatening liver damage. Hepatitis C was formerly called "non-A/non-B" hepatitis.

HEPATOMEGALY: An enlargement of the liver.

HEPATOSPLENOMEGALY: Enlargement of the liver and spleen.

HEREDITARY MATERIAL: Material responsible for the transmission of qualities from ancestor to descendant through genes.

HERPES SIMPLEX VIRUS 1 (HSV-1): A virus that can cause painful "cold sores" or blisters on the lips ("fever blisters") or in the mouth or around the eyes. The symptomatic disease stage occurs at unpredictable intervals of weeks, months or years. The latent (inactive) virus can reactivate due to emotional stress, physical trauma, other infections, or suppression of the immune system. HSV-1 responds well to treatment with acyclovir.

HERPES SIMPLEX VIRUS 2 (HSV-2): A virus closely related to HSV-1 that causes similar lesions. However, HSV-2 is usually transmitted sexually, and its lesions generally are in the anogenital area.

HERPES VARICELLA ZOSTER VIRUS (HVZ/VZV): The varicella virus causes chicken pox in children and may reappear in adulthood as herpes zoster. Herpes zoster, also called shingles, consists of very painful blisters on the skin that follow nerve pathways.

HERPES VIRUS: A family of viruses including Herpes simplex I and II, Herpes zoster, Epstein-Barr virus, Cytomegalovirus, and the newly discovered Kaposi's Sarcoma-associated herpes virus.

HISTOPLASMOSIS: A fungal disease resulting from infection with *Histoplasma capsulatum*. Histoplasmosis is geographically limited, generally appearing only in the Mississippi River Valley.

HORMONE: An active regulatory chemical substance formed in one part of the body and carried by the blood to another part of the body, where it signals the coordination of cellular functions.

HOST: A cell or organism that supports the growth of a parasite or virus.

HOT VIRUS: A virus that has the potential ro spread rapidly and therefore must be handled with extreme care. Examples include the Ebola virus and Hantavirus.

HUMAN IMMUNODEFICIENCY VIRUS or HIV INFECTION: Infection with the retrovirus that causes the acquired immunodeficiency syndrome (AIDS).

HUMAN IMMUNODEFICIENCY VIRUS TYPE 1 (HIV-1): The probable cause of AIDS. A human retrovirus of the lentivirus family, notable for long duration of asymptomatic infection, often followed by progressive deterioration of cell mediated immune function, and eventual opportunistic infection and neoplasm. The median time for an HIV infected individual to develop full clinical AIDS is probably over ten years.(formerly called HTLV III or LAV).

HUMAN PAPILLOMAVIRUS (HPV): A member of the papova family of viruses. HPV causes warts or nipplelike protrusions. HPV has also been associated with cervical cancer in women and anal cancer.

HUMATIN: Brand name of Paromomycin.

HUMORAL IMMUNE RESPONSE: The immune response that is mediated by B-cells and involves production of antibodies. Humoral immunity (also known as TH-1 immune response) is associated with the production of the cytokines interleukin-4 and interleukin-10.

HUMORAL IMMUNITY: The branch of the immune system that relies primarily upon antibodies.

HYPERGAMMAGLOBULINEMIA: Abnormally high levels of antibodies in the blood. Common in persons with HIV.

HYPERPLASIA: Excessive cell growth.

HYPERSENSITIVITY: Abnormal sensitivity; medically, when the body responds in an exaggerated manner to a foreign agent; allergy.

HYPERTENSION: High blood pressure.

HYPOTHESIS, ALTERNATIVE: Ha: The investigator's initial supposition that the study will demonstrate that "something is going on," *i.e.,* results observed are more than the outcome of chance. Common examples include that there is a difference in outcome measures between groups, correlation between factors of interest, or association between exposure and disease, *i.e.,* observations are the result of real differences or correlations.

HYPOTHESIS, NULL: Ho: The investigator's initial supposition that the study will demonstrate that "nothing is going on," *i.e.,* results observed are the outcome of chance. Common examples include that there is no difference in outcome measures between groups, no correlation between factors of interest, or no association between exposure and disease, *i.e.*, observations are the result of random variation. Generally, the goal is to "falsify" the null hypothesis.

IDIOTYPES: The unique and characteristic parts of an antibody's variable region, which can themselves serve as antigens.

IMMUNE COMPLEX: Clusters formed when antigens and antibodies bind together.

IMMUNE DEFICIENCY: A breakdown or inability of certain parts of the immune system to function, thus making a person more susceptible to certain diseases to which the person would not ordinarily be subject. In disease associated with HIV, cell mediated immunity related to the function of T-helper lymphocytes deteriorates, increasing the likelihood of disease from a number of pathogens, many of which are ubiquitous (opportunistic infection).

IMMUNE RESPONSE: The activity of the immune system against foreign substances.

IMMUNE SYSTEM: The complex functions of the body that recognize foreign agents or substances, neutralize them, and recall the response later when confronted with the same challenge.

IMMUNE THROMBOCYTOPENIC PURPURA (ITP): An HIV-related loss of platelets in the blood. Its exact cause is unclear.

IMMUNITY: A natural or acquired resistance to a specific disease. Immunity may be partial or complete, long lasting or temporary.

IMMUNOCOMPETENT: Having a normally functioning immune system.

IMMUNOCOMPROMISED: Having a deficient or damaged immune response.

IMMUNOGLOBULIN: A protein that acts as an antibody to help the body fight off disease. There are 5 classes: IgG, IgA, IgD, IgM and IgE. Recombinant and pooled immunoglobulins from blood donations have been used successfully to help HIV-infected children and some adults resist bacterial infections.

IMMUNOMODULATOR: Any substance that influences the immune system. Generally the term "immunomodulator" is used.

IMMUNOSUPPRESSION: A state of the body in which the immune system is damaged and does not perform its normal functions. Immunosuppression may be induced by drugs or result from certain disease processes (such as HIV infection).

IMMUNOTHERAPEUTIC: Aiming at reconstituting an impaired immune system.

INACTIVATED VACCINE: Dead microorganisms used as antigens to produce immunity.

INCLUSION BODIES: Unusual structures occasionally found inside a host cell during virus replication.

INDINAVIR: An experimental protease inhibitor drug made by Merck pharmaceuticals

INFECTION: The state produced by the presence of an infective agent in or on a suitable host.

IN VIVO: Latin for "in life": Studies conducted within a living organism, *e.g.*, animal or human studies.

IN VITRO: Latin for "in glass": An artificial environment created outside a living organism, *e.g.*, a test tube or culture plate, used in experimental research to study a disease or process.

INACTIVE CONTROL TREATMENT: A control treatment that is not considered to have any pharmacological or physiological effect. A placebo treatment or sham procedure.

INCUBATION PERIOD: The time interval between the initial infection and appearance of the first clinical symptom or sign of disease.

INDICATION: Sign or symptom. Also, in terms of drug approval, the exact cause or purpose for which a drug is approved by the FDA to be prescribed. Also called "label indication"

INDOLENT: At rest, or in a quiescent state.

INDUCED SPUTUM: A test where saline mist is breathed to induce a cough. Resultant sputum is then cultured or stained to look for microorganisms, often *pneumocystis carinii.*

INDUCTION: The initiation of a particular therapy.

INFECTION: Condition in which virulent organisms are able to multiply within the body and cause a response from the host's immune defenses. Infection may or may not lead to clinical disease.

INFECTIOUS: Capable of being transmitted by infection, with or without actual contact.

INFILTRATES: Something seeping or filling in a space or cavity.

INFLAMMATION: Redness, swelling, soreness of tissues.

INFORMED CONSENT: The voluntary consent given by a patient to participate in a study after being informed of its purpose, method of treatment, procedure for assignment to treatment, benefits and risks associated with participation, and required data collection procedures and schedule.

INFUSION: Administration of treatment in a dilute form by slow injection into a vein.

INJECTION DRUG USER (IDU, IVDU, AVDA): Also known as intravenous drug user (IVDU), or intravenous drug abuser (IVDA). None of these terms are very precise: often IVDU will be used to describe someone who injects IM or Sub-Q, and the distinction between "user" and "abuser" is a controversial and emotionally charged one. Nevertheless, a working definition for an IDU is anyone who regularly injects any substances, whether pharmaceutically or illicitly made, not under medical prescription and supervision.

INSTITUTIONAL REVIEW BOARD (IRB): A committee of physicians, statisticians, community advocates, and others which ensures that a clinical trial is ethical and that the rights of the study participants are protected. All clinical trials in the United States must be approved by an IRB before they begin.

INTENTION-TO-TREAT ANALYSIS: A method of data analysis in which the primary tabulations, summaries and comparisons of patient outcome data are by assigned treatment, regardless of compliance to therapy or the protocol.

INTEGRASE: The HIV enzyme that governs the insertion of HIV's proviral genetic material into the host genome. Integrase is a target for a new generation of HIV drugs.

INTERACTION: A situation in which the magnitude of the test-control treatment difference for the outcome of interest depends upon the value assumed by a third factor, such as age or prior disease state of the study patients.

INTERACTION EFFECT: Treatment interaction effect.

INTERCURRENT: Occurring at the same time, or accompanying.

INTERFERONS: A family of secreted proteins (lymphokines) in the body with the ability to induce an antiviral state in most cell types. They are secreted by infected host cells to protect uninfected cells from viral infections. There are 3 main classes of interferon: alpha, beta and gamma. The interferons have been synthesized by genetic engineering, and are being tested as treatments for HIV infections and other diseases. Alpha interferon is FDA-approved for treatment of HIV-related Kaposi's sarcoma, chronic hepatitis B and genital warts.

INTERLEUKIN: A chemical hormone messenger (cytokine) secreted by and affecting many different cells in the immune system.

INTERLEUKIN-1 (IL-1): A natural cytokine released by monocytes, macrophages, T-cells and other immune cells that fights infection.

INTERLEUKIN-2 (IL-2): A cytokine that is produced by both T-helper and suppresser lymphocytes, IL-2 increases the expression of natural killer and other cytotoxic cells. IL-2 is associated with a cell-mediated or TH-2 immune response. A recombinant IL-2 is under study as a treatment for HIV disease (immunomodulator).

INTERLEUKIN-4 (IL-4): A cytokine released by lymphocytes (the TH-2 subset of T-helper lymphocytes) that enhances the humoral response, increasing antibody production.

INTERLEUKIN 6 (IL-6): A cytokine whose production affects many different cells in the immune system.

INTERLEUKIN-10 (IL-10): A cytokine released by lymphocytes (the TH-2 subset of T-helper lymphocytes) that enhances the humoral response, increasing antibody production.

INTERLEUKIN-12 (IL-12): A cytokine that induces the production of natural killer and other cytotoxic immune cells. IL-12 is associated with a cell-mediated or TH-1 immune response. A recombinant IL-12 is under study as a treatment for HIV disease (immunomodulator).

INSTERSTITIAL: A space or gap in a tissue: in the context of "interstitial infiltrates", means between the air passage in a lung

INTOLERANT: Unable to take a drug because of toxicity.

INTRADERMAL: Within the layers of the skin.

INTRALESIONAL: Injected directly into a lesion

INTRAMURAL: Research that is done at the National Institutes of Health, by one or more institute, often within the NIH clinical research facility.

INTRAMUSCULAR: Into the muscle: frequently in reference to injections.

INTRAPARTUM: During childbirth.

INTRATHECAL: Injection of a substance through the theca of the spinal cord into the subarachnoid space.

INTRAVENOUS (IV): Within or into the veins. Intravenous drugs are injected directly into the veins.

INTRAVITREAL: Injected in to the vitreous humor of the eye.

INVESTIGATIONAL NEW DRUG APPLICATION: (INDA, also IND) An application directed to the Food and Drug Administration (made by submitting a Notice of Claimed Investigational Exemption for a New Drug) for permission to evaluate a drug (new or old) for a new indication in humans.

ISONIAZID: An antibiotic that is one of the most common components in treatment for tuberculosis, either alone or in combination.

ISOSPORIASIS: A Protozoal infection, isosporiasis is usually restricted to the lower gastrointestinal tract. Symptomatically similar to cryptosporidiosis, isosporiasis usually responds well to treatment.

ISOTONIC: Refers to a solution whose salinity is the same as human blood.

ITRACONAZOLE (SPORANOX): An antifungal drug that is used to treat fungal infections in persons with HIV, including candida, cryptococcus, histoplasmosis, and others.

JAUNDICE: Yellowish discoloration of the skin and eyes due to bile. Usually associated with some form of liver damage or malfunction.

JC PAPOVAVIRUS: Suggested as a possible cause of progressive multifocal leukoencephalopathy.

KAPLAN-MEIER CURVE: A way of graphing patient progress (how many are still alive or free of infection) against time: A lifetable curve showing the percent of people free of a specific event at times following randomization. The Kaplan-Meier method is especially well suited to situations with censored data, such as those encountered in clinical trials, where patients are enrolled over a period of time and followed to a common calendar time point.

KAPOSI'S SARCOMA (KS): A tumor of the wall of blood vessels, or the lymphatic system. Usually appears as pink to purple, painless spots on the skin but may also occur internally in addition to or independent of lesions.

KARNOFSKY SCORE: A subjective score between 0-100, assigned by a physician to describe a patient's ability to function and perform common tasks.

KINASE: An enzyme that mediates the addition of phosphorus to chemical groups. The AZT nucleoside must be phosphorylated to become a genetically active nucleotide.

KETACONAZOLE (NIZORAL): An antifungal drug that is used to treat fungal infections in persons with HIV, including candida, cryptococcus, histoplasmosis, and others.

KILLER T CELL: Cytotoxic lymphocyte. A T-cell that directly kills of infected cells of the body.

LAMIVUDINE (3TC): A nucleoside analogue drug, in the same family as AZT, that is used as an antiretroviral to treat HIV disease.

LANGERHANS CELL: A type of dendritic cell found in the skin.

LAMPRENE: Brand name of Clofazimine

LATENCY: The period when an organism is in the body, shows no symptoms, but is in an inactive state (also known as incubation period).

LATENT INFECTION: Viral infection in which the virus responsible is able to avoid the hosts immune system and defenses.

LD50: Short for "Lethal dose 50"; the dosage of a drug or poison that is toxic to 50 per cent of the organisms it is tested against. Often applies to does of drug against microorganisms.

LEGIONELLA PNEUMOPHILAE: An ubiquitous bacteria causing pneumonia, or "legionnaire's disease". Recognized fairly recently, L. pneumophilae may be the cause of many pneumonias of indeterminate etiology. Most exposure is from tap water, often aerosolized by a shower head or air-conditioner

LENTIVIRUS: A sub-family of retrovirii that is cytopathic and causes chronic diseases. HIV is a lentivirus.

LESION: A change in tissue caused by disease; a point or localized patch of a skin disease. Other words that mean approximately the same thing as lesion are "sore" or "spot."

LEUCOVORIN: A form of the B vitamin, folate that is preferentially taken up by human cells. Leucovorin thereby protects from the effect of folate-antagonist antibiotics.

LEUKINE: Brand name of Sargramostim

LEUKOCYTES: White blood cells which generally fulfill immune functions, as opposed to red blood cells, which are primarily involved in oxygen transport. Leukocytes may be classified as granular or agranular.

LEOKOCYTOSIS: An abnormal high number of white blood cells in the circulating blood.

LEUKOPENIA: An abnormally low number of white blood cells in the circulating blood.

LIFETABLE ANALYSIS: A method of analysis that relies on a count of the number of events observed and the time points at which those events occurred, relative to some zero point. The event may be death or some other event. In clinical trials, the time to an event for a patient is usually measured from the time of randomization. Treatment effects are assessed by comparing event rates in the different treatment groups.

LIPID: A fatty and oily compound used by cells as energy reserves and material for structure.

LIPOSOME: Microscopic globules of fat used to encapsulate drugs, and ensure their delivery to the proper targets, reducing toxicity, and hopefully increasing activity.

LIVER FUNCTION TEST (LFT): Any of a number of tests that measure the health of the liver by checking the levels of various liver-secreted chemicals in the blood. See "Alkaline phosphatase", "Aspartate aminotransferase", and "Alanine aminotransferase".

LUMBAR PUNCTURE: A procedure in which fluid from the sub-arachnoid space in the lumbar region is tapped for examination; also known as a spinal tap.

LYMPH NODES: Small bean-sized organs of the immune system, distributed widely throughout the body. Lymph fluid is filtered through the lymph nodes in which all types of lymphocytes take up temporary residence. Antigens which enter the body find their way into lymph or blood and are filtered out by lymph nodes or the spleen respectively, for attack by the immune system.

LYMPHADENOPATHY: Swollen, firm and possibly tender lymph nodes. The cause may range from an infection such as HIV, the flu, mononucleosis, or lymphoma (cancer of the lymph nodes).

LYMPHATIC VESSELS: A bodywide network of channels, similar to the blood vessels, which transport lymph to the immune organs and into the bloodstream.

LYMPHOCYTE: White blood cells that mature and reside in the lymphoid organs, and are responsible for the acquired immune response.

LYMPHOID TISSUE: Tissue made up white blood cells and lymphatic vessels.

LYMPHOKINES: Cytokines (Chemical messengers) produced by lymphatic cells.

LYMPHOMA: Cancers of the lymphatic system, often of T- or B-lymphocytes. There are many categories of lymphoma, including lymphoblastic, cleaved, non cleaved, Burkitt's, and Hodgkin's disease. Many lymphomas count as an AIDS diagnosis

LYSE: To rupture or destroy a cell.

LYSIS: The process of lysing or destroying cells.

MACROLIDES: A family of antibiotics with wide spectrum activity, including Clarithromycin, Azithromycin, and erythromycin. Some macrolides are sometimes used as parts of combination therapy for *mycobacterium avium*.

MACROPHAGE: A large scavenger cell that ingests degenerated cells, blood tissue and foreign particles, and secretes messenger proteins (monokines) involved in inflammatory reactions, lymphocyte activation and acute systemic immune responses. Macrophages exist in large numbers throughout the body and are key to the development of immunity to a variety of organisms. Along with their precursor blood cell, the monocyte, macrophages are a major reservoir of HIV infection.

MAGNETIC RESONANCE IMAGING (MRI): A diagnostic imaging procedure that uses magnetic fields and radio waves, instead of X-radiation. MRI produces very accurate three dimensional computer generated images.

MAJOR HISTOCOMPATIBILITY COMPLEX (MHC): Two classes of molecules of the surfaces of antigen presenting cells. Antigen is complexed with MCH I or MHC II, and presented to an effector cell. Antigen without MHC is ignored. Discordant MHC type is the source of graft versus host disease and other rejection phenomena. MHC Class I is used for presentation to CD8 cells. Class II is used to present antigen to CD4 cells.

MALAISE: A generalized nonspecific feeling of discomfort and/or fatigue.

MALIGNANT: Cells or tumors growing in an uncontrolled fashion.

MALNUTRITION: Faulty nutrition resulting from poor diet, under eating or abnormal absorption of nutrients from the gastrointestinal tract.

MAST CELL: A type of immune cell that features prominently in allergies.

MATCHING PLACEBO: A pill (capsule or tablet) that is designed to resemble in shape, texture, size, taste, etc., a therapeutically active drug and that is used as the control treatment.

MEAN CORPUSCULAR VOLUME (MCV): The average volume of erythrocytes, conventionally expressed in cubic micrometers per red blood cell.

MEDICAL DEVICE: A diagnostic or therapeutic contrivance that does not interact chemically with a person's body. Includes diagnostic tests, kits, pacemakers, arterial grafts, intraocular lens and orthopedic pins.

MEDLINE: Medical Literature Analysis Retrieval System on Line. A computer searchable database of published medical literature.

MEGESTROL ACETATE: A synthetic hormone used in patients with wasting to increase appetite and weight gain

MEMBRANE: A thin sheet or layer of pliable tissue (or lipids in the case of cells) that serves as a semi-permeable covering.

MEMORY T CELL: Persistent T cells that bear a receptor for a specific antigen that was previously encountered in the course of illness or vaccination. Memory T cells allow a rapid response to pathogens that the body has been previously exposed to.

MESSENGER RNA (mRNA): Used as the carrier of genetic codes and information directly from DNA to cell structures.

METABOLIC: Refers to the process of building the body's molecular structures from nutrients (anabolism) and of breaking them down for energy production and excretion (catabolism).

MICROBES: Microscopic living organisms, including bacteria, protozoa, and fungi.

MICRONUTRIENT: A trace element; an organic compound like a vitamin that is essential but only in small amounts for physical health, growth and metabolism.

MICROSPORIDIOSIS: Disease resulting from infection with a protozoal pathogen from the *Microsporidia* order; similar in symptoms (diarrhea, cramps) and often misdiagnosed as cryptosporidiosis.

MITOCHONDRIA: Organelles within the cytoplasm of the cells, mitochrondia have their own independent DNA, and serve as a source of energy for the cell.

MITOGEN: A substance that induces cell division

MINIMUM INHIBITORY CONCENTATION: The smallest amount of a substance, when diluted, which kill pathogens or stop them from reproducing. Usually means concentration required to achieve ED90.

MOLLUSCUM CONTAGIOSUM: An infectious skin condition characterized by small whitish papules, generally on the face or the trunk.

MONOCLONAL ANTIBODY: An artificially produced antibody, made in the lab by use of an immortalized cell line. Monoclonal antibodies bind to one unique epitope.

MONOCYTE: An antigen presenting white blood cell, monocytes mature into macrophages.

MORBIDITY: The condition of being diseased or sick.

MORTALITY: The condition of being dead.

MRNA (MESSENGER RNA): Used as the carrier of genetic codes and information directly from DNA to cell structures.

MUCOCUTANEOUS: Pertaining to the mucous membranes and the skin, *e.g.*, mouth, vagina, lips, anal area.

MUCOSAL: Pertaining specifically to the mucous membranes

MUTATION: A rearrangement of genes or change in base pairs so they produce different effects within their environment.

MYCOBACTERIUM AVIUM COMPLEX (MAC): A common opportunistic infection caused by two very similar mycobacterial organisms, *Mycobacterium avium* and *M. intracellulare*. In PWAs, it can spread through the bloodstream to infect lymph nodes, bone marrow, liver, spleen, spinal fluid, lungs and intestinal tract. Symptoms of MAC include prolonged wasting, fever, fatigue and enlarged spleen. MAC infection is one of the disease making up the AIDS definition.

MYCOPLASMA: A class of microorganism, simpler than a bacteria, but more complex than a virus. Some mycoplasma may play a role in HIV pathogenesis, although the evidence is unclear.

MYCOBUTIN: Brand name of Rifabutin.

MYCOSIS: Any disease caused by a fungus.

MYCELEX:Brand name of Clotrimazole.

MYELITIS: Inflammation of the spinal cord, or of the bone marrow.

MYELOPATHY: A degenerative process involving the spinal cord.

MYELOTOXIC: Toxic to the bone marrow.

MYOPATHY: A degenerative condition of the muscles. Can be caused by both HIV and AZT.

NAIVE T CELL: A newly formed T cell that has not yet been exposed to antigen.

NANOMETER: One-millionth of a millimeter.

NATIONAL INSTITUTES OF HEALTH (NIH): A group of institutes and related support structures located in Bethesda, Maryland, that is part of the United States Public Health Service. Responsible for funding basic and applied research in the health field. Also initiates and carries out medical research on an intramural and extramural basis.

NATURAL KILLER (NK) CELL: A type of lymphocyte (white blood cell) that lyses infected or cancerous cells. NK response does not require antigen presentation to lymphocytes.

NCI: National Cancer Institute (part of the NIH).

NDA: New drug application. A package of information submitted by the sponsor of a treatment to the FDA, containing all of the information in support of the treatment's approval

NEBULIZER: A device used to reduce liquid medication to extremely fine cloudlike particles; useful in delivering medication to deeper parts of the respiratory tract (*e.g.*, into the lungs).

NEBUPENT: Brand name of pentamidine.

NECROSIS: Cell death and decay.

NECROTIZING: Causing necrosis.

NEUPOGEN: Brand Name of Filgrastim

NEUROLOGIC: Pertaining to the brain or nervous system.

NEUROPATHY: An abnormal and degenerative state of the nervous system. HIV, some treatments, and other diseases can cause a peripheral neuropathy marked by burning tingling sensations in the extremities, loss of deep tendon responses, and decrease in sensitivity to touch stimulation.

NEUTRALIZING ANTIBODY: An antibody that neutralizes (renders harmless) the infectivity of microorganisms, particularly viruses.

NEUTREXIN: Brand name of Trimetrexate.

NEUTROPENIA: An abnormally low number of neutrophils (white blood cells) in the circulating blood.

NEUTROPHILS: One of the white blood cells called granulocytes, filled with granules of toxic chemicals that can digest microorganisms. Neutrophils are comprised of two kinds of cells, polymorpho- nuclear cells ("polys" or PMNs), and bands, which are immature polys. Neutrophils are one of the key components of the immune response, especially to bacterial infection, and the neutrophil count is often elevated in acute infection.

NEVARIPINE: An antiretroviral drug of the non-nucleoside reverse transcription inhibitor family. These drugs target the same enzyme as AZT and the other nucleoside drugs, but are chemically different.

NEW DRUG APPLICATION (NDA): An application submitted by the manufacturer of a drug to the Food and Drug Administration for a license to market the drug for a specified indication.

NIAID: The National Institute of Allergy and Infectious Diseases. One of the institutes of the National Institutes of Health (NIH), which is part of the U.S. Public Health Service of the federal government. The NIAID is responsible for most of the federally funded AIDS research.

NIH: A collection of institutes covering all of the medical specialties and subspecialties, NIH conducts or coordinates the bulk of the biomedical research funded by the united states. A Branch of the Public Health Service of the department of Health and Human Services.

NONCOMPLIANCE: Not in compliance with a designated procedure. Usually in reference to some treatment or data collection procedure

NON NUCLEOSIDE REVERSE TRANSCRIPTASE INHIBITOR (NNRTI): A class of drugs including nevaripine and delaviridine that targets the same reverse transcriptase enzyme of HIV as the nucleoside drugs like AZT, ddI, etc. NNRTIs have been characterized by strong antiviral activity, followed by extremely rapid acquisition of resistance by exposed virus.

NUCLEIC ACID: An organic compound made up of a phosphoric acid, a carbohydrate and a base of purine or pyrimidine; formed in helical chains.

NUCLEOSIDE: A precursor to the cellular building blocks, nucleotides, from which new DNA is constructed. Nucleosides are phosphorylated to produce nucleotides.

NUCLEOSIDE ANALOGUE REVERSE TRANSCRIPTASE INHIBITORS: A family of antiviral compounds including AZT, ddI, ddC, d4t, 3tc, FLT, PMEA, and others. These compounds are phosphorylated into nucleotide analogues, and interfere with the activity of the viral enzyme reverse transcriptase.

NUCLEOTIDE: The basic building blocks genetic material is made of. Nucleotides are produced by the phosphorylation (adding of phosphorus groups) of nucleosides.

NUCLEUS: A cellular organelle that is the essential control mechanism for cell function; contains the DNA and genetic material.

NULL HYPOTHESIS: 1. (statistics) A hypothesis that postulates no underlying difference in the populations or groups being compared with regard to the factor, trait, characteristic, or condition of interest. 2. Null treatment hypothesis.

NULL TREATMENT HYPOTHESIS: A hypothesis that states that the true underlying effect of the test treatment, as expressed by a specified outcome measure, is no more or less than for the control treatment.

NUTRIENT: Any item of food that nourishes or promotes growth and metabolism; may be essential or non-essential.

NUTRITION: The processes involved in the taking in and metabolism of food material by living plants and animals.

NYSTATIN: A brand name of Clotrimazole.

OCULAR: Pertaining to the eye.

OPPORTUNISTIC INFECTIONS: An infection in an immune compromised person caused by an organism that does not usually cause disease in healthy people. Many of these organisms are carried in a latent state by virtually everyone, and only cause disease when given the opportunity of a damaged immune system.

ORIGIN: Location where the process of replication in a nucleic acid begins.

ORPHAN DRUG: A category created by FDA for medications used to treat diseases that occur rarely, so there is little financial incentive for industry to develop them. Orphan drug status gives the manufacturer specific financial incentives to provide the drug.

OXIDATIVE STRESS: Physiological effects of increased levels of free radicals and oxidating molecules, associated with disease and aging; effects include cell membrane damage and cell death.

p24 ANTIGEN: A core protein making up the nucleocapsid of the HIV Virus, p24 was thought at one time to be a surrogate marker for disease progression. Now it is recognized that some long term asymptomatics have relatively high elevations of p24, while others die never having been positive (anything less than 10 picograms/mole is effectively negative)

P VALUE: The probability of obtaining a given outcome due to chance alone. For example, a study result with a significance level of $p \leq 0.05$ implies that 5 times out of 100 the result could have occurred by chance.

PALLIATIVE: A treatment which provides symptomatic relief, but not a cure.

PANCREATITIS: Inflammation of the pancreas. Pancreatitis is often characterized by abdominal pain, nausea/vomiting and elevated triglyceride and amylase levels. Pancreatitis, which can be fatal, is a known side effect of ddI

PANDEMIC: Referring to an-epidemic disease of widespread prevalence.

PAP SMEAR (PAPANICOLOU SMEAR): A specimen of vaginal or cervical cells placed on a slide and examined under the microscope for abnormal development.

PAPILLOMAVIRUS: A family of papova viruses associated with sexually transmitted diseases, including condylomata. Certain papillomavirus variants have also been associated with cervical cancer, particularly in HIV infected women.

PAPULE: A small raised bump or protrusion on the skin

PARALLEL TRACK: A system of distributing experimental drugs to patients who are unable to participate in ongoing clinical efficacy trials and have no other treatment options.

PARASITE: A plant or organism that lives on or in the host, deriving nourishment from it. Some cause inflammation, but others cause infection and destroy tissue. Human parasites include fungi, yeast, bacteria, protozoa, worms and viruses.

PARENTERAL: Not through the mouth. Intravenous, intramuscular, and intradermal administration are all parenteral

PARESTHESIA: Abnormal sensations: numbness, tingling, burning.

PAROMOMYCIN (HUMATIN): An antibiotic used for treating intestinal infections.

PASSIVE IMMUNOTHERAPY/PASSIVE IMMUNIZATION: Infusion of antibodies from another individual. AS distinguished from actively stimulating an immune response in the recipient.

PATHOGEN: Any disease-producing microorganism or material.

PATHOGENESIS: The natural evolution of a disease process in the body without intervention (*i.e.*, without treatment); Description of the development of a particular disease, especially the events, reactions and mechanisms involved at the cellular level.

PCR (POLYMERASE CHAIN REACTION): A highly sensitive test that can detect and/or DNA fragments of viruses or other organisms in blood or tissue. PCR works by repeatedly copying genetic material using heat cycling, and enzymes similar to those used by cells.

PEDIATRIC: Relating to the medical specialty concerned with the development, care and treatment of children from birth through adolescence.

PELVIC INFLAMMATORY DISEASE (PID): A painful gynecological condition usually caused by infection, where disease spreads upward from the vagina into the pelvic cavity.

PENTAM: Brand name of Pentamidine.

PENTAMIDINE: An antibiotic effective in an IV formulation against PCP. An aerosolized form is used to prevent PCP.

PEPTIDE: A short string of amino acids. Peptides are small proteins.

PERINATAL: Relating to the period around the time of birth.

PERIPHERAL: Toward the outside, or at the extremities.

PERIPHERAL NEUROPATHY: A disorder of the nerves that usually involves the feet or hands and sometimes the legs, arms and face. Symptoms may include numbness, a tingling or burning sensation, sharp pain, weakness and abnormal reflexes.

PEYER'S PATCH: Lymphatic tissue lining the intestines. One of the largest areas of lymphatic tissue in the body. May take over some of the role of the thymus in adults

pH: A term used to describe the acidity or alkalinity of a solution. It directly measures the hydrogen concentration of a solution. If the pH is low the solution is acidic; if high, alkaline.

PHAGOCYSTOSIS: The consumption and destruction of foreign materials by white blood cells like macrophages

PHARMACOKINETIC: Concerning the study of how a drug is processed by the body, with emphasis on the time required for absorption, duration of action, distribution in the body and method of excretion.

PHARMACEUTICAL AND REGULATORY AFFAIRS BRANCH: That office at NIAID responsible for coordinating communication with FDA and industry. In particular, PRAB is concerned with drugs under IND status, and with prompt reporting, summary, and analysis of adverse events in NIAID sponsored trials.

PHASE I TRIAL: The first stage in testing a new drug in humans. Performed as part of an approved Investigational New Drug Application under Food and Drug Administration guidelines. The studies are usually done to generate preliminary information on the chemical action and safety of the drug using normal healthy volunteers. Usually done without a comparison group.

PHASE II TRIAL: The second stage in testing a new drug in humans. Performed as part of an approved Investigational New Drug Application under Food and Drug Administration guidelines. Generally carried out on patients with the disease or condition of interest. The main purpose is to evaluate activity, and possibly provide preliminary information on treatment efficacy and to supplement information on safety obtained from phase I trials. Usually, but not always, designed to include a control treatment and random allocation of patients to treatment.

PHASE III TRIAL: The third and usually final stage in testing a new drug in humans. Performed as part of an approved Investigational New Drug Application under Food and Drug Administration guidelines. Concerned primarily with assessment of dosage effects and efficacy and safety. Usually designed to include a control treatment and random allocation to treatment. Once this phase is completed the drug manufacturers may request permission to market the drug by submission of a New Drug Application to the Food and Drug Administration, assuming the results of the phase I, II and III trials are consistent with such a request.

PHASE IV TRIAL:Generally, a randomized controlled trial that is designed to evaluate the long-term safety and efficacy of a drug for a given indication and that is done with Food and Drug Administration approval. Usually carried out after licensure of the drug for that indication.

PICO: Prefix meaning a one one-trillionth, as in picogram, a trillionth of a gram.

PILOT STUDY: A preliminary study designed to indicate whether a larger study is practical. See "Feasibility Study."

PHOSPHORYLATION: The process of adding phosphorus to a compound. Often done within the body by enzymes named "kinases". The nucleoside drugs, like AZT, ddI, ddC, etc. all need to be phosphorylated in the body before they become active.

PID: Patient identification number. A unique identifier that refers to a particular patient, yet preserves confidentiality for record keeping

PLACEBO: A pharmacologically inactive agent given to a patient as a substitute for an active agent and where the patient is not informed whether he is receiving the active or inactive agent.

PLACEBO EFFECT: The effect produced by a placebo due to the expectations of the patient. The effect in placebo-controlled clinical trials is generally measured by comparison of the effect observed in patients receiving the placebo treatment with the effect observed in patients receiving the active treatment.

PLACENTA: A combination of fetal and maternal cells that serves as the organ of exchange for nutrients and other chemicals between mother and fetus during pregnancy.

PLATELET: Blood cells that are essential to clotting.

PNEUMOCYSTIS CARINII PNEUMONIA: An opportunistic pneumonitis often seen in HIV-infected patients. PCP generally produces a dry, hacking cough. Although previously thought to be a protozoa, and responsive to anti-protozoal treatment, recent genetic analysis suggest that *p carinii* is closer to the funguses.

POLYNEUROPATHY: Neuropathy involving a number of different nerves

POST-HERPETIC NEURALGIA (PHN): Literally "pain following herpes." The term is usually applied to the severe pain that sometimes follows the healing of herpes zoster lesions.

POST HOC ANALYSES: Analyses conducted after the results are available that were not defined before the start of the trial. Such analyses are particularly prone to false-positive claims or type I error.

POSTHUMOUS: After Death

POST-MARKETING SURVEILLANCE: Term used by the Food and Drug Administration to characterize any procedure, implemented after licensure of a drug for a given indication, that is designed to provide information on the actual use of the drug for that indication and on the occurrence of related side effects. The surveillance usually involves survey techniques rather than controlled trials.

POST-MORTEM: After death, or after the event.

POST-PARTUM: After giving birth.

POST-STRATIFICATION: The process of classifying patients into strata after they have been enrolled in the study‹usually for data analysis purposes.

PREMATURE TERMINATION: Early termination of a trial before data are sufficiently strong to be convincing.

PRENATAL: Relating to the period before birth.

PRERANDOMIZATION EXAMINATION: Any examination that is part of the evaluation process of a patient for enrollment into a trial and that is carried out before the randomization examination.

PRERANDOMIZATION VISIT: Any visit made to the clinic by a potential study patient for the purpose of evaluation for enrollment into the trial and that takes place prior to the randomization visit.

PRESUMPTIVE: Presumed. In the context of diagnosis, one that is not definitely made, but where signs and symptoms make it exceedingly likely to be the proper one, even without confirmatory evidence. Diagnosis of CMV retinitis by characteristic appearance, without using a biopsy is an example if a presumptive diagnosis.

PRIMARY DRUG RESISTANCE (PDR): Resistance of bacteria or other pathogens to drugs which exists prior to the beginning of treatment.

PRION: An infectious crystallizing protein, which affects the brain.

PROCRIT: Brand name of Erythropoietin

PRODRUG: A chemical precursor of a drug that is converted into the desired substance in the body.

PRODROMAL: Pertaining to symptoms indicating the onset of a disease. May include symptoms prior to those adequate for accurate diagnosis.

PROGNOSIS: The probable future course of a disease.

PROGRESSION OF DISEASE: A common endpoint used in AIDS clinical trials. The entity "progression of disease" usually consists of new occurrences of non-recurrent AIDS-defining illnesses or death.

PROGRESSIVE MULTIFOCAL LEUKOENCEPHALOPATHY: A rapidly degenerative neurological condition associated with HIV, characterized by diffuse gray-matter pallor on CT, and no focal lesions. Thought to be associated with JC Papovavirus.

PROKINE: Brand name of Sargramostim

PROPHYLACTIC TRIAL: A trial that is designed to assess the efficacy of a treatment procedure aimed at preventing the development or progression of a specific disease or condition.

PROPHYLAXIS: Treatment intended to preserve health and prevent the occurrence or recurrence of a disease.

PROTEASE: An enzyme that cleaves proteins. HIV protease is required to separate the long gag-pol polyprotein into it's constituent parts during the process of viral replication.

PROTEASE INHIBITOR: A new class of experimental antiretroviral drugs that work by inhibiting the HIV protease. Some examples include Saquinavir, Indinavir, and Ritonavir.

PROTEIN: A large group of substances made up of amino acids that are formed naturally by plants and all living organisms. An essential human nutrient, proteins provide the structures essential for the growth and repair of living cells and tissue.

PROTOCOL: A detailed plan for studying a treatment for a specific condition.

PROTOZOA: A family of unicellular organisms including amoebas, that are the simplest form of animal life. Protozoa can be a cause of parasitic disease. In AIDS, Toxoplasmosis, Pneumocystis carinii, Giardia, and Cryptosporidium belii are among the most harmful protozoa.

PROVIRUS: Viral genetic material, in the form of DNA, that has been integrated into the host genome. HIV, when it is dormant in human cells, is in a proviral form

PRURITIC: Causing itching

PRURITUS: An itching rash

PULMONARY: Pertaining to the lungs.

PURIFIED PROTEIN DERIVATIVE (PPD): The most common test for exposure to *m. tuberculosis*, the bacteria that causes TB. In the PPD test, a small amount of protein from TB is injected under the skin. If the patient has been previously infected, they will mount a delayed type hypersensitivity reaction, characterized by a hard red bump called an induration.

PYRIMETHAMINE (DARAPRIM): An antibiotic used in treating toxoplasmosis, and rarely for the treatment or prophylaxis of PCP

QUALITY ASSURANCE: Any procedure, method, or philosophy for collecting, processing, or analyzing data that is aimed at maintaining or improving the reliability or validity of the data and the associated procedures used to generate them.

QUALITATIVE: Of, relating to, or expressed in relative or subjective terms‹impossible to precisely quantify.

QUANTITATIVE: Of, relating to, or expressed in terms of quantity.

RADICULOPATHY: Infection or other damage of the peripheral nerves and spinal roots, accompanied by weakness, numbness, and eventual paralysis. Radiculopathy is distinguished from peripheral neuropathy frequently by its asymmetric presentation

RADIOLOGY: The science of diagnosis and/or treatment using radiant energy. Includes X-rays, MRI, destruction of tumors by radiation, etc.

RADIOGRAPHY: The use of X-rays in diagnosis

RALES: Sounds in the lungs, often indicative of disease.

RECEPTOR: A structure on a cell which joins with proteins to produce changes in cellular function.

RECOMBINANT: Produced by genetic engineering in the laboratory.

RECORD: A paper or electronic document that contains or is designed to contain a set of facts related to some occurrence, transaction, or the like

RECTUM: The very bottom-most portion of the lower intestine, including the anus.

RECURRENCE: The return or flare up of a condition thought cured or in remission

REFRACTORY: Resistant to treatment, as of a disease.

REGIMEN: Any particular treatment plan specifying which drugs are used, in what doses, according to what schedule, and for how long.

REGRESSION: Any of several statistical techniques concerned with predicting some variables by knowing others. Regression is used to answers such questions as "how well can I predict the values of one variable, (such as survival) by knowing the value of another variable (such as treatment assignment)

REGRESSION ANALYSIS: Methods of explaining or predicting the variability of a dependent variable using information about one or more independent variables, or techniques for establishing a regression equation. The equation indicates the nature and closeness of the relationship between two or more variables, specifically the extent to which you can predicted some by knowing other, the extent to which some are associated with others. THe equation is often represented by a regression line, which is the straight line that comes closest to approximating a distribution of points in a scatter diagram.

REMISSION: A reduction of the severity or duration of a condition, or the abatement of symptoms altogether over a period of time.

REPLICATION: The action or process of reproducing exact copies of one's self.

RESISTANCE (BACTERIAL, FUNGAL, or VIRAL): Refers to the ability of some pathogens to grow and multiply even in the presence of certain drugs which normally kill them. (Such strains are referred to as "drug resistant strains.")

RESPONSE VARIABLE: Outcome variable.

RETINITIS: Inflammation of the retina, linked in AIDS to CMV infection. Untreated, it can lead to blindness.

RETROVIRUS: A class of viruses, which copy genetic material using RNA as a template for making DNA (HIV is a retrovirus).

REVERSE TRANSCRIPTASE: A retroviral enzyme that is capable of copying RNA into DNA, an essential step in the life-cycle of HIV. AZT, ddI and ddC act against reverse transcriptase.

RIBAVIRIN: An antiviral drug, used for treatment of Respiratory Syncytial Virus infection. May have some activity against HIV.

RIBONUCLEIC ACID (RNA): A single stranded molecule that carries genetic information. In the human body, DNA is transcribed to RNA, which creates proteins. In HIV and other retroviruses, the process is reversed, and RNA (The native form of the virus) is transcribed to DNA for insertion in the human genome.

RIBOSOME: The spherical structure that assembles proteins after being fed the genetic instructions by mRNA.

RIFABUTIN (MYCOBUTIN): An antibiotic used as a component in the combination treatment of *mycobacterium avium* complex. Rifabutin is also the only drug currently approved for the prevention of MAC.

RIFADIN: Brand name of Rifampin.

RIFAMATE: A combination of Rifampin and Isoniazid in one pill.

RIFAMPIN (RIFADIN, RIMACTANE): An antibiotic used alone, or as a component in the combination treatment of tuberculosis.

RIMACTANE: Brand name of Rifampin

RISK FACTOR: Anything in the environment, personal characteristics, or events that make it more or less likely one might develop a given disease or experience a change in health status.

RISK FACTOR ANALYSIS: Any analysis, usually involving regression or subgroup analyses, that is aimed at identifying risk factors for a given disease or condition.

Rx: Common abbreviation for "prescription"

SALINE: Salt water

SALINITY: The amount of salt in a solution.

SALMONELLA: A ubiquitous family of bacteria, salmonella can cause serious disseminated disease in HIV positive patients.

SALVAGE THERAPY: A treatment used when the usual treatment(s) have failed. A last ditch option.

SAMPLE SIZE: The number of patients required for a trial. In planning an interventional trial, a sample size is calculated, taking into account the frequency of the condition to be prevented or delayed, the anticipated effectiveness of the treatment thought to be clinically significant, and variables such as predicted drop outs and cross overs. The sample size is a calculated value that takes all of the above into account, and makes a statistically significant result likely within the prescribed amount of time.

SAQUINAVIR (INVIRASE): A protease inhibitor drug made by Hoffman-LaRoche.

SARCOMA: A malignant tumor of the skin or soft tissues.

SARGRAMOSTIM (GRANULOCYTE-MACROPHAGE COLONY STIMULATING FACTOR, GM-CSF): A recombinant version of an endogenous cytokine that stimulates the production of white blood cells. Can be used for treatment of neutropenia attendant upon chemotherapy, or treatment with AZT, but

Filgrastim is usually used instead, because of concerns that Sargramostim may increase HIV replication.

SAS: Statistical Analysis System (a package of data analysis programs.

SEPSIS: The presence of harmful microorganisms or associated toxins in the blood

SEPTICEMIA: Disease due to sepsis

SEPTRA: Brand name of trimethoprim-sulphamethoxazole

SEROCONVERSION: The development of antibodies detected by blood testing. In HIV, seroconversion is the time when you first test positive.

SEROLOGIC TEST: Any of a number of tests that are performed on the clear, liquid portion of blood (serum). Often refers to a test which determines the presence of antibodies to antigens such as viruses.

SEROSTATUS: Refers to the presence or absence of antibodies in the serum portion of blood.

SERUM: The clear, non-cellular, fluid portion of the blood.

SHAM: Something false presented to be genuine; a spurious imitation. Derived from the word *shame*, meaning trick or fraud.

SHINGLES (HERPES ZOSTER): A skin condition characterized by painful blisters in a linear distribution on one side of the body that generally dry and scab, leaving minor scarring. Shingles is caused by the re-activation of a previous infection with the varicella-zoster virus that causes chickenpox, usually early in life. Shingles may be a symptom of HIV disease progression. Shingles may recur in people with poor immunity.

SIDE EFFECTS: The action or effect of a drug beyond what it is supposed to do. The term usually refers to undesired or negative effects, such as headache, skin irritation, or liver damage‹although side effects can be expected or unexpected, desired or undesired. Experimental drugs must be evaluated for both immediate and long-term side effects.

SIGNIFICANCE, CLINICAL: Statistical significance does not necessarily imply that the results have practical implications or are clinically significant from the clinician's point of view, and vice versa.

SIGNIFICANCE LEVEL: (statistics) 1. The permissible type I error level for a test of the null hypothesis with a specified test statistic. The null hypothesis not rejected if the test statistic yields a p-value which is larger than the specified level and is rejected if it is equal to or less than this value. 2. p-value.

SIGNIFICANCE, STATISTICAL: Infers that an observation was unlikely to have occurred by chance alone. Statistical significance is often based on a p value ≤ 0.05. Below this level, the smaller the p value, the greater the statistical significance.

SIMIAN IMMUNODEFICIENCY VIRUS (SIV): A retrovirus closely related to HIV, which causes disease in monkeys

SIMPLE RANDOMIZATION: Unrestricted randomization.

SINUSITIS: An infection of the sinus cavities in the head, often bacterial.

SPARFLOXACIN: An experimental flouroquinolone antibiotic active against *mycobacterium avium* complex.

SPLEEN: A lymphoid organ in the abdominal cavity that is an important center for immune system activities.

STANDARD DEVIATION: A Statistic that shows the spread or dispersion of scores in a distribution of scores. in other words, a measure of dispersion. The more widely the scores are spread out, the greater the standard deviation. the standard deviation is the square root of the variance. For example: imagine trued groups of patients. In one CD4 cell counts range from 310-650, in the other they range from 43-700. The standard deviation would be much greater for the second group.

STANDARD ERROR: Short for "standard error of the mean", or "standard error of the estimate". A statistic indicating how greatly the mean score of a single sample is likely to differ from the mean score of the population. It is the standard deviation of a sampling distribution of the means. The Standard error of the mean indicates how much the sample mean differs from the expected value. The standard error of estimate is how much you are off when using a regression line to predict particular scores. In either case, the smaller the standard error, the better the sample statistic is an estimate of the population parameter. The standard error is a measure of sampling error, due to random fluctuations in the samples.

STATISTICAL TEST, ANALYSIS OF VARIANCE (ANOVA): A statistical technique used to test the difference among the means of two groups or more regarding continuous data. The purpose of ANOVA is to test whether the observed differences among the means of these groups is significant or due to chance.

STATISTICAL TEST, CHI-SQUARE: A statistical technique used to test the difference between groups regarding discrete (categorical) data. The purpose of chi square is to test whether differences between observed and expected frequencies are significant or due to chance.

STATISTICAL TEST, MULTIPLE REGRESSION: A statistical technique used to assess the relationship between one dependent variable and two or more input variables. The purpose of multiple regression analysis is to measure the degree of association between variables or predict the dependent variable as a function of the input variables.

STATISTICAL TEST, T-TEST: A statistical technique used to test whether the means of two groups differ significantly regarding continuous data. The purpose of the t-test is to infer whether the difference between the means of two groups on one variable is statistically significant or due to chance.

STATISTICS, DESCRIPTIVE: The intent of descriptive statistics is to summarize and present data, *e.g.*, measures of central tendency (mean, mode, median) and measures of variability (standard deviation, variance, standard error of the mean).

STATISTICS, INFERENTIAL: Inference refers to making generalizations about a population on the basis of a sample from that population. The intent of inferential statistics is to determine whether differences between groups are real or due to chance.

STAVUDINE: A nucleoside antiretroviral drug recently approved for treatment of advanced HIV disease.

STEERING COMMITTEE (SC): 1. A committee responsible for directing the activities of a designated project. 2. One of the key committees in the organizational structure of a multicenter clinical trial. Committee responsible for conduct of the trial and to which all other committees report, except the adverse experience committee and Data and Safety Monitoring Board or advisory-review and treatment effects monitoring committee.

STEM CELLS: Cells from which all blood cells derive. Bone marrow is rich in stem cells.

STEROIDS: A large family of structurally similar chemicals. Various steroids have sex determining, anti-inflammatory, and growth-regulatory roles.

STEVEN JOHNSON REACTION: A severe allergic drug reaction, Steven Johnson is characterized by high fevers and desquamating rash, and is frequently fatal.

STOMATITIS: Inflammation of the upper gastric tract, including the mouth.

STRAIN: A specific type, quality, or disposition of a material.

STRATA: (pl. of stratum) A series of distinct levels or layers. In this book, generally subgroups of patients formed by classification on some variable or set of variables, usually baseline variables.

STRATIFICATION: 1. The process of classifying observation units into strata. 2. The process of classifying patients into strata as part of the randomization process or for purposes of data analysis.

SUBCLINICAL INFECTION: An infection, or phase of infection, without readily apparent symptoms or signs of disease.

SUBCUTANEOUS: Beneath or introduced beneath the skin (*e.g.*, subcutaneous injections).

SUBTYPE: A genetic variant.

SULFADIAZINE: A sulfa antibiotic, sometimes used in the treatment of Toxoplasmosis, especially in combination with pyrimethamine

SULPHAMETHOXAZOLE: A sulfa antibiotic, and half of the combination "trimethoprim-sulphamethoxazole", it is rarely used alone.

SULFADOXINE: A sulfa antibiotic, only occasionally used in AIDS

SUSCEPTIBLE: Vulnerable or predisposed to a disease. Also refers to bacteria which can be killed or inhibited by the drugs used against them.

SYMMETRIC: Equally divided, and identical on both sides of a middle line; mirror image

SYMPTOMATIC: Showing clinical pathology or changes indicative of disease, such as HIV disease. Often symptomatic HIV disease means clinical evidence of HIV infection in absence of an AIDS diagnosis, indicative of, relating to or constituting the aggregate of symptoms of a disease.

SYMPTOMS: Any perceptible, subjective change in the body or its functions that indicates disease or phases of disease, as reported by the patient.

SYNCYTIA: A mass of cells which fuse together to form one "giant cell." In HIV infection this condition leads to direct cell-to-cell infection and continued HIV replication.

SYNDROME: A group of symptoms and diseases that together are characteristic of a specific condition.

SYNERGISM/SYNERGISTIC: An interaction between two or more agents (drugs) that produces or enhances an effect which is greater than the sum of the individual agents.

SYSTEMIC: Throughout the body. Sometimes applies to medications that are taken orally or parenterally that saturate the entire body.

T CELLS (T LYMPHOCYTES): A thymus derived white blood cell that precipitates a variety of cell mediated immune reactions. Three fundamentally different types of T cells are recognized: helper, killer, and suppresser (each has many subdivisions).

T HELPER CELLS: Lymphocytes responsible for assisting other white blood cells in responding to infection, processing antigen, and triggering antibody production (also known as T4 cells, CD4 cells).

T KILLER CELLS: A major component of cytotoxic lymphocyte response (CTL), responsible for lysing infected or cancerous cells, T killer cells (not to be confused by natural killer cells) are a subset of CD8+ lymphocytes.

T SUPPRESSER CELLS: T lymphocytes responsible for turning the immune response off after infection is cleared, a subset of CD8+ lymphocytes.

TERATOGENICITY: The production of physical defects in offspring in utero.

TEST OF SIGNIFICANCE: 1. The evaluation of observed data by calculating a specified test statistic and then deriving the associated p-value. 2. Test statistic. See "Significance"

TEST TREATMENT: The drug, device, or procedure to be evaluated in a particular trial.

TH1/TH2 CELLS: Subsets of T-helper lymphocytes, involved in cell-mediated immune responses. TH1 cells secrete IL-1 and gamma interferon, which enhance cell-mediated responses and inhibit both TH2 subset cell activity and the humoral immune responses. TH2 cells, the other subset of T-helper cells, are also involved in cell-mediated immune responses. TH2 cell activity and secretions are thought to inhibit cell-mediated responses and to enhance the humoral response. TH2 cells secrete IL-4 and IL-10.

THERAPEUTIC TRIAL: A trial designed to test the safety and efficacy of a particular drug, device, or procedure that is considered to have therapeutic value.

THYMUS: A lymphoid organ in the upper chest cavity; site of T-lymphocyte differentiation and hormone secretion.

TID: Common abbreviation for "three times a day"

TIME OF ENROLLMENT: The time point at which a patient (treatment unit) is regarded as having officially entered the trial and after which is regarded as a part of the study population. Operationally, the time point at which the treatment assignment is revealed to clinic staff, or when treatment is initiated when assignments are known in advance of enrollment.

TITER ("TITRE"): A laboratory measurement of the amount (or concentration) of a given component in solution.

T LYMPHOCYTE CELLS: Thymus dependent cells, coordinate the cell-mediated immune system.

TOPICAL: Applied to the skin, or other external area.

TORULOPSIS GLABRATA: A fungal infection similar to *Candida albicans*, but less susceptible to many of the treatments used for candida

TOTAL PARENTAL NUTRITION (TPN): A type of nutritional feeding that delivers all nutrients in liquid form through a plastic tube into a vein.

TOXICITY: The extent, quality, or degree of being poisonous or harmful to the body.

TOXIN: A harmful or poisonous agent.

TOXOPLASMOSIS: A life-threatening opportunistic infection caused by a microscopic parasite (*Toxoplasma gondii*) found in raw or undercooked meat and cat feces. Symptoms may be so mild as to be barely noticeable or may be more severe with headache, lymphadenopathy, malaise, muscle pain, fever and dementia. Toxoplasmosis may lead to brain swelling, coma and death in people with suppressed immune systems.

TRANSCRIPTION: Constructing a mRNA molecule using a DNA molecule as a template; results in the transfer of genetic information to the mRNA.

TRANSCUTANEOUS: The passage of substances through unbroken skin, as in absorption.

TRANSIENT: Short-lived; passing; not permanent.

TREATMENT: 1. The act of treating, as in caring for a patient. 2. The specific regimen, method, or procedure being tested in a clinical trial.

TREATMENT ALLOCATION: 1. The process of assigning patients to treatment. 2. The treatment assignment of a particular patient.

TREATMENT ASSIGNMENT: The treatment to be administered to a patient, but may also mean treatment assigned to some other larger unit such as all members of a family or members of a hospital ward, as indicated in the treatment allocation schedule.

TREATMENT ASSIGNMENT PROBABILITY: The probability associated with a specified treatment assignment. The value is fixed over the course of patient enrollment in trials with fixed allocation designs. It changes in trials using adaptive allocation designs.

TREATMENT SIDE EFFECT: A by-product of treatment, either expected or unexpected, desired or undesired.

TRIMETHOPRIM-SULFAMETHOXAZOLE: A combination of two antibiotics that is the preferred treatment for the prophylaxis and treatment of PCP. Active against a wide range of other organisms, possibly including *Toxoplasma gondii*. Generally administered Intravenously or (less often) orally for treatment of PCP, and orally for prophylaxis

TRIMETREXATE: A drug used for "salvage" therapy of *pneumocystis carinii* pneumonia, when other treatments have failed. Trimetrexate is enormously toxic unless administered with leucovorin, which "rescues" human cells from it's effects.

TRIPLE-BLINDED: Sometimes used in a jocular fashion to characterize a situation in which neither the patient, physician, nor statistician knows how the trial is designed or operated.

TUBERCULOSIS: The disease caused by *M. tuberculosis* (or rarely *M. bovis*). Condition in which tuberculous infection has progressed so that the individual typically has signs and symptoms of illness, an abnormal radiograph, a "positive" bacteriological examination (smear and/or culture), as well as a positive tuberculin reaction. Individuals with disease may be infectious.

TUMOR NECROSIS FACTOR: A macrophage produced cytokine that helps activate T-cells. It is also thought to upregulate HIV replication, and may contribute to the pathogenesis of wasting.

UNBLIND: To reveal the treatment assignment of an individual patient or group of patients to an individual or group of individuals associated with the trial (*e.g.,* patients, study physicians, treatment effects monitoring committee) who have heretofore been denied this information.

VACCINATION: The act of administering a vaccine.

VACCINE: A substance that contains antigenic components from an infectious organism. By stimulating an immune response (but not disease), it leads to immunity to a certain microorganism and protects against subsequent infection by that organism.

VARIANCE: The degree to which a set of quantities vary: A measure of the spread of scores in a distribution of scores, that is, a measure of dispersion. The larger the variance, the further the individual cases are from the mean. The smaller the variance, the closer the individual score are to the mean. Specifically, the population variance is the mean of the sum of the squared deviations from the mean score. The Sample variance is computed by dividing the sum of squared deviations by the number in the sample minus 1. The Standard Deviation is the square root of the variance.

VARIANT: A variation of a particular strain of virus or infective agent; slightly different in form or function.

VARICELLA-ZOSTER VIRUS (VZV): A virus in the herpes family that causes chickenpox during childhood and may reactivate later in life to cause herpes zoster (shingles) in immunosuppressed individuals.

VARIX: A painfully enlarge blood vessel; plural is varices

VECTOR: Anything capable of moving or transferring genetic material.

VERTICAL TRANSMISSION: Transmission of HIV from mother to fetus.

VIRAL BURDEN/LOAD: The concentration of a virus in the body.

VIREMIA: The presence of virus in the blood.

VIROID: A very exotic type of virus-like particle that only infects plant cells and consists of a group of membraneless circular RNAs that neither code for nor contain any structural protein. They replicate without other viruses being present.

VIRION: A virus particle existing freely outside a host cell.

VIROLOGY: The study of viruses and viral disease.

VIRULENCE: Refers to the ability of a microorganism to produce serious disease. *Tuberculosis* is a virulent organism. Some nontuberculous mycobacteria are virulent (*e.g., M. kansasii*), while others (*e.g., M. gordonae*) are not. (PATHOGENICITY is a related–though not identical–concept.)

VIRUS: A group of infectious agents characterized by their inability to reproduce outside of a living host cell. Viruses may subvert the host cells' normal functions, causing the cell to behave in a manner determined by the virus.

VISCERAL: Pertaining to the major internal organs.

VITREOUS HUMOR: The gel-like substance that fills the eyeball.

WASTING SYNDROME: A condition among HIV-infected individuals characterized by involuntary weight loss of more than 10 per cent of baseline body weight. Other symptoms may include chronic diarrhea or chronic weakness and fever for more than 30 days; a CDC AIDS-defining condition.

WESTERN BLOT: A laboratory test of blood for specific antibodies; more accurate than the ELISA test, the Western blot is used as a confirmatory test if an HIV ELISA test is positive.

WILD-TYPE VIRUS: The customary type of a virus before genetic manipulation or mutation; virus isolated from an individual, as opposed to from a lab culture

ZITHROMAX: Brand name for Azithromycin

ZOVIRAX: Brand name for Acyclovir.

References

Abedon, S. T. 2000. The murky origin of Snow White and her T-even dwarfs. Genetics 155:481-486. *(historical description of the isolation of the T4-like phages T2, T4, and T6)*

Acebedo G, Hayek A, Klegerman M, Crolla L, Bermes E, Brooks M (1975). "A rapid ultramicro radioimmunoassay for human thyrotropin". *Biochem. Biophys.*

Ackermann, H.-W., and H. M. Krisch. 1997. A catalogue of T4-type bacteriophages. Arch Virol 142:2329-2345. *(nearly complete list of then-known T4-like phages).*

Air, G.M., Coulson, A.R., Fiddes, J.C., Friedmann, T., Hutchison, C.A., Sanger, F., Slocombe, P.M. and Smith, A.J., Nucleotide sequence of the F protein coding region of bacteriophage PhiX174 and the amino acid sequence of its product, J. Mol. Biol., 125, 247-254, 1978.

Alcamí A, Symons JA, Smith GL (December 2000). "The vaccinia virus soluble alpha/beta interferon (IFN) receptor binds to the cell surface and protects cells from the antiviral effects of IFN". *J. Virol.* 74 (23): 11230–9. doi:10.1128/JVI.74.23.11230-11239.2000. PMID 11070021. PMC: 113220.

Alwine JC (2008). "Modulation of host cell stress responses by human cytomegalovirus". *Curr. Top. Microbiol. Immunol.* 325: 263–79. PMID 18637511.

Amrine and Stansy, *Catalogue of the Eriophyoidea (Aceria: Prostigmata) of the World*, Indira Publishing, Bloomfield, MI, USA. 1186 pp., 1994.

Baldacci, Belli, Betto and Refatti, *Annali della Facoltà di Agraria dell' Università di Milano* 10: 23, 1962.

Barbanti-Brodano G, Sabbioni S, Martini F, Negrini M, Corallini A, Tognon M (2004). "Simian virus 40 infection in humans and association with human diseases:

results and hypotheses". *Virology* 318 (1): 1–9. doi:10.1016/j.virol.2003.09.004. PMID 15015494.

Barbier, Demangeat, Perrin, Cobanov, Jaquet and Walter, *Extended Abstracts of the 12th Meeting of ICVG, Lisbon 1997*: 131, 1997.

Barozzi P, Potenza L, Riva G, Vallerini D, Quadrelli C, Bosco R, Forghieri F, Torelli G, Luppi M (December 2007). "B cells and herpesviruses: a model of lymphoproliferation". *Autoimmun Rev* 7 (2): 132–6. doi:10.1016/j.autrev.2007.02.018. PMID 18035323. Retrieved on 20 December 2008.

Bass, Vuittenez and Legin, *Proceedings of the 6th Meeting of ICVG, Cordoba 1976*: 325, 1978.

Benfey, P.; Protopapas, A.D. (2004). *Essentials of Genomics*. Prentice Hall.

Brown, Terence A. (2002). *Genomes 2*. Oxford: Bios Scientific Publishers. ISBN 978-1859960295.

Bhatti Z, Berenson CS (2007). "Adult systemic cat scratch disease associated with therapy for hepatitis C". *BMC Infect Dis* 7: 8. doi:10.1186/1471-2334-7-8. PMID 17319959.

Boscia *et al.*, In *Sanitary Selection of the Grapevine. Protocols for the Detection of Viruses and Virus-like Diseases*, p. 129, ed. B Walter, Les Colloques INRA n°86,

Cabrera, C.M. *et al.* (2006) Identity tests: Determination of cross contamination. *Cytotechnology* 51, 45–50. DOI: 10.1007/s10616-006-9013-8

Chang HW, Watson JC, Jacobs BL (June 1992). "The E3L gene of vaccinia virus encodes an inhibitor of the interferon-induced, double-stranded RNA-dependent protein kinase". *Proc. Natl. Acad. Sci. U.S.A.* 89 (11): 4825–9. doi:10.1073/pnas.89.11.4825. PMID 1350676. PMC: 49180.

Chatterjee, R. (2007) Cell biology. Cases of mistaken identity. *Science* 315, 928–931 PMID: 17303729.

Cheng S, Fockler C, Barnes WM, Higuchi R (1994). "Effective amplification of long targets from cloned inserts and human genomic DNA". *Proc Natl Acad Sci.* 91: 5695–5699. doi:10.1073/pnas.91.12.5695. PMID 8202550.

Chibani-Chennoufi, S., C. Canchaya, A. Bruttin, and H. Brussow. 2004. Comparative genomics of the T4-Like *Escherichia coli* phage JS98: implications for the evolution of T4 phages. J. Bacteriol. 186:8276-8286. *(characterization of a T4-like phage)*.

Desplats, C., C. Dez, F. Tetart, H. Eleaume, and H. M. Krisch. 2002. Snapshot of the genome of the pseudo-T-even bacteriophage RB49. J. Bacteriol. 184:2789-2804. *(overview of the RB49 genome, a T4-like phage)*.

Desplats, C., and H. M. Krisch. 2003. The diversity and evolution of the T4-type bacteriophages. Res. Microbiol. 154:259-267. *(characterization of T4-like phages)*.

Dickinson, M. 2003. *Molecular Plant Pathology*. BIOS Scientific Publishers.

Dinant, Lot, Albouy, Kuziak, Meyer and Astier-Manifacier, *Archives of Virology* 116: 235, 1991.

Dinant, Maisonneuve, Albouy, Chupeau, Chupeau, Bellec, Gaudefroy, Kusiak, Souche, Robaglia and Lot, *Molecular Breeding* 3: 75, 1997.

Drexler, H.G. *et al.* (2001) Cross-contamination: HS-Sultan is not a myeloma but a Burkitt lymphoma cell line. *Blood* 98, 3495–6. PMID: 11732505.

Duijsings *et al.*, *In vivo* analysis of the TSWV cap-snatching mechanism: single base complementarity and primer length requirements. The EMBO Journal, Vol. 20 pp. 2545–2552.

Dunham, J.H. and Guthmiller, P. (2008) Doing good science: Authenticating cell line identity. *Cell Notes* 22, 15–17.

Drexler, H.G., Dirks,W.G. and MacLeod, R.A.F. (1999) False human hematopoetic cell lines: cross-contaminations and misinterpretations. *Leukemia* 13, 1601–1607. PMID: 10516762.

Eddy, S. R. 1992. Introns in the T-Even Bacteriophages. Ph.D. thesis. University of Colorado at Boulder. *(chapter 3 provides overview of various T4-like phages as well as the isolation of then-new T4-like phages)*

Eibl RH, Kleihues P, Jat PS, Wiestler OD (1994) A model for primitive neuroectodermal tumors in transgenic neural transplants harboring the SV40 large T antigen. Am J Pathol. 1994 Mar;144(3):556-64.

Fiers W *et al.*, Complete nucleotide-sequence of SV40 DNA, Nature, 273, 113-120, 1978.

Fiers, W., and R. L. Sinsheimer, "The structure of the DNA of bacteriophage PhiX174. III. Ultracentrifuge evidence for a ring structure", *J. Mol. Biol.* 5:424-434, 1962

Fiers W, *et al.* (1976). "Complete nucleotide-sequence of bacteriophage MS2-RNA– primary and secondary structure of replicase gene". *Nature* 260: 500–507. doi:10.1038/260500a0. PMID 1264203.

Fiers W, Contreras R, Haegemann G, Rogiers R, Van de Voorde A, Van Heuverswyn H, Van Herreweghe J, Volckaert G, Ysebaert M (1978). "Complete nucleotide sequence of SV40 DNA". *Nature* 273 (5658): 113–120. doi:10.1038/273113a0. PMID 205802.

Filee, J., F. Tetart, C. A. Suttle, and H. M. Krisch. 2005. Marine T4-type bacteriophages, a ubiquitous component of the dark matter of the biosphere. Proc. Natl. Acad. Sci. USA 102:12471-12476. *(indication of prevalence and T4-like phages in the wild)*

Fleischmann R, Adams M, White O, Clayton R, Kirkness E, Kerlavage A, Bult C, Tomb J, Dougherty B, Merrick J (1995). "Whole-genome random sequencing and assembly of Haemophilus influenzae Rd". *Science* 269 (5223): 496–512. doi:10.1126/science.7542800. PMID 7542800.

Frederick R. Blattner, Guy Plunkett III, *et al.* (1997). "The Complete Genome Sequence of *Escherichia coli* K-12". *science* 277: 1453–1462. doi:10.1126/science.277.5331.1453. PMID 9278503.

Gibson, Greg; Muse, Spencer V. (2004). *A Primer of Genome Science* (Second Edition ed.). Sunderland, Mass: Sinauer Assoc. ISBN 0-87893-234-8.

Goodpasture, E.W., Woodruff, A.M., Buddingh, G.J. (1931) "The cultivation of vaccine and other viruses in the chorioallantoic membrane of chick embryos" *Science* 74, pp. 371–372 PMID 17810781

Gregory, T. Ryan (ed) (2005). *The Evolution of the Genome*. Elsevier. ISBN 0-12-301463-8.

Greilhuber, J., Borsch, T., Müller, K., Worberg, A., Porembski, S., and Barthlott, W. (2006). "Smallest angiosperm genomes found in Lentibulariaceae, with chromosomes of bacterial size". *Plant Biology* 8: 770–777. doi:10.1055/s-2006-924101. PMID 17203433.

Hunter and Bowyer, *Journal of Phytopathology* 140: 11, 1994.

Hunter and Bowyer, *Journal of Phytopathology* 145: 521, 1997.

Holmes, *Handbook of phytopathogenic viruses*, p. 221, Minneapolis: Burgess Publishing Company, 1939.

Huttinga and Mosch, *Netherlands Journal of Plant Pathology* 80: 19, 1974.

Joseph Sambrook and David W. Russel (2001). *Molecular Cloning: A Laboratory Manual* (3rd ed. ed.). Cold Spring Harbor, N.Y.: Cold Spring Harbor Laboratory Press. ISBN 0-87969-576-5. Chapter 8: In vitro Amplification of DNA by the Polymerase Chain Reaction.

Joshua Lederberg and Alexa T. McCray (2001). "'Ome Sweet 'Omics – A Genealogical Treasury of Words". *The Scientist* 15 (7).

Jordan MC, Jordan GW, Stevens JG, Miller G (June 1984). "Latent herpesviruses of humans". *Ann. Intern. Med.* 100 (6): 866–80. PMID 6326635.

Karam, J. D. *et al.* 1994. Molecular Biology of Bacteriophage T4. ASM Press, Washington, DC. *(the second T4 bible, go here, as well as Mosig and Eiserling, 2006, to begin to learn about the biology T4 phage)*

Karam, J., Petrov, V., Nolan, J., Chin, D., Shatley, C., Krisch, H., and Letarov, A. The T4-like phages genome project. http://phage.bioc.tulane.edu/. *(the T4-like phage full genomic sequence depository).*

Kneller et all. Cap-independent translation of plant viral RNAs. Virus Research, Volume 119, Issue 1, July 2006, pp. 63–75.

Konstantin Kanyuka, Elaine Ward and Michael J. Adams. Polymyxa graminis and the cereal viruses it transmits: a research challenge. Molecular Plant Pathology, Vol 4 pp. 393–406.

Krause-Sakate, Fakhfakh, Peypelut, Pavan, Zerbini, Marrakchi, Candresse and Le Gall, *Archives of Virology*, 149: 191, 2004.

Krause-Sakate, Le Gall, Fakhfakh, Peypelut, Marrakchi, Varveri, Pavan, Souche, Lot, Zerbini and Candresse, *Phytopathology* 92: 563, 2002.

Kuida, Inouye and Inouye, *Nogaku Kenkyu* 56: 33, 1977.

Kutter, E., K. Gachechiladze, A. Poglazov, E. Marusich, M. Shneider, P. Aronsson, A. Napuli, D. Porter, and V. Mesyanzhinov. 1995. Evolution of T4-related phages. Virus Genes 11:285-297. *(comparison of the genomes of various T4-like phages).*

Lin RJ, Liao CL, Lin E, Lin YL (2004 aaa). "Blocking of the alpha interferon-induced Jak-Stat signaling pathway by Japanese encephalitis virus infection". *J. Virol.* 78 (17): 9285–94. doi:10.1128/JVI.78.17.9285-9294.2004. PMID 15308723.

Liscovitch, M. and Ravid, D. (2007) A case study in misidentification of cancer cell lines: MCF-7/AdrR cells (re-designated NCI/ADR-RES) are derived from OVCAR-8 human ovarian carcinoma cells. *Cancer Lett.* 245, 350–352. PMID: 16504380

Liu YJ (2005). "IPC: professional type 1 interferon-producing cells and plasmacytoid dendritic cell precursors". *Annu Rev Immunol* 23: 275–306. doi:10.1146/annurev.immunol.23.021704.115633. PMID 15771572.

Lowe DB, Shearer MH, Jumper CA, Kennedy RC (2007). "SV40 association with human malignancies and mechanisms of tumor immunity by large tumor antigen". *Cell. Mol. Life Sci.* 64 (7-8): 803–14. doi:10.1007/s00018-007-6414-6. PMID 17260087.

Lwoff A, Horne R, Tournier P (1962). "A system of viruses". *Cold Spring Harb. Symp. Quant. Biol.* 27: 51–5. PMID 13931895.

MacLeod, R.A.F. *et al.* (1999) Widespread intraspecies cross-contamination of human tumor cell lines arising at source. *Int. J. Cancer* 83, 555–563 PMID: 10508494

Madigan M, Martinko J (editors) (2006). *Brock Biology of Microorganisms* (11th ed. ed.). Prentice Hall. ISBN 0-13-144329-1.

Masters, John R. (2002): HeLa cells 50 years on: the good, the bad and the ugly. *Nature Reviews Cancer* 2:315-319. See List of contaminated cell lines.

Mathews, C. K., E. M. Kutter, G. Mosig, and P. B. Berget. 1983. Bacteriophage T4. American Society for Microbiology, Washington, DC. *(the first T4 bible; not all information here is duplicated in Karam et al., 1994; see especially the introductory chapter by Doermann for a historical overview of the T4-like phages)*

Milton Zaitlin and Peter Palukaitis (2000), *Advances in Understanding Plant Viruses and Virus Diseases.* Vol. 38: 117–143 (doi:10.1146/annurev.phyto.38.1.117)

Milton Zaitlin (1998), *Discoveries in Plant Biology*, New York 14853, USA. Pp.: 105-110. S.D Kung and S. F. Yang (eds).

Miller, E. S., E. Kutter, G. Mosig, F. Arisaka, T. Kunisawa, and W. Ruger. 2003. Bacteriophage T4 genome. Microbiol. Mol. Biol. Rev. 67:86-156. *(review of phage T4, from the perspective of its genome).*

Miller JE, Samuel CE (September 1992). "Proteolytic cleavage of the reovirus sigma 3 protein results in enhanced double-stranded RNA-binding activity: identification of a repeated basic amino acid motif within the C-terminal binding region". *J. Virol.* 66 (9): 5347–56. PMID 1501278. PMC: 289090.

Minks MA, West DK, Benvin S, Baglioni C (October 1979). "Structural requirements of double-stranded RNA for the activation of 2′,5′-oligo(A) polymerase and protein kinase of interferon-treated HeLa cells". *J. Biol. Chem.* 254 (20): 10180–3. PMID 489592. http://www.jbc.org/cgi/pmidlookup?view=long&pmid=489592.

Moiseeva O, Mallette FA, Mukhopadhyay UK, Moores A, Ferbeyre G (2006). "DNA damage signaling and p53-dependent senescence after prolonged beta-interferon stimulation". *Mol. Biol. Cell* 17 (4): 1583–92. doi:10.1091/mbc.E05-09-0858. PMID 16436515.

Monod, C., F. Repoila, M. Kutateladze, F. Tétart, and H. M. Krisch. 1997. The genome of the pseudo T-even bacteriophages, a diverse group that resembles T4. J. Mol. Biol. 267:237-249. *(overview of various T4-like phages from the perspective of their genomes)*

Mosig, G., and F. Eiserling. 2006. T4 and related phages: structure and development, R. Calendar and S. T. Abedon (eds.), The Bacteriophages. Oxford University Press, Oxford. *(review of phage T4 biology)*

Nakabachi A, Yamashita A, Toh H, *et al.* (October 2006). "The 160-kilobase genome of the bacterial endosymbiont Carsonella". *Science (journal)* 314 (5797): 267. doi:10.1126/science.1134196. PMID 17038615.

Parfrey, L.W.; Lahr, D.J.G.; Katz, L.A. (2008). "The Dynamic Nature of Eukaryotic Genomes". *Molecular Biology and Evolution* 25 (4): 787. doi:10.1093/molbev/msn032. PMID 18258610.

Pavlov AR, Pavlova NV, Kozyavkin SA, Slesarev AI (2004). "Recent developments in the optimization of thermostable DNA polymerases for efficient applications". *Trends Biotechnol.* 22: 253–260. doi:10.1016/j.tibtech.2004.02.011. PMID 15109812.

Pershouse M, Heivly S, Girtsman T (2006). "The role of SV40 in malignant mesothelioma and other human malignancies". *Inhal Toxicol* 18 (12): 995–1000. doi:10.1080/08958370600835377

Poulin DL, DeCaprio JA (2006). "Is there a role for SV40 in human cancer?". *J. Clin. Oncol.* 24 (26): 4356–65. doi:10.1200/JCO.2005.03.7101. PMID 16963733. http://jco.ascopubs.org/cgi/content/full/24/26/4356.

Reece, Richard J. (2004). *Analysis of Genes and Genomes*. Chichester: John Wiley and Sons. ISBN 0-470-84379-9.

Rosen, F.S.(2004) "Isolation of poliovirus—John Enders and the Nobel Prize" *New England Journal of Medicine*, 351,pp. 1481–83 PMID 15470207

Roulston A, Marcellus RC, Branton PE (1999). "Viruses and apoptosis". *Annu. Rev. Microbiol.* 53: 577–628. doi:10.1146/annurev.micro.53.1.577. PMID 10547702. Retrieved on 20 December 2008.

Rudin, D., Shah, S.M., Kiss, A., Wetz, R.V., Sottile, V.M. (2007) "Interferon and lamivudine vs. interferon for hepatitis B e antigen-positive hepatitis B treatment: meta-analysis of randomized controlled trials." *Liver Int.* 9, pp. 1185–93. PMID 17919229.

Saccone, Cecilia; Pesole, Graziano (2003). *Handbook of Comparative Genomics*. Chichester: John Wiley and Sons. ISBN 0-471-39128-X.

Saiki, RK; Scharf S, Faloona F, Mullis KB, Horn GT, Erlich HA, Arnheim N (December 20 1985). "Enzymatic amplification of beta-globin genomic sequences and restriction site analysis for diagnosis of sickle cell anemia". *Science* 230 (4732): 1350–4. doi:10.1126/science.2999980. PMID 2999980. http://sunsite.berkeley.edu/cgi-bin/ebind2html/pcr/034.

Saiki, RK; Gelfand DH, Stoffel S, Scharf SJ, Higuchi R, Horn GT, Mullis KB, Erlich HA (1988). "Primer-directed enzymatic amplification of DNA with a thermostable DNA polymerase". *Science* 239: 487–91. doi:10.1126/science.2448875. PMID 2448875. http://sunsite.berkeley.edu/cgi-bin/ebind2html/pcr/009.

Sanger, F., Air, G.M., Barrell, B.G., Brown, N.L., Coulson, A.R., Fiddes, C.A., Hutchison, C.A., Slocombe, P.M. and Smith, M., Nucleotide sequence of bacteriophage PhiX174 DNA, Nature, 265, 687-695, 1977.

Sanger, F., Coulson, A.R., Friedmann, T., Air, G.M., Barrell, B.G., Brown, N.L., Fiddes, J.C., Hutchison, C.A. III., Slocombe, P.M. and Smith, M, The nucleotide sequence of bacteriophage PhiX174, J. Mol. Biol., 125, 225-246, 1978.

Sen GC (2001). "Viruses and interferons". *Annu. Rev. Microbiol.* 55: 255–81. doi:10.1146/annurev.micro.55.1.255. PMID 11544356.

Sinclair J (March 2008). "Human cytomegalovirus: Latency and reactivation in the myeloid lineage". *J. Clin. Virol.* 41 (3): 180–5. doi:10.1016/j.jcv.2007.11.014. PMID 18164651. Retrieved on 20 December 2008.

Sissons JG, Bain M, Wills MR (February 2002). "Latency and reactivation of human cytomegalovirus". *J. Infect.* 44 (2): 73–7. doi:10.1053/jinf.2001.0948. PMID 12076064. Retrieved on 20 December 2008.

Smith, Hamilton O.; Clyde A. Hutchison, Cynthia Pfannkoch, J. Craig Venter (2003-12-23). "Generating a synthetic genome by whole genome assembly: {phi}X174 bacteriophage from synthetic oligonucleotides". *Proceedings of the National Academy of Sciences* 100 (26): 15440-15445.

Stanley, W.M., Loring, H.S., (1936) "The isolation of crystalline tobacco mosaic virus protein from diseased tomato plants" *Science*, 83, p.85 PMID 17756690

Stanley, W.M., Lauffer, M.A. (1939) "Disintegration of tobacco mosaic virus in urea solutions" *Science* 89, pp. 345–347 PMID 17788438

Stenger, Hall, Choi and French, *Phytopathology* 88: 782, 1998.

Stewart M. Grayl and Nanditta Banerjee–Mechanisms of Arthropod Transmission of Plant and Animal Viruses. Microbiology and Molecular Biology Reviews, Volume 63, pp. 128–148.

Subramanya D, Grivas PD (November 2008). "HPV and cervical cancer: updates on an established relationship". *Postgrad Med* 120 (4): 7–13. doi:10.3810/pgm.2008.11.1928. PMID 19020360

Takaoka A, Hayakawa S, Yanai H, *et al* (2003). "Integration of interferon-alpha/beta signalling to p53 responses in tumour suppression and antiviral defence". *Nature* 424 (6948): 516–23. doi:10.1038/nature01850. PMID 12872134.

Wang, Daowen and Andrew J. Maule (June 1994). "A Model for Seed Transmission of a Plant Virus: Genetic and Structural Analyses of Pea Embryo Invasion by Pea Seed-Borne Mosaic Virus" *The Plant Cell*, Vol 6, 777–787.

Werner, E. (2003). "In silico multicellular systems biology and minimal genomes". *Drug Discov Today* 8 (24): 1121–1127. doi:10.1016/S1359-6446(03)02918-0. PMID 14678738.

Witzany, G. (2006). "Natural Genome Editing Competences of Viruses". *Acta Biotheoretica* 54 (4): 235–253. doi:10.1007/s10441-006-9000-7. PMID 17347785.

Witthoft, T., Moller, B., Wiedmann, K.H., Mauss, S., Link, R., Lohmeyer, J., Lafrenz, M., Gelbmann, C.M., Huppe, D., Niederau, C., Alshuth, U. (2007) "Safety, tolerability and efficacy of peginterferon alpha-2a and ribavirin in chronic hepatitis C in clinical practice: The German Open Safety Trial." *J Viral Hepat.* 14, pp. 788–796. PMID 17927615.

Index

www.ingramcontent.com/pod-product-compliance
Lightning Source LLC
Chambersburg PA
CBHW050508190326
41458CB00005B/1476